G...
S...
C... .... –
Firefighter Safety

Euro Firefighter

Paul Grimwood FIFireE
London Fire Brigade (retired)

*Compartment Fire Behavior
(CFBT) Instructor
European Edexcel® CFBT
Training Syllabus*

EURO FIREFIGHTER

Paul Grimwood

Published by Jeremy Mills Publishing Limited
113 Lidget Street, Lindley, Huddersfield,
West Yorkshire HD3 3JR

www.jeremymillspublishing.co.uk

This edition first published 2008

ISBN: 978–1–906600–25–9

A CIP catalogue record for this book is available from the British Library

Typeset in 9.5/11pt Plantin
Concept Huddersfield

Contents

Introduction

There was a fire in 2007 on a European fire-ground that represented a typical 'routine' approach faced by firefighters all over the world. The fire involved a small single-story abandoned warehouse but there were surrounding buildings that posed some moderate exposure risks from a developing fire. The initial response consisted of just five firefighters on a single engine, with limited back-up for the next twelve minutes on-scene.

As this book will show, a lot can happen in twelve minutes on the fire-ground. In this case two firefighters died. What went wrong at this 'routine' fire? Or was it just one of those fires where the hazardous nature of the profession took its inevitable course?

Firstly, please understand that there is nothing 'routine' about firefighting! If firefighters get into the habit of making routine approaches to fires then they will become complacent. Rule number one – Complacency is the firefighter's worst enemy! Secondly, there are nearly always things we can do to reduce a firefighter's exposure to risk. For example, we can pre-plan more effectively and we can communicate more effectively, as well as taking actions on-scene that will secure team safety and save lives. Thirdly, we should provide safe operational 'systems of work' based on clear directives and protocols (Standard Operating Procedures or SOPs), with an overall objective to optimize the tactical deployment of on-scene resources. Finally, we must effectively train both firefighters and commanders from differing perspectives and across a broad range of operational issues – most specifically, to observe, 'read' and understand changing fire conditions. This point is absolutely critical and provides the underpinning knowledge needed to stabilize and control fire development within a structure whilst securing the safety of crews on the fire-ground.

In the fire described above, the incident commander (IC) of the primary response immediately formed a two-pronged plan of attack, utilizing three firefighters manning hose-lines at two points of entry. His initial approach was to cut off and control fire spread in the derelict warehouse from a defensive (exterior) stance. However, two firefighters were directed to advance their hose-line in 'a few feet' to enable them to get a better angle on the fire. The incident commander's intention was to site these firefighters just inside the street entry doorway to enable them to hit the fire. However, this directive served to change the tactical mode of attack from 'defensive' to 'offensive', although it appeared this was never truly the intention.

The two firefighters' interpretation of 'a few feet' became twenty feet. As the two firefighters advanced further into the structure, the IC was assisting another firefighter in cutting through a steel street door to allow a better hit on the fire with the other hose-line. At this stage we have to question roles and assigned tasks, in linewith any declared (or non-declared) tactical mode of attack. If all firefighters on-scene are outside the structure, in relatively 'safe' positions, the IC may effectively take part in operational tasks, as needs dictate (even this is arguable according to structural safety hazards). However, from the moment firefighters step a foot inside the structure, this now becomes an interior offensive operation. At this point critical strategic concerns would include:

- A tactical mode (offensive or defensive) was never declared;
- As firefighters have entered 'a few feet' into the structure, an offensive tactical mode of operations should now be declared and communicated to all on-scene via the command system;
- With firefighters deployed to the interior, the IC must now take a command position and locate himself effectively in order to observe and 'read' fire conditions, looking out for changing circumstances and recognizing warning indicators of hazardous situations developing. This position may ideally be at a corner of the structure to allow visual contact with at least two sides of the building, which will also provide vital information as to structural safety;
- Crew briefings must be clear as to their objectives – the reference to advancing in 'a few feet' may be interpreted in different ways, as was the case here.

With difficulties being experienced in gaining access via the steel door, the IC decided to create additional openings by breaking some windows for the purpose of creating points of access for the second hose-line. At this stage the two firefighters on the first line had advanced some way into the structure and although the intention may have been to break windows to create access points, they also clearly served as ventilation points.

Over the next two minutes the IC (who was also reportedly a trained CFBT instructor) was not in a position to see the changing fire conditions as smoke became darker and started to push out of the eaves of the roof under great pressure. Suddenly there was an 'event' of rapid fire progress and the two firefighters inside were tragically caught and trapped.

At another fire in the same year, this time on a US fire-ground, nine firefighters were to tragically lose their lives. Again, the 'routine' approach to a seemingly 'minor' fire demonstrated how things can take a turn for the worse in a few brief moments. A 'trash' fire situated against the exterior wall of a large furniture superstore spread into the large volume structure, suddenly trapping firefighters as the interior fire intensified.

Again, the command structure reportedly failed to establish 'control' from a command perspective and progressive ICs, within the first few minutes, were tied up at the operational level, or were micro-managing the scene without stepping back and gaining a wider perspective of what was occurring. It will be seen time and again, throughout this book, that 'tunnel vision' is a result of 'command without control'. It is clearly possible to have command of a fire-ground but at the same time, not be in control of operational aspects. Any opportunities to break the tactical 'error chain' are sadly missed because of this inability to 'control' the fire-ground, and any exposure to risk is therefore frequently increased beyond acceptable limits.

In both of the above fire-ground tragedies, as in so many cases, we must look closely at:

- Pre-planning
- Effective command and control
- Adequate communications
- Adequate training of all staff
- Adequate equipment provision and maintenance
- Effective tactical deployments, ensuring available staffing and on-scene resources are optimized to their best effect
- Clearly defined directives and protocols (SOPs) that provide safe systems of work at fires

You might honestly believe that all these issues have been effectively addressed in your own brigade or department, or perhaps you work within a reckless culture of simply 'ticking boxes'. However, experience demonstrates time and again that vital links in the above chain are seen to fail regularly and serve as causal factors in the deaths of both firefighters and building occupants alike. Of course, in many situations the available budgets and resources will limit us. Even so, we still have the opportunity to optimize our resources and ensure they are used most effectively by careful analysis of critical roles and operational tasking needs.

In the UK, firefighters experience over fifty backdrafts and 600 events of 'rapid fire' progress every year. On average, that is once in every 187 working fires. With this in mind, one area of tactical firefighting operations that has influenced the training of firefighters from a global perspective more than any other is that of Compartment Fire Behavior Training (CFBT). This book covers the new 2007 European (Edexcel BTEC) CFBT instructor syllabus and provides guidelines and information for those studying for these awards. In the USA this form of training has been termed 'flashover' training. In fact, the European CFBT training approach goes much further than this and delivers training in a wide range of firefighting techniques, fire phenomena, compartment entry procedures, attack techniques, tactical ventilation and tactical deployment procedures, using various configurations of ISO shipping containers.

When firefighters occupy these Fire Development Simulator (FDS) units they work at very close quarters with the fire but under extremely safe and controlled conditions. They experience a wide range of fire phenomena and are able to practice a range of nozzle techniques to deal with fire in the gaseous-phase. They also learn how the creation of ventilation openings, or anti-ventilation techniques, are likely to affect compartment fire development. They will also learn safety techniques used to gain entry to fire-involved compartments. One common factor that makes CFBT so effective is that each student gets the very same experience, which is impossible where acquired structure burns are used for such training.

However, as effective as CFBT has been in teaching firefighters basic skills in 'real fire' environments, the training has also created a false sense of security amongst the British Fire Service, and practitioners should become aware of these serious issues. The author has demonstrated in his survey of fifty-eight UK fire brigades that 89% of brigades are under-flowing their attack hose-lines simply because they have no basic understanding of the importance of flow-rate. There are several accounts of firefighters being unable to suppress fires or rescue trapped colleagues because the hose-line they are advancing is flowing at less than 230 liters/min (60 gallons/min) into rapidly developing fires.

One of the misconceptions about CFBT is that container fires are 'real' fires. In fact they are only 1.5 MW simulations that produce pure gaseous flaming combustion. In 'real' compartment fires, energy release rates may commonly be between 5 to 15 MW and the fuel-load will be more concentrated, requiring deeper penetration and cooling of the fuel-base. As CFBT instructors learn to deal with vast amounts of flaming combustion in FDS units using flow-rates as low as 40 liters/min (10 gallons/min), they may take on a false sense of security and become over-confident in the ability of low-flow hose streams to deal with actual room fires verging on, or surpassing, the flashover stage. This misunderstanding can cost lives!

The biggest learning curve here is that you cannot identify a 'good' fire stream by simply 'looking' at it! Modern nozzles are designed to trade flow-rate for reach and may provide a totally false impression of true water content.

It is not just firefighter safety issues that this book deals with, but also the safety of the very people we serve. There are many instances where we may have accepted greater levels of exposure to risk to ourselves and yet through some tactical error or miscommunication, we may have fallen short in our responsibility towards building occupants. Therefore, it is not just the firefighter who is affected by inappropriate tactics or error chains but also the occupants of fire-involved buildings.

We must train fire commanders and firefighters to deal more effectively with incidents. More importantly, we must establish a clearer appreciation of what is 'acceptable risk' under specific circumstances and reflect this through our SOPs. We can increase a firefighter's chances of survival by taking simple precautions. We can improve efficiency and safety through the provision of simple checklists in Standard Operating Procedures that will encourage a risk-based culture through a selective operational thought process. We can ensure that fire commanders and firefighters have a more in-depth appreciation of *when to ventilate and when not to*. We can also educate firefighters in the most effective methods of opening up a structure and advancing in with greater safety, whilst anticipating and recognizing potential hazards through the general dynamics of air and smoke movements.

This book will discuss simple guidelines used to establish safe but effective tactical approaches into fire-involved structures. It is written in such a way to assist the instructor, fire chief, or firefighter to learn from key points provided in bullet lists. Through a series of simple SOPs covering a wide variety of fire-ground situations, you will learn to apply basic concepts using a more pro-active approach, whilst effectively balancing 'risk' versus 'gain' and implementing Risk Control Measures, that may one day save the lives of you or your crew.

Always make it your personal objective to:

Learn from the past, seek out new information, gain new knowledge and use that new knowledge to challenge assumptions and conventional wisdom, to stimulate and share new ideas.

But in doing this always remember, respect and honor those who have gone before us, as you study their own experiences through the various accounts and case histories:

It is not the critic who counts, not the man who points out how the strong man stumbled, or where the doer of deeds could have done them better. The credit belongs to the man who is actually in the arena; whose face is marred by dust and sweat and blood; who strives valiantly; who errs and comes short again and again; who knows the great enthusiasms, the great devotions, and spends himself in a worthy cause; who, at the best, knows in the end the triumph of high achievement; and who, at worst, if he fails, at least fails while daring greatly, so that his place shall never be with those cold and timid souls who know neither victory nor defeat.

Theodore Roosevelt 1910

Reader's LINK CODE

Website downloads and updates
http://www.euro-firefighter.com/w977p631wz.htm

Foreword

Paul Grimwood is a consummate professional who, while passionate and holding strong opinions, has an open mind an insatiable curiosity. These personal characteristics serve as a powerful foundation for this text. Paul is a student of the firefighter's craft and throughout his adult life has not only worked tirelessly to improve his own mastery, but to add to the body of fire service knowledge and share this information with others in our community.

In *Fire Protection: A Complete Manual of the Organization, Machinery, Discipline and General Working of the Fire Brigade of London* (1876, page v), Massey Shaw wrote:

> *From the remotest periods of antiquity to the present time, the business of extinguishing fires has attracted a certain amount of attention; but it is a most curious fact that, even now, there is so little method in it that it is a very rare circumstance to find any two countries, or even any two cities in one country, adopting the same means ...*

The same holds true today. Paul's work in this text is a first rate effort to integrate best practice from the world's fire services. *Euro Firefighter* serves as an excellent source for fire service practitioners seeking a reference that will challenge their assumptions and stimulate their interest in improving mastery of their craft.

Battalion Chief Ed Hartin, MS, EFO, MIFireE, CFO
Gresham Fire and Emergency Services
Gresham, Oregon USA

Author biography

Paul Grimwood is a thirty-five-year veteran of the British Fire Service, having served most of his time as a firefighter in London Fire Brigade's busy West End district. In the mid 1970s he also served an eighteen-month detachment into New York's South Bronx 7th Division, during the busiest period in FDNY's history. From 1976–77 he further served as a volunteer firefighter/EMT on Long Island's south shore.

For more than thirty years Paul has been undertaking global research into structural firefighting strategy and tactics and has contributed in excess of 200 technical articles since 1979, in an effort to advance firefighter safety. During this time he has served and responded out of more than 100 fire stations around the world, working alongside some of the finest firefighters you could ever wish to meet. He has also presented papers on fire service operations at international conferences in several countries since 1993.

His other books include *Fog Attack* (1992) and *3D Firefighting* (2005), the latter of which he joint authored with firefighting colleagues Battalion Chief Ed Hartin (USA), and Station Officers John McDonough and Shan Raffel (Australia).

From 1984 he served eleven years as a London Fire Brigade fire investigator and was part of the six-person team that investigated the tragic King's Cross fire in 1987 where thirty-one lives were lost, including a colleague (Station Officer Colin Townsley) from London's Soho fire station.

He is a trained USAR instructor (EMT) and was deployed on operational disaster relief assignments into Iraq (1991) and Bosnia (1993). He is also a CFBT and tactical ventilation specialist (1984-2008) and a Tactical Deployment (command and control) and High-rise Firefighting instructor.

Paul is the founder and principal of Firetactics.com, a website which has provided in excess of 14,000 pages of structural firefighting SOGs in six languages FREE to over 2.5 million visitors from more than seventy countries since July 1999 (source Webstat.com).

He is an advisor to several UK Government Task Groups including ODPM Compartment Fire Behavior Training; BDAG High-rise Firefighting; CLG High-rise Firefighting, as well as an editorial reviewer for the Fire Safety Journal (the official journal of the International Association of Fire Safety Science). He is also an established 'expert' technical witness and advisor in fire service operations, having worked on several high-profile cases in the USA and Europe.

In 2008 Paul was awarded the Institute of Fire Engineers' (IFE) highest academic status (FIFireE) in recognition of his thirty-year professional commitment to firefighter safety.

Acknowledgments

As with any work such as this, the author is generally blessed by some major commitment from a large body of people who have provided a vast amount of research, advice and information that goes to make up the completed text. I do not accept the entire credit for this book but would wish to mention a few colleagues here whose own work and endeavors have contributed, inspired, motivated, and assisted my own professional development and objectives.

The late Assistant Chief Officer Roy Baldwin	London Fire Brigade
Station Officer Tom Stanton (retired)	London Fire Brigade
Deputy Chief William Bohner (retired)	City of New York Fire Department
Deputy Chief Vince Dunn (retired)	City of New York Fire Department
The late Battalion Chief William Clark	City of New York Fire Department
Assistant Commissioner Jon Webb	London Fire Brigade
Deputy Assistant Commissioner Terry Adams	London Fire Brigade
Commandant Frederic Monard	Sapeurs Pompiers de Paris
Battalion Chief Ed Hartin	Gresham Fire and Emergency Services, Oregon, USA
Station Officer Shan Raffel	Queensland Fire and Rescue, Australia
Station Officer John McDonough	New South Wales Fire Brigades, Australia
Chief Fire Officer Peter Holland	Lancashire Fire and Rescue Service, UK
Deputy Chief Fire Officer Paul Richardson	Lancashire Fire and Rescue Service, UK
Chief Fire Officer Barry Dixon	Greater Manchester County Fire Service UK
Chief Fire Officer John Craig (retired)	Wiltshire Fire and Rescue Service, UK
Matt Beatty (Rescue One)	City of New York Fire Department
Nate DeMarse (Engine 68 Bronx)	City of New York Fire Department
Lt. Daniel McMaster	Alexandria (Virginia) Fire Department, USA
Major Stéphane Morizot	Versailles, Paris
Chief Jan Südmersen	City of Osnabrück Fire Service, Germany
Fire Officer Tony Engdahl	City of Gothenburg Fire Service, Sweden
Captain Juan Carlos Campaña	City of Madrid Fire Brigade, Spain
Captain Jose Gomez Antonio Milara	City of Madrid Fire Brigade, Spain
Pierre Louis Lamballais	www.flashover.fr
Frank Gaviot Blanc	Fire Engineer, France
Jesper Mandre	Fire Engineer, Sweden
Mr. Khirudin bin Drahman @ Hussaini	Malaysia Fire and Rescue Service (Bomba)
Stefan Svensson	Swedish Rescue Services Agency, Revinge
Cliff Barnett	Fire Engineer, New Zealand
Dietmar Kuhn	Fire Engineer, Germany
Dave Dodson	*Reading the Smoke* – Firefighter and author
Chief Billy Goldfedder	www.firefighterclosecalls.com
Adrian Ridder	Fire Engineer, Germany
Station Officer Jürgen Ernst	Böblingen, Germany
Station Officer John Chubb	Dublin Fire Brigade, Eire
Cas Seyffert (retired)	City of Johannesburg Fire Department, SA
Nigel (Snowy) Kind	South Yorkshire Fire and Rescue, UK

Plus all my dear friends and close family who have put up with me whilst I toil away for hours on end engrossed in research; to my two dear sons Richie and Paul Jnr who are now following their own successful careers; to my best friend and partner in life Lorraine; and to each and every firefighter I have had the pleasure to meet – you are a special breed, stay safe!

Finally, to anyone I have forgotten to mention – Thank you!

Institution of Fire Engineers (IFE)

The Institution of Fire Engineers now has approaching 10,000 members in over twenty countries, who represent a complete cross-section of the fire engineering discipline.

The objectives of the IFE are to encourage and improve the science and practice of Fire Extinction, Fire Prevention and Fire Engineering and all operations and expedients connected therewith, and to give an impulse to ideas likely to be useful – in relation to such science and practice – to the members of the Institution and to the community at large.

The global appeal of the organization assists international networking, which means you are able to tap into a vast membership, from around the globe, in efforts to share common aims, interests and knowledge.

What academic opportunities can the IFE offer you?

- The opportunity to further your career and achieve higher academic standards by progressing through the range of examinations set by the Institution.
- A number of educational establishments run courses such as Fire Engineering and Fire Safety, which are accredited by the Institution.
- Scholarships are available, funded by the Fire Service Research and Training Trust.
- The Institution supports the concept and practice of Continuing Professional Development (CPD), believing it to be essential to effective performance as a professional fire engineer.

The Institution of Fire Engineers offers global opportunities to assist your professional development, experience and qualifications, if you are a fire chief, fire engineer, firefighter, scientist, or fire service management professional.

US branch of the IFE	http://www.ife-usa.org
UK branch of the IFE	http://www.ife.org.uk
Canadian branch of the IFE	http://www.ife.ca
Australian branch of the IFE	http://www.ifeaustralia.org.au/
International branches	http://www.ife.org.uk/branches/

Chapter 1

Guiding principles and managing risk at fires

Fire chiefs and chief officers are not the ones to generally go into a burning building to save someone or save someone's property. We are not the ones that have to deal with a shooting at three o'clock in the morning when the whereabouts of the perpetrator are unknown. We are not the ones who are generally laying our lives on the line each and every day to protect our communities, so the least we can do is damn well make sure that those brave men and women who are, have the best equipment, the best PPE, the best training, the best policies and procedures, the best safety practices, the best management, and the best leadership . . .

    If you can look in the mirror without any hesitation and say, 'Yes, I have done all that I can,' then you should have no trepidation or concern about a task force coming into your community following a line-of-duty death. If you can't say yes, then you need to turn in your badge . . . today!

Brian Crawford – National Fire Academy

No matter what the questions are, the answers are in the mirror.

Fire Chief Magazine Editorial – September 2007

1.1 INTRODUCTION

London 1971 and 1985

Things are done a lot differently in the fire service these days. When I first joined in 1971 we had plastic gloves, turnout trousers and smart woolen fire tunics. We would tackle most fires without the protection of self-contained breathing apparatus (SCBA), as this was considered a sign of weakness amongst the brothers. Often, a guvnor's (captain's) boot, placed gently up the backside (polite word for arse), was all the back-up you were likely to get inside a fire – literally! However, in the days before occupational health and safety and managing risk on the fire-ground became commonplace, we seemed to get things done just fine. Didn't we?

I remember once crawling over three firemen lying prone in a hallway. Dark black smoke was rolling over our backs and heading for the entry doorway behind us. I had crawled into this basement apartment fire feeling I had just taken a clear lead over my three brothers in a race of destiny. After all, I had my plastic protective clothing and woolen jacket with bright silver buttons keeping me safe, so who the hell needed breathing apparatus? I crawled further on into the apartment and managed to find some glow in the darkness up ahead. I could hear my brothers coughing behind me and that assured me I was going to win this one outright. Sure enough, I reached the kitchen ahead of them. As I crawled around the corner I became mesmerized by the awesome power of the flames roaring up across the ceiling over my head. The fire was loud, it was hot and it was bright. It gave me a clear vision of everything that was around me and I was confident that we were in control of the situation. Despite the heavy black smoke layer hanging in the overhead, we continued to search throughout the apartment and were happy to find it clear of any occupants. By the time we had returned to the entry doorway, the attack hose-line was just being brought in through the hallway. 'It's in the kitchen to your left,' I told the crew. All went well after that and we were home and dry!

Another time, in 1985, I remember an oxy-acetylene cylinder trolley had fallen over whilst in use. It had turned over itself as it fell upside down on a steel sliding roller belt, into the basement of one of London's top West End hospitals. Flames were roaring up out of the tiny street opening where the roller belt went down into the basement. We all huddled together behind a piece of street furniture – perhaps a telephone company junction box – spraying our water in the general direction of the basement opening.

The 'guvnor' asked for a volunteer to crawl down the roller belt and close the valve on the cylinder. I conjured up a glamorous image of First World War trenches, and being asked to volunteer for a mission to run ahead and single-handedly take out a machine gun nest! 'I'll go guv,' I shouted, and up I leapt, reaching for the additional protection that might save my life if the cylinder were to explode – a pair of fire resisting gloves and a flash-hood!

As I crawled headfirst down the roller belt I came face to face with a sight I will never forget. The cylinder set was totally enveloped in flames and I was lying upside down and almost on top of it. I reached over and placed the cylinder key into position. I thought to myself, 'This might be the last moment you'll ever spend on this earth.' I recall that as clearly as if it were yesterday! The thought of lying down with my mates outside in the fresh air, behind that junction box, suddenly seemed a far better option! Then I turned the key and the flames disappeared. Hey – we got things done in those days right?

I look back now – all these years later – and shudder! What the hell were we doing? We could have achieved virtually the same results whilst using 'safer systems of work' (nice buzz words from the 1980s). But really, we just needed to take a step back in these situations, balance the 'risks' against the potential for 'gains' and apply some simple Risk Control Measures. We might have achieved the same outcomes but with a lot less exposure to risk.

Traumatic operational fatalities amongst firefighters

Statistically, the UK Fire Service incurs traumatic operational fatalities at a fairly consistent average of about one firefighter per year per 100,000 structure fires.[1] Given the inherent dangers of firefighting operations, frequency of exposure to risky situations and a firefighter population of 50,000, this would indicate that the risk is generally well managed in this domain.[2]

In the USA the traumatic death rate amongst firefighters is twice as high, where currently around 1.9 firefighters are killed per year per 100,000 structure fires (a rate only slightly lower than that observed in the early 1980s). However, this rate was at its highest (3.0 per 100,000 structure fires) over a thirty-year period during the 1990s.[3]

The main causes of these line-of-duty deaths (LODDs) are smoke inhalation, burns, crushing injuries and related trauma. Most importantly, both the UK and US statistics provided above are strictly related to firefighting operations and exclude all other causes of death, such as heart attacks and road accidents en route etc.

The author has served in both a professional and voluntary capacity in the UK and USA across a thirty-year career; working on assignment in three UK metropolitan brigades and across eight US states. It is clear that fires are generally fought with similar tactics, based on aggressive interior firefighting approaches common to both countries. However, the implementation of 'risk management' principles at fires is clearly viewed from different perspectives and the author believes this may be a prime reason why US LODD statistics are disproportionately higher in relation to such causes.

Operational risk management refers primarily to the risk of death or injury to firefighters and other emergency responders that could result from the performance of their duties. In a broader sense, it applies to other types of accidents and undesirable events that could occur during emergency operations. Emergency responders knowingly subject themselves to elevated levels of risk in the performance of their duties. Some of those risks are unpredictable and unavoidable. On the other hand, many are well-known and can be effectively limited or avoided through the application of operational risk management practices.

Firefighters' reputations are frequently associated with courage and bravery. That perception often suggests that firefighters are willing to accept any risk to their personal safety when performing their duties. Blind acceptance of risk used to be virtually unlimited and unquestioned in the fire service. It was not unusual as recently

1. Office of the Deputy Prime Minister, *UK Fire Statistics*
2. Tissington, P. & Flin, R., (2004), *Assessing Risk in Dynamic Situations: Lessons from Fire Service Operations*, RPO432, Universities of Birmingham and Aberdeen
3. Fahy, R.F., LeBlanc, P.R. & Molis, J.L., (2007), *Firefighter Fatality Studies 1977–2006: What's Changed over the Past 30 Years?*, NFPA Journal

as twenty years ago for firefighters to be exposed to very high levels of risk, with very little concern for their personal safety. Firefighters were expected to follow any order without question and to accept any risk to accomplish the mission. The most respected firefighters were often those with the most obvious disregard for their own safety – those who demonstrated the attitude that the fire must be defeated 'at any cost'.[4]

Today, we are moving toward a different perception of the relationship between bravery and risk. Without question, we still respect, value, and honor bravery and courage – particularly when a situation involves saving lives. Even so, a contemporary sense of values requires a very different assessment of appropriate and inappropriate risks. In many cases, that calls for **limiting the exposure** of personnel to risks that they might be willing to accept for themselves. A fire department's definition of acceptable risk might be more conservative than the level of risk an individual firefighter might willingly accept. In the current value system, higher-level officers are often more responsible for limiting risk exposure than for demanding courage from their forces.

It is not acceptable for fire departments to risk the lives of their members because they are not adequately **trained** or **equipped** or because they do not apply appropriate judgment in conducting emergency operations.

Every incident commander (IC) should anticipate that the authority having jurisdiction for occupational safety and health laws will thoroughly review any incidents in which injuries or fatalities occur – using *NFPA 1500* and other applicable standards as benchmarks – to consider if actions taken were reasonable under the circumstances. A fire department should expect that an investigation would seek to determine if its members were provided with every appropriate form of protection, including training and Standard Operating Procedures (SOPs).

The move to 'risk-managed' fire-grounds in the UK has been gradually, but strictly, enforced through national occupational health legislation since 1974,[5] although the basic framework for firefighter safety is clearly rooted in national practices that were adopted at least two decades earlier. Risk-managed concepts associated with fire-ground 'accountability', SCBA air management and Rapid Intervention Teams (RITs), became part of the UK Fire Service culture following several multiple LODDs in London in the 1940–50s.[6] The US approach is legislated by federal (Occupational Safety and Health Administration or OSHA) regulations and NFPA guides that serve as established industry 'standards'. However, it was the mid 1980s before risk management principles and fire-ground safety standards (i.e. *NFPA 1500* and other OSHA regulations) for the US fire service were seriously addressed.

Risk assessment is a powerful tool for informing, but not dictating, decisions on the management of risk. The implication is that a fire commander, having assessed that a particular course of action may involve exposure to risk, would not necessarily abandon it. As in many other industrial settings, some level of risk is accepted and has to be managed. Indeed, as will be seen later in this book, more recent guidance specifically encourages controlled, deliberate risk-taking in certain circumstances.

---

4. FEMA, (1996) *Risk Management Practices in the Fire Service*, FA-166, United States Fire Administration
5. Health & Safety at Work Act 1974 (UK)
6. Fires at Covent Garden and Smithfield Market (see Chapter Fourteen)

The firefighter's risk management model

- Establish what the risks are.
- Select a safe system of work (**mode of attack**).
- Implement Risk Control Measures.
- Monitor the dynamic processes on the fire-ground.
- Are the risks proportional to the benefits or gains?

This list clearly shows the need to actively involve the strategic and task levels of command and operation.

So how is 'risk management' defined? One common definition interprets it as, 'The systematic application of principles, approaches and processes to the tasks of identifying and assessing risks, and then planning and implementing risk responses'. Introducing risk management principles on the fire-ground seems a surprisingly simple process. Firstly we establish what the risk is and then we select and implement control measures to reduce or remove the risk. We subsequently document clear directives (SOPs) as to how various fire-ground risks and hazards shall be managed and controlled. If you were crossing a busy road you would (probably without thinking) implement typical Risk Control Measures in a way that reduces the risks to an acceptable level. These measures might include looking several times each way before crossing, listening for vehicles, looking to use a purpose-built crossing or a pedestrian bridge, or waiting for the 'green man' or 'walk' sign. Such control measures commonly guide our general safety and well being in life. To cross a busy road without looking is reckless and would obviously increase the chances of an accident occurring. We are going to pursue our objective anyway but it is sensible to take reasonable precautions in the process to increase our chances of success.

This approach to a risk-managed fire-ground provides direction and guidance for firefighters, through carefully worded protocols, ensuring both that the employer is covered legally and that responders are protected personally, from exposure to unnecessary risks. Such a process further ensures personnel may be held accountable where deviations from Standard Operating Procedures occur. However, the applied definitions, applications to all situations and wording in our SOPs are absolutely critical to their effective implementation on the fire-ground, and a single word inappropriately placed may detract from what would have been a sound basis for providing a risk-based approach to firefighting. It is also critical that firefighters and fire commanders possess the knowledge and ability to apply risk-based concepts and undertake dynamic risk assessments at fires. Without the necessary levels of fire-ground experience, and/or practical training that ensures core firefighting skills are regularly and effectively updated, we cannot expect personnel to implement even the most basic principles of risk-based tactical approaches.

In his book *Fire Officer's Handbook of Tactics*,[7] FDNY Deputy Assistant Chief John Norman proposes 'Five General Principles of Firefighting' upon which he bases his tactical approach theories. He further suggests these five guiding rules, or principles, are so important that they should never be broken unless under the most unusual of circumstances. It is well known that many fire departments around the USA have even structured their entire primary response tactics and SOPs around Chief Norman's 'famous five' basic rules.

---

7. Norman, J., *Fire Officer's Handbook of Tactics*, Fire Engineering/Penwell Publishing

1.2 CHIEF JOHN NORMAN'S 'FIVE GUIDING PRINCIPLES OF FIREFIGHTING'

- When sufficient manpower isn't available to effect both rescue and extinguishment at the same time, rescue must be given priority.
- When you don't have sufficient manpower to perform all of the needed tasks, first perform those that protect the greatest number of human lives.
- Remove those in the greatest danger first.
- When sufficient personnel are available to perform both functions they must carry out a coordinated attack.
- When there is no threat to occupants, the lives of firefighters shouldn't be unduly endangered.

1.2.1 RECEO/REVAS

Furthermore, there also exist some well-known and simple acronyms that are widely used by firefighters to assist the prioritization of critical tasking at a structural fire. The first of these is known as **RECEO** – this strategic approach was provided as far back as the 1940s by Chief Lloyd Layman:

R – Rescue
E – Exposures
C – Confinement
E – Extinguish
O – Overhaul

Later training texts also add:

V – Ventilation
S – Salvage

Another well-known acronym is **REVAS:**

R – Rescue
E – Evacuate
V – Ventilate
A – Attack
S – Salvage

A review of John Norman's excellent 'Five Guiding Principles of Firefighting' suggests that the most important primary action on arrival at a fire-scene is obviously the *rescue of those in immediate peril*. This does **not** account for occupants who may be trapped inside the structure but rather prompts an immediate rescue action to remove visible occupants who are at windows or on balconies, or offers approval towards attempts to locate and rescue 'known' life risk. This may entail the urgent placement of a ladder or an exterior access by firefighters, using rescue ropes from an upper level, or from the roof itself.

The risk-based approach applied to the concept of **'known life hazard'** is one that is well established and defined under *OSHA 29 CFR 1910.156* as in any Immediate Danger to Life or Health (IDLH) environment where:

- Immediate action could prevent loss of life
- For a 'known' life risk only
- Not for standard 'search and rescue' of 'possible' or 'suspected' life risk

Any such deviations from the regulations must be exceptions and not de facto standard practices. When the exception becomes the practice, OSHA citations are authorized *(29 CFR 1910.134[g][4][Note 2])* (see notes in Chapter Five).

In the UK, the term 'known life hazard' refers to a definition provided by *Technical Bulletin 1/97 – Safe Practice for SCBA Air Management*, where the Rapid Deployment Procedure provides an adequate but minimal level of safety and accountability when staffing and resources may be restricted during the initial stages of fire service response. This level of control is only for use in exceptional circumstances where **persons are at great risk requiring very urgent assistance, or where dangerous escalation of the incident can be prevented.** The 'known life hazard' in this case must be either within view, or 'known' to be within a short distance of the entry point to the risk area. Although not accounted for in the wording of the bulletin's definition, it may also be argued that exceptional circumstances include cries for help from within the fire-involved structure.

Chief Norman goes on to suggest that where staffing is restricted on arrival, simple actions might serve to save a large number of lives and these should be implemented as a matter of urgency where possible. Such actions may include the closing of a door to confine the fire, the placement of a primary hose-line to protect an escape route, or a primary attack made to suppress the fire itself – all prior to interior searches taking place. He goes on to say that where staffing permits, both 'fire attack' and 'interior search' of the building should occur at the same time, under a coordinated approach. These are simple guidelines borne out of the extensive experiences of literally thousands of inner-city firefighters over decades of fire response and yet, an annual review of LODD incidents clearly demonstrates how firefighters are repeatedly being killed, simply because they fail to follow these basic principles of firefighting which clearly promote risk-based concepts.

1.3 MANAGING 'RISK' ON THE FIRE-GROUND

The entire concept of risk analysis at fires is based upon industry standards provided through occupational safety guidelines for employees whilst they are at work. In fire service terms these guides generally state that fire departments must be well trained, adequately staffed and effectively equipped to deal with fire-ground emergencies and firefighting in general. These guidelines further state that the basic principles of effective risk management at fires rely on the ability of firefighters to *recognize hazards, implement Risk Control Measures,* and *monitor their success.* They of course must be **effectively** trained to do this.

Effective risk management will include such issues as careful and safe deployments, fire-ground accountability, SCBA air management, incident command, tactical operations, and water provisions – amongst other things.

The risk to fire department members is the most important factor considered by the incident commander in determining which strategy will be employed in each situation. The management of risk levels includes all of the following examples (**Risk Control Measures**) as a means of reducing the hazards faced by firefighters:

- Routine evaluation of risk in all situations
- Well-defined strategic options (mode of attack)
- Standard Operating Procedures
- Effective training

- Accountability and SCBA air management
- Full protective clothing ensemble and equipment
- Effective incident management and communications
- Safety procedures and safety officers
- Back-up crews for interior attack
- Back-up crews for rapid intervention
- Covering hose-lines
- Adequate resources
- Rest and rehabilitation
- Regular evaluation of changing conditions
- Experience based on previous incidents and critiques

There are three main guiding principles (*NFPA 1500*)[8] upon which effective risk-managed tactical operations are founded, and these are as follows:

- Actions that present a **high level of risk** to the safety of firefighters are justified only where there is a **potential** to save lives.
- Only a **limited level of risk** is acceptable to save valuable property.
- It is **not acceptable** to risk the safety of firefighters when there is **no possibility whatsoever** to save lives or property.

The section of *NFPA 1500* that specifically refers to operational risk management was introduced in the 1992 edition.

The *acceptable level of risk* is directly related to the potential to save lives or property. Where there is no potential to save lives, the risk to fire department members should be evaluated in proportion to the ability to save property of value. When there is no ability to save lives or property, there is no justification to expose fire department members to any avoidable risk, and defensive fire suppression operations are the appropriate strategy.

However, various definitions and interpretations of terms such as 'potential life risk' exist and there is further disagreement between firefighters, many arguing that such guides cannot be applied to the dynamic processes involved in fighting fires. In fire service terms, what might be considered an *acceptable risk* in line with *achievable objectives*, is also open to personal definition. Put more simply, many firefighters are quite willing to accept higher levels of risk as part of the very nature of firefighting and will look at things differently when balancing risk versus gain. There is no situation where this approach becomes more obvious than the interior search of buildings for 'suspected', 'potential' or 'known' occupants, where the variable interpretations associated with what is acceptable risk are played out.

1.4 WHAT IS CONSIDERED AN 'ACCEPTABLE RISK'?

This is a most critical question that leads to widespread debate and diverse opinions. An acceptable level of risk, on the fire-ground, is something that remains difficult to define.

The quantitative engineering definition of risk is:

$$Risk = (probability\ of\ accident) \times (losses\ per\ accident)$$

8. National Fire Protection Association, NFPA 1500: Standard on Fire Department Occupational Safety & Health

However, firefighters are all influenced by definitions of their primary roles (to save life and property), which are often ensconced in a long history of fire service tradition and the way firefighters are perceived. They are often seen as the 'last line of defense' or 'heroes', and these cultural and traditional concepts strongly influence how firefighters, as individuals, are likely to perceive and accept risk themselves.

An acceptance of some risk is a necessary trait of a good firefighter whose personality is strongly driven by challenge. The desire and determination to serve and succeed is what makes firefighters who they are. Without these personality traits the team concept would be weaker and less likely to succeed. Following the tragic fires and collapses that killed 343 New York City firefighters when the World Trade Center buildings were subjected to terrorist attack in 2001, FDNY Chief of Safety Al Turi put these personality traits into perspective when he said:[9]

When I reflect back on the whole thing, what I think is really important to bring out is that the courage and the bravery of the firemen was more outstanding than I thought it possibly could have been. You could look in their faces and you could see the fear. They knew what they were getting into. They knew what they were going to. They knew they were going to have the worst firefight of their lives, yet they all went, without question. You could almost see the relief on some of the people that we didn't send in, put in the staging area. You could almost sense the relief in their faces that we weren't sending them across the street at that time. All we had to do was say, 'You're up next, you're on your way' and they would have gone in.

There's been such tremendous talk about how many firemen came to the WTC on their own and these did contribute to our fatalities because they themselves became fatalities. The answer is, Yes it's a shame and it's unfortunate that we didn't have better discipline within the department, where we would have assured they would have all reported to a staging area or a central location, but when you think about it, it's part of our culture as firefighters to do exactly what they did. That's why they did it. It's that mental attitude that enables a normal person, which is what a firefighter is, just a normal person, male or female, to go into a burning building. That's what keeps the fire department running, that mental attitude. The same thing that caused those people to leave what they were doing when they were off and report to that site, that's the attitude that enables them to enter burning buildings on an everyday basis. Obviously, in the future, the department is going to have to demand more discipline from people, but somehow not stifle that attitude that enables them to do their job.

Chief Turi's words were strikingly clear in stating that we must find the right balance between accepting risk without stifling the attitude that makes firefighters do their job well.

The question, 'What is an acceptable level of risk?' is one with which firefighters will always have conflict. We might examine this from two angles:

- Property conservation
- Life hazard

9. www.firetactics.com /FDNY-TRIBUTE-11SEPT2001.htm

1.4.1 PROPERTY CONSERVATION

Where property is concerned, effective risk management recognizes that no (or few) structures are worth risking/losing firefighters' lives, and this belief is normally easier for firefighters to acknowledge when assessing acceptable levels of risk. However, even here there are issues of some concern! *(The author mentions 'few' as some firefighters consider specific occupancies of sacred religious worship or those which are important by their very nature, as worthy of much higher levels of risk)*.

In an effort to estimate or effectively define the term 'acceptable risk' in simple firefighting (property conservation) terms, let's take as an example fires in vacant or abandoned buildings. Many firefighters see such fires as opportunities to enhance their personal development, skills and experience. The author spent many years in socially deprived areas, where a vast number of buildings were either abandoned or unoccupied. In fact many of these buildings were actually occupied by homeless types that brought with them a wide range of other problems. These structures presented a good learning ground for probationary firefighters and interior firefighting was a daily event. Gradually, these 'ghetto' firefighters would become very experienced at their job and were able to mold themselves into some of the finest practical firefighters the author has ever worked with. However, it is arguable that in many situations, these buildings – which were due for demolition – should never have been entered in the first place, as the risks clearly outweighed the benefits.

However, if the 'benefit' is perceived as advancing the general performance levels of firefighters working in these areas, there may be some argument that they will approach all fires with a greater experience level and wider understanding of risks and hazards. Therefore their actual exposure to risk is somewhat reduced.

City of Flint – Fires in vacant and abandoned buildings

A recent study of fires[10] in Flint (Detroit) involving vacant and abandoned buildings demonstrated that:

- Out of the 767 total structure fires dispatched, 443 resulted in a report of an actual structure fire occurring. The 443 actual structure fires involved 264 occupied structures and 179 vacant structures.
- Vacant structure fires represented 40% of the department's structure fire volume.
- The department's injury rate at vacant structure fires is more than triple the national average reported by the National Fire Protection Association.
- 62% of the department's fire-ground injuries occurred at vacant structures fires.
- 79% of the cost from fire-ground injuries resulted from fires at vacant structures.
- 93% of the cost of injuries at fires in vacant structures occurred in buildings that were unsecured when firefighters arrived.
- Fire-ground operations produced twenty-one injuries at vacant buildings. Thirteen injuries occurred during fires at occupied buildings – whilst most injuries were minor by nature, the potential for serious injury or LODD clearly exists.

10. Graves., A., (2007), *Vacant Structure Fires and Firefighter Injuries In The City Of Flint,* Flint (Detroit) Fire Department

The National Fire Protection Association (NFPA) reported a national average of 5.6 firefighter injuries per 100 'special structure' fires and 1.9 firefighter injuries per 100 structure fires in general. The NFPA defines vacant buildings and buildings under construction as 'special structures'.

The rate of injury for Flint firefighters was, alarmingly, higher than the findings in the NFPA reports. During the survey period, Flint firefighters incurred an injury rate of 11.7 per 100 *vacant* structure fires. (An injury rate of 4.9 per 100 *occupied* structures was incurred and the rate of injury for structure fires in general was 7.6 per 100).

One important consideration here was that the fire department in question were somewhat restricted in staffing levels, and their excessively high injury rates were in alignment with national staffing studies that demonstrated how low staffing may influence the projected injury rate. Another reason for an abnormally high rate of injury might be the typically aggressive approach (characteristic of firefighters in busy fire areas) undertaken by Flint firefighters during offensive structural fire-fighting operations. This saw them entering buildings earlier and more often in comparison to the generally less aggressive – and perhaps more common – national firefighting approaches.

As a result of this research the City of Flint Fire Department responded with a new risk-based SOP for structure fires. They based their tactical approaches on the *NFPA 1500* standard rules of engagement (see above) and provided clear documented directives as follows:

Normally occupied buildings

- Highest level of risk taken to save savable life
- Acceptable level of risk to preserve savable property may be taken based on *NFPA 1500* rules of engagement

Vacant buildings

- Highest level of risk taken to save savable life
- Acceptable level of risk to preserve savable property may be taken based on *NFPA 1500* rules of engagement

Abandoned buildings

- Highest level of risk taken to save savable life
- No level of unacceptable risk may be taken to attempt to save abandoned property of little or no value based on *NFPA 1500* rules of engagement
- Defensive strategies shall be used to minimize risks and protect exposures
- Defensive strategies can be used transitionally to control fire from the exterior, followed by interior extinguishments and overhaul if structural and hazard conditions permit *safe entry*
- Interior attacks should not be initiated unless there is a *known life* in jeopardy or unless fire conditions are *incipient* or *minimal* and structural and hazard conditions permit safe entry

In forming these protocols the City of Flint Fire Chief stated that:

- The risk to fire department members is the most important factor considered by the incident commander in determining the strategy that will be employed in each situation.

- Fire-ground strategy decisions cannot be made due to peer pressure, tradition, public perception or any other non-safety related factor.

The fire department were basing their protocols on *NFPA 1500* 'rules of engagement' but, again, we come back to an undefined or controversial range of terms. These terms include:

1. 'Known' life risk
2. 'Savable' life
3. Highest level of risk
4. Acceptable level of risk
5. Structural and hazard conditions permitting safe entry
6. Incipient or minimal fire conditions

The definitions applied to some or all of these terms may be open to debate and individual interpretation, as well as legal argument.

It is worth noting that their new protocols opted to approach 'vacant' structures with the same level of commitment as normally 'occupied' buildings and this decision was based on the fact that vacant structures were still under ownership as opposed to 'abandoned' structures, which were not. Hence, there was a property conservation issue. Flint firefighters made 136 offensive attacks into 124 vacant/abandoned buildings during the survey period.

1.4.2 LIFE HAZARD

The City of Flint had two issues of civilian life safety at vacant structure fires during the survey period. The two incidents involving civilian life safety represented 1.1% of a total 179 'working' vacant structure fires.

In the first incident, firefighters arrived to find a vacant two-story house fully involved in fire. Defensive operations were initiated. Reports were then received from bystanders that a vagrant might be inside the building. Despite the appearance that the fire would be non-survivable for anyone inside, fire crews then made an **interior attack** into poor conditions. They encountered structural instability on the stairway and noted the fire was growing despite their suppression efforts. Crews were withdrawn from the building and defensive operations were resumed. Several hours after extinguishment, the remains of a civilian were discovered amidst collapsed debris in the basement.

In the second incident, firefighters arrived to find a vacant two-story house with fire emanating from one room on the second floor. No reports of persons trapped within the building were made to fire crews or the 911 center. Fire crews initiated an **interior attack** and successfully rescued two injured civilians, one conscious and one unconscious, who were found on the second floor and whose means of escape down the stairway had been blocked by the fire.

Considering that statistically, only one in every 100 vacant/abandoned structure fires in Flint involved occupants, it might be well argued there is a 'risk versus gain' reason to control interior offensive approaches in situations where occupant status is unknown. However, even though the 11.7 per 100 injury rate was dramatically higher in this type of occupancy, the injuries were normally very minor in nature and the risks to firefighters when operating offensively may therefore have been justified. Despite the fact there were no LODDs during the study, the potential for such a tragedy remains.

Deputy Assistant Chief John Norman (FDNY) reflects upon his own experiences[11] as follows:

As a young firefighter, I confess to having enjoyed the challenge of fires in vacant buildings. I regarded them as occasions where I could sharpen my skills and test myself without civilians being endangered. It was something like a trip to an amusement park, where I could experience all of the thrill and excitement without any of the distractions posed by concern for the occupants. This attitude was extremely common in the fire departments in which I served.

Then a string of tragedies occurred that started changing the firefighters' thinking. Probably none of them individually would have succeeded in effecting this change, but the combined weight of their loss awakened a number of the members. The death of a lieutenant; the crippling of two firefighters in a vacant building, followed rapidly by the death of a chief and severe injury to other firefighters at yet another vacant building; the narrow escape of two firefighters when a collapsing wall of an unoccupied building sheared the bucket off of their platform, carrying them to the ground – all of these incidents served to change the attitude of our members toward vacant buildings.

Now firefighters, at least in the New York area, display an attitude of caution when operating at vacants. They no longer rush headlong into aggressive interior attack. More often than not, they assume a defensive mode, using an outside stream in conjunction with a careful survey of the stability of the structure. The officers in command must exercise tight control over their subordinates to ensure that they don't unnecessarily expose themselves to dangerous conditions. Otherwise, the lessons that these firefighters paid for with their lives will have been wasted. The real shame is that the lesson has only been learned locally, for it is still common in some areas for (firefighter) casualties to occur in buildings that are in such poor condition that they were barely standing prior to the fire and shouldn't have been entered in the first place.

So now we might ask ourselves, are we honestly able to justify subjecting our firefighters to varying levels of risk or fire-ground hazard without addressing the management of such risk in a way that effectively balances the 'risk versus benefit' conundrum? Is it acceptable to account for risk on the basis that, 'We haven't suffered locally from any serious injury or LODD in 40 years so we must be doing something right'?

- Even though there was life risk at only one in 100 fires (above) can we justify applying the same SOP to both occupied and vacant buildings?
- Are fires in vacant or abandoned buildings presenting a greater risk to our firefighters than normally occupied structures?
- Is it possible that fires in such buildings will burn through more rapidly due to a lack of compartmentation and removal of window glass?
- Are firefighters more likely to become disoriented in structures where windows are boarded up?
- Should we consider establishing – and might we be legally liable in doing so – a more defensive approach for vacant buildings even though there is savable property?

11. Norman, J., *Fire Officer's Handbook of Tactics*, Fire Engineering/Penwell Publishing

- If there is no 'known' life hazard at a vacant or abandoned building, can we justify an interior search without implementing additional Risk Control Measures?

Effective risk management of the fire-ground relies on recognizing the risks, grading the risks, and monitoring the risks, but perhaps most importantly the implementation of **Risk Control Measures**. Where we grade risk in proportion to the potential benefits or gains and take remedial actions to lessen the risk, surely we are taking the logical course in protecting our own safety.

Take for example the argument that all buildings must be considered occupied until searched. This is a belief held by many of the most aggressive fire departments,

Life Priority	Life Hazard	Control Measures
HIGH (Known)	• Occupants **seen** at windows or on balconies etc., or very close to exit • Confirmed reports of occupants seen, heard, or known inside	Immediate rescue attempt required; isolate or attack the fire where possible; consider Vent-Enter-Search (VES) tactics (see Chapter Two)
MEDIUM (Suspected)	• Potential occupants believed possibly involved because of the time of day; the type of occupancy; insecurity of the building etc. • Hotels or large residential homes	Isolate or attack the fire before interior search and rescue is attempted
LOW (No Reason to Suspect)	• Secure residences during wakeful hours • Large volume structures without reported occupants • Vacant (unoccupied) buildings	Search team to take a hose-line as protection
VERY LOW (Unlikely)	• Vacant or abandoned (dangerous) buildings	Search team to take a hose-line as protection; but consider a defensive operation where necessary

Fig. 1.1 – An example of grading risk and documenting directives (SOPs) to implement effective Risk Control Measures or strategy during interior search and rescue operations can be seen above. In adopting these priority levels for search and rescue assignments we can implement effective Risk Control Measures to deploy more safely, whilst effectively balancing 'risk versus gain'.

Note: In all cases, coordinate fire attack with search and rescue where on-scene staffing permits.

but is it in any way justifiable in a court of law? Some might argue from this very view that there is a legal onus upon the fire service to deploy for the purpose of interior search at every incident. In the case of vacant or derelict properties it is worth noting[12] that the International Association of Fire Chiefs (IAFC) supports the view that interior offensive operations should not be undertaken where there is a reasonable belief that the *structure is unoccupied*. This is somewhat in opposition to the common notion that the very same structures should be entered at any stage where there is reasonable belief that they *may be occupied*. However, legal arguments and case histories suggest that decisions to commit firefighters into dangerous and hostile environments must be made on a reliable assumption that the 'risk versus benefit' conundrum has been addressed in our tactics, and based on information known or reasonably believed at any particular time.

Would it not be more logical to suggest that where there is 'known' life risk our firefighters will do all they can to save life, but where life risk is only a possibility we should perhaps temper our approach in a more controlled manner. One way we can do this is by ensuring that unless there is a 'known' life risk, interior search is always undertaken with the direct and personal protection of a hose-line. This may slow operations, but in some situations we may come across trapped or downed occupants whilst advancing the line – as occurred in both of the recorded cases in Flint (see above). In occupied structures during the early hours, where we can reasonably suspect occupants may still be inside, interior search may justifiably take place ahead of, or on floors above, the primary attack hose-line, provided that sensible precautions are taken and there is no reasonable alternative. However, where there is no sound reason to 'know' or **strongly** 'suspect' a life risk, we should take greater care with our tactical approach and implement more effective Risk Control Measures, just as we would/should with the OSHA Two In/Two Out ruling[13].

1.5 COMMAND AND CONTROL

Whilst the origins of the Incident Command System (ICS) were being developed through California wild fires during the 1970s, Chief Alan Brunacini in Phoenix was developing the Fire-ground Command system (FGC). As both systems (ICS and FGC) matured and improved over the years, Chief Brunacini once again took the leadership role, along with others, to merge the ICS and FGC systems into model procedures and guidance for the fire service industry. One thing is clear, that ICS/FGC concepts have raised the standard of how emergency incidents may be managed and resolved safely and effectively and further demonstrated how lives may be saved through careful implementation.

It is absolutely critical that an efficient Incident Command System is implemented from the moment the first vehicle/firefighter arrives on scene. The tactical approach and deployment of crews that take place primarily within the first sixty seconds – and secondarily during the initial five minutes following arrival at an incident – generally lay the foundation for the outcome of the incident. Where critical decisions are made within those vital first few minutes following arrival on-scene, a chain of

12. International Association of Fire Chiefs, (2001), *The 10 Rules of Engagement for Structural Firefighting and the Acceptability of Risk*
13. OSHA 29 CFR 1910.134 (Occupational Safety & Health Administration USA)

events may unfold that is either to the advantage or disadvantage of the firefighting operation. Where inappropriate decisions have been made, a period of 'catch-up' may ensue. It is during this period that an organized command system, supported by a powerful culture of leadership, may be able to assert some redirection over the path this chain of events has laid. Without such a system in place, and without strong leaders, there may be utter chaos and a negative outcome!

A fire department may sincerely believe they have a command system in place. They have a rank structure, they have a system of staged response, they have a document that says 'Incident Command' as its heading. But if they don't have coordination, if they don't have command from the first unit arriving on-scene, if they don't have the most vital parts of the ICS in place, if they don't have trained and knowledgeable leaders and an effective system of communication, they will fail somewhere. If the 'leader' simply turns up on-scene, says, 'I am now in command,' and begins to shout orders at the top of his/her voice in a micro-managed style of operational management – moving around the fire-ground without any logical purpose, direction or clear objectives – then failure is inevitable.

Effective command relies on control. That's why we call it 'command and control,' for you may be in command but unless you are also in control, your system is destined to fail when tested under the most extreme circumstances. The effective control of a fire-scene can only come from an organized and disciplined structure of command that provides practical channels of communication. The concept of 'fail-safe' operations must also be inherent throughout the system to ensure that where things may go wrong, there is always a back-up to check and counter errors, or deal with changing circumstances.

There are countless examples of communication failure at fires and these have often led to fatalities. On occasions these failures are due to technology limitations or malfunctions. In other situations these breakdowns in communication are due to the human 'error chain'. There are many situations where **maydays** have been called over the fire-ground radio but were never heard by on-scene commanders or safety chiefs who were too busy shouting orders and taking care of 'business'. There have been situations where a critical message has been passed by radio to an incident commander by a chief's aides or field communications staff who were actually within view and walking distance of the IC. However, because this critically important message was never received or acknowledged (according to standard radio procedure), and the sender simply 'assumed' it had reached its intended target, it is clear and well documented that lives have been lost in this way on numerous occasions.

The **NIOSH**[14] **(USA)** five most common factors associated with firefighter deaths are:

1. Lack of incident command from first response onwards
2. Inadequate risk assessment
3. Lack of firefighter accountability or SCBA air management
4. Inadequate communication
5. Lack of adequate or effective SOPs

14. National Institute for Occupational Safety & Health (NIOSH), Firefighter Fatality Investigation & Prevention Program

Is it beginning to hit home yet? It should! We may sincerely believe we all have firefighter safety covered, and an effective ICS in place, but be honest with yourself – is that true?

- Who establishes command on the primary response, prior to a chief arriving?
- How do they establish and communicate a 'mode' of command?
- How does any subsequent transfer of command occur effectively?
- Do you frequently carry out 'table-top' exercises to test the ability of command functions to recover, where various peripheral events and situations might arise to throw a normal routine approach astray?

1.5.1 RISK ASSESSMENT OR SIZE-UP?

But we have always carried out a 'risk analysis' – it's called size-up!

It is a commonly held view that a fire-ground commander's responsibility to address 'risk assessment' is already part of his/her 'size-up'. It is something that we have always done! But is this correct? What does size-up consist of? Is size-up the same as risk assessment?

What is the primary purpose of a size-up? It is something that is, or should be, carried out by the incident commander on arrival. Whilst every firefighter should also be carrying out his/her own size-up from the moment they arrive on-scene, it is the IC's size-up we are analyzing here.

The primary objectives of the basic size-up are:

- Gather information – as much as possible
- Get a view of at least three sides of the structure if possible
- Complete a 360 degree walk around, where viable
- Form a plan, initiate a mode of attack and transmit a status message based on the following:
  1. Are the staffing and resources adequate at this stage?
  2. Occupancy type
  3. Structure (construction)
  4. Floors (number of)
  5. Dimensions (area)
  6. Fire involvement (percentage estimate)
  7. Life hazard (known)
  8. Deployments (rescue/fire attack/interior search/exposures)

The main objective of a size-up is to answer the questions, 'How can I most effectively deploy my forces to achieve the objectives of life and property protection?' and 'Have I got sufficient resources on-scene?' In contrast, the purpose of a risk assessment is to establish the level and types of exposure to risk that personnel may encounter, and to decide how these hazards might be managed, controlled, prevented or *balanced against the potential for gains*. There is undoubtedly some crossover here but, answer this – Can you complete a size-up without addressing the risk factors? Yes of course! In fact, that is very common indeed. If firefighters are climbing ladders or operating in potential hazard zones without full PPE or SCBA, you may have sized-up the fire effectively but failed to address their exposure to risk!

In his book[15], Chief Michael Terpak of Jersey City USA refers to an acronym – **COAL TWAS WEALTHS,** which serves as a useful reminder of how to undertake a very advanced size-up. In this version of a size-up we can see how a more complex analysis of a structural fire situation might be reviewed during the first five minutes of an incident.

Fire-ground SIZE-UP	Fire-ground RISK ASSESSMENT
Construction	
Occupancy	
Apparatus and staffing	
Life hazard	
Terrain	
Water supply	
Auxiliary appliances and aides	
Street conditions	
Weather	
Exposures	
Area	
Location and extent of fire	
Time	
Height	
Special considerations	

Fig. 1.2 – Chief Terpak's Acronym COAL TWAS WEALTHS is representative of a very detailed size-up. It took an entire book to explain it and yet this analysis of a fire-scene needs to be undertaken during the first few seconds following arrival.

Here's another size-up acronym[16] – **WALLACE WAS HOT:**

• Water
• Area

15. Terpak, M.A., (2002), *Fireground Size-up*, Penwell Corporation, USA
16. Montgomery County FRS In-Service Training Program, Fire-ground Decision Making

- Life hazard
- Location
- Apparatus
- Construction
- Exposures
- Weather
- Auxiliary appliances
- Special matters
- Height
- Occupancy
- Time

As an exercise, see if you can pick out what's missing (above) in terms of risk assessment. Or, add in where you can, how risk assessment needs addressing. For example, there is no mention of building utilities (electric/gas supply etc.) in the above lists. This is a risk assessment issue: find them and shut them down where needed. This is an example of recognizing a risk and managing the hazard by removing it as far as possible.

Remember:

- Establish what the risks are
- Select a safe system of work (**mode of attack**)
- Implement Risk Control Measures
- Monitor the dynamic processes on the fire-ground
- Are the risks proportional to the benefits or gains?

As well as the following examples of Risk Control Measures:

- Routine evaluation of risk in all situations
- Well-defined strategic options
- Standard Operating Procedures
- Effective training
- Accountability and SCBA air management
- Full protective clothing ensemble and equipment
- Effective incident management and communications
- Safety procedures and safety officers
- Back-up crews for interior attack
- Back-up crews for rapid intervention
- Covering hose-lines
- Adequate resources
- Rest and rehabilitation
- Regular evaluation of changing conditions
- Experience based on previous incidents and critiques

1.5.2 MODE OF ATTACK
Possibly one of the most critical decisions made by an incident commander is that of mode of attack: deciding from the outset whether on-scene staffing and resources will allow you to implement that aggressive interior attack, or to go 'defensive' from the exterior, protecting the most threatened exposures or surrounding the fire. This is a clear strategic decision that may reflect a risk-based approach. The priority is

to ensure a safe 'system of work' is selected, based on the staffing and resources immediately available. Is there a 'known' or 'reasonably suspected' life hazard? Is the fire simply compartmental or has it breached structural boundaries? What is the level of fire involvement? What is the potential fire load? What are the indicators for smoke build-up and transport to voids and other areas far removed from the fire? Can this be safely removed? Can you assure 'interior attack' or 'search team' security? These are all questions that have an element of firefighter safety attached.

An incident commander has a wide span of discretionary authority for making risk management decisions. A strategic plan must not needlessly place the lives of firefighters or emergency responders in danger, but it should not be so over-cautious that it allows a fire to destroy property that could be saved – or keeps other valuable functions from being performed. The ultimate test of a risk management decision is whether or not a reasonable, well-informed person would find the decision appropriate under the circumstances.

Safety officer
The role of an incident safety officer does not relieve an incident commander of the responsibility for managing risk at an incident. By the same token, an incident commander should be able to rely on the incident safety officer to provide a balancing perspective on the situation. An incident commander should look at a situation as: **'How to get the job done and operate safely.'** The incident safety officer should look at the situation as: **'How to operate safely and still get the job done.'**[17]

1.5.3 MODES OF COMMAND
NFPA 1561[18] states that the incident commander shall be responsible for the overall coordination and direction of all activities at an incident and that this role should be clearly assigned through SOPs, from the beginning of operations, at the scene of each incident. Following the initial stages of an incident the IC shall establish a stationary command post. To effectively coordinate and direct firefighting operations on the scene, it is essential that *adequate staff* is available for immediate response to ensure that the incident commander is not required to become involved in firefighting efforts.

According to many common Fire-ground Command systems, an initial arriving company officer must assume either a mobile command or a stationary command.

A. Mobile command

1. *Nothing Showing Mode:* These situations generally require investigation by the first arriving unit. The officer can go with his/her company to check while utilizing a portable radio to maintain mobile command.
2. *Fast Attack Mode:* Circumstances which call for immediate action to stabilize the situation – such as interior fires in residences, apartments, or small commercial occupancies – require that the officer quickly decide how to commit his/her company.

17. FEMA, (1996) *Risk Management Practices in the Fire Service*, FA-166, United States Fire Administration
18. National Fire Protection Association, NFPA 1561: Standard on Emergency Services Incident Management System

This mode should transition to one of the following as quickly as possible:

- Situation is stabilized; or
- Situation is not stabilized and the officer/company withdraws to set up a command post, or transfers command to an arriving company or chief officer who shall establish a stationary command.

If a company officer assumes mobile command and elects to join his/her company in action/investigation, he/she should announce to alarm, ?Command will be operating in the mobile command mode.'

Whenever the mobile command mode is chosen, it should be concluded as soon as possible with one of the following outcomes:

- The situation is quickly stabilized by the initial offensive attack or the pre-liminary investigation reveals no problem requiring the incident commander's active participation. In either case, the company officer should then return to a fixed command location and continue to discharge his/her command responsibilities; or
- The situation is not likely to be quickly stabilized, or initial investigations indicate possible long-term involvement. The company officer should recognize these situations and assign command of his/her company to a company member or another company officer, return to a fixed command location and continue to function as the incident commander until relieved of this responsibility ; or
- Command is passed to the next arriving company or officer.

Note: The 'passing of command' must occur only once during any given incident and should not be passed to an officer who is not yet on the scene.

B. Stationary (or fixed) command

Incidents that require stationary command are situations that by virtue of the size, complexity, or potential of the incident, require strong and direct overall command from the outset. In such cases, the officer will initially assume a command position (exterior command post) and maintain that position until relieved by a ranking officer.

This should not preclude the option of the first arriving company officer having another company officer arriving with him/her and taking command. This may be by pre-arrangement or may be necessitated by circumstances; in either case it must be confirmed by both parties via radio. Command should not be transferred to an officer who is not yet on the scene.

If a first arriving company officer assumes command and elects not to join his/her company in action, the officer may operate within the following options with regard to the assignment of his/her crew:

- The officer may assign a 'move up' within his/her company and place the company into action with the personnel available. The individual and collective experience and capability of the crew will regulate this action.
- The officer might assign company members to perform non-hazard zone functions such as reconnaissance or intelligence gathering.

- The officer might assign company members to another company to work under the direction of the officer of that company. In such cases, the officer **must** communicate with the receiving officer and **confirm** the assignment of personnel.

While the company officer assuming command has a choice of modes and degrees of personnel involvement in the attack, that officer continues to be fully responsible for the identified tasks assigned to the command function.

In all cases, the initiative and judgment of the officer are of great importance. The command modes identified are not strict rules, but general guidelines to assist the officer in planning his/her actions.

Command modes (summary)

1. **Nothing showing (mobile mode)**
   - Investigate situation
   - Hold/stage other companies
2. **Fast attack (mobile mode)**
   - Command on the move with crews
   - Assign crew command to experienced member, or
   - Pass command to another company officer on-scene
3. **Command mode (stationary or fixed mode)**
   - Assume command position at exterior command post

Passing command

In some SOPs, fire authorities allow for the initial (first response) command responsibility to be 'passed' over to another on-scene company officer. This may occur in situations where the first company officer is deeply involved in a life safety issue, for example, and is unable to effectively take command. In this situation the command is 'passed' by radio but this passing of command is only allowed to occur once during any incident and then, only under extenuating circumstances. The passing of command is not to be confused with the 'transfer' of command.

Transfer of command

NFPA 1561 states that Standard Operating Procedures shall define the circumstances and procedures for transferring command as well as to whom any such command will be transferred. As an incident becomes larger or more complex, the transfer of command has historically been one of the most dangerous phases of incident management. A briefing that captures all essential information for continuing effective command of the incident and provides for firefighter and public safety must occur *prior* to transfer of command. This information should be recorded and displayed for easy retrieval and subsequent briefings (command board or even audio taped).

During the transfer of command, the following information should be handed over and acted upon:

- Assume command
- Confirm existing tactics and tactical priorities (strategic plan)
- Confirm the tactical mode as 'offensive' or 'defensive'

- Ensure 'safety' is reviewed or assigned as a command function
- Review whether the resources on-scene are adequate for the needs
- Ensure communications to all parts of the fire-ground are effective
- Ensure adequate situation reports have (a) been received from crews and (b) transmitted to alarm control
- Evaluate accountability, air management and rehab requirements
- Establish effective spans-of-control

Past experience has demonstrated that in some situations – normally in metropolitan areas where command staff are geographically closely spaced – where an incident is escalating rapidly during the initial stages, there may be several transfers of command initiated as additional alarms are struck. These transfers of command are often hurried and prevent any opportunity for a commander to actually take effective control. It has been common to see anything up to three or four transfers of command within the first fifteen minutes of an incident and during this period the operational effects of command suffers badly. With this in mind it is now recognized that an incident management 'team' approach serves the need for a more fluid transition in the command function as additional chiefs arrive on-scene.

Incident Advisory Team
NFPA 1561[19] effectively deals with this approach by introducing the Incident Advisory Team (IAT) concept. An IAT consists of three individuals (preferably command officers) located at the strategic level of the incident management system that have specific roles and responsibilities for the management of a fire (or other major incident). The Incident Advisory Team consists of:

- Incident commander
- Support advisor
- Incident advisor

In general (see *NFPA 1561* for guidance) the roles of support and incident advisors are to assist and mentor the IC in his/her role. Additionally there are specific responsibilities assigned to each role. The support advisor is more tactical, reviewing the strategy employed and assigning logistics and safety responsibilities. The incident advisor will liaise with other agencies, where necessary, and provide strategic support, but will not become involved at the tactical level. A local officer may well fill this role most effectively where possible.

An Incident Advisory Team is **not** incident management by committee. Each of the team members has a specific set of roles and responsibilities and the IC role is not necessarily adopted by the senior ranking chief in the team but rather the first arriving chief. The Incident Advisory Team process is designed to increase the effectiveness of command and firefighter safety during the most critical stages of the incident. This 'front-end loading' of the command organization allows the team to effectively manage the first hour of an incident, which is statistically the most dangerous period for firefighters. It is also the most critical time for decision-making and it is almost impossible to recover from poor operations on the front-end of an

---

19. National Fire Protection Association, NFPA 1561: Standard on Emergency Services Incident Management System, (2008 Version)

incident. Accountability within the team should apply to all three members but ultimate responsibility should lie with the IC, even though he/she may not be the senior ranking officer.

Advantages of the Incident Advisory Team concept

- Fewer transfers in command during critical stages of an incident
- Ideal for local domestic incidents
- Three officers are better than one
- New command officers learn more quickly and effectively
- Strong command presence during the critical first hour

The use of an Incident Advisory Team on the front-end of an incident allows for both the expansion of command organization and the continued focus on the incident tactics, strategy, and risk management assessments. The command staff always (and rightfully so) says that the system they are going to use on the 'Big One' should be the same system they use on a daily basis. If it is not the same system they will probably not use it when the 'Big One' happens. One of the many advantages of an Incident Advisory Team is that it transitions very smoothly from small-scale incidents to large incidents that require the use of a complete incident management system, while providing the incident commander with the support he or she needs.

Another distinct advantage is the ability of the system to allow a new command officer to manage an incident from start to finish. New command officers get to run major incidents with a support advisor sitting next to them providing guidance, experience, and expertise. The only reason to transfer command is to improve it. Any time command is transferred the overall operation loses vital information and previous planning efforts. The Incident Advisory Team process prevents this loss of important information and strengthens the role of command by adding support to command instead of transferring it to a ranking officer. This process clearly allows for better decision-making on the front-end and provides a safe and effective learning opportunity for young officers. Since the Phoenix Fire Department started using this process in the early 1990s, command has never been transferred from IC-2, except when the incident escalated and required a transition to a full incident management system.

Span-of-control

Span-of-control is perhaps the most fundamentally important management principle of ICS. It applies to the management of individual responsibilities and response resources. The objective is to limit the number of responsibilities being handled by, and the number of resources reporting directly to, an individual. ICS considers that any single person's span-of-control should be between three and seven, with five being ideal. In other words, one manager should have no more than seven people working under them at any given time.

When span-of-control problems arise around an individual's ability to meet responsibilities, they can be addressed by expanding the organization in a modular fashion. This can be accomplished in a variety of ways. An incident commander can delegate responsibilities to a deputy and/or activate members of the command staff. Members of the command staff can delegate responsibilities to assistants, etc.

There may be exceptions, usually in lower-risk assignments or where resources work in close proximity to each other.

The four levels of incident command

- Unified command level – large-scale incidents (UK **Gold** Command)
- Strategic level – incident command (UK **Silver** Command)
- Tactical level – sector command (UK **Bronze** Command)
- Task (operational) level – company command

The task and tactical levels of incident command are basically assigned to company and sector commands respectively and the IC, or IAT normally undertakes the strategic level of command. However, in the UK system the **Bronze** Command is assigned to sector management, coordinating companies at the operational level, whilst the **Silver** Command position will coordinate the sector commands, functioning as the IC. The UK **Gold** Command level represents unified command status and is the most senior in the organization, rarely coming into play in pure fire service operations. However, it can often feature in multi-service operations such as major incidents, large-scale civil disorder, wide area flooding or other protracted and serious incidents, broadly similar to a US commissioner's role. Whereas Gold does not directly influence operations on the ground, at the tactical or Silver level, it can often involve political considerations and policy level decisions that extend beyond a single organization. Gold, or strategic command is invariably exercised at a distance from the scene of the incident. It is intended to take the longer view of the situation; the time frame of Gold, or strategic command, is normally in days rather than hours or minutes. Each of the primary agencies will have pre-designated and trained Gold commanders.

1.6 CREW RESOURCE MANAGEMENT (CRM) – THE ERROR CHAIN

On December 28, 1978 United Airlines Flight #173 was traveling to a destination of Portland, Oregon[20]. The crew that day consisted of a pilot, first officer and flight engineer to handle the operations of the DC-8 aircraft. The journey along the way was uneventful and routine, until the plane was made ready for landing. Instead of the usual '3 down and green' (landing gear down and locked into place) indication, the nose gear green light did not illuminate. The following series of events that occurred in the cockpit was unbelievable.

The captain radioed the landing gear problem to the air traffic controller and requested to remain aloft, circling in a holding pattern to buy more time to resolve the issue with the landing gear light problem. The captain went through the checklists and procedures to ensure that all steps were properly taken to prepare for landing. However, the indicator light still showed red (nose gear not locked). The plane continued in a holding pattern around the Portland (Oregon) Airport until more trouble visits the cockpit. The plane had only 58 minutes of fuel remaining when they started circling and unbelievably ran empty with the airport six miles out. One by one the four mighty aircraft engines sputtered and flamed out from being fuel starved. Ironically, the flight engineer and the first officer had warned the pilot on several occasions that the fuel supply was running low without the proper action being taken by the captain. Because of miscommunications, lack of teamwork leadership, improper task allocations, and poor critical decision making, the (perfectly capable, but fuel starved) jet aircraft fell from the air, killing ten people and injuring 23 others. It was later discovered that

20. Rubin, D., (Chief), *Crew Resource Management 1/2/3*, www.firehouse.com

the nose landing gear operated correctly, but the 59 cents indicator green light bulb was at fault. In the wake of the needless death and destruction, the Crew Resource Management (CRM) program was developed and implemented by the commercial airline industry.

There are **five critical components** that comprise the basis of the Crew Resource Management program (CRM). These are:

- Communication under stress
- Teamwork
- Leadership
- Task allocation
- Critical decision making

The error chain

The 'error chain' is a concept that describes human error accidents as the result of a sequence of events that culminates in death or serious injury. Typically there is usually a chain of mistakes, or omissions, inactions, or failings, that all contribute to the final outcome. Often, none of these errors on their own are seen as a single overpowering cause in a tragedy but they combine as causal factors. It is generally the case that one of the five critical components of CRM (above) is at fault.

The links of these error chains are generally identifiable by means of ten 'clues' divided into *operational* and *human behavioral* factors. Recognizing and preventing any one link in the chain from failing, offers the potential to ensure the entire error chain remains intact, thus avoiding a situation where firefighter injuries or fatalities are likely to result. This entails a proactive analysis of where these 'links' might exist or evolve.

More than fifty fire-ground incidents where firefighter fatalities or serious injuries have occurred have been examined in developing and testing the concept of the error chain. Each case-study was examined from the following perspective: 'If this fire crew had been trained to recognize the links in the error chains that were present, would this knowledge have increased the probability of a different crew response and outcome (specifically to avoid the fatality or serious injury)?' In most of the events considered, the answer was, 'Yes'.

The fewest error chain links discovered in any one accident was four and the average number was seven. Yet, recognizing and responding to only one link may be all that is necessary to prevent a negative outcome.

Familiarizing firefighters with the concept of recognizing and eliminating the error chain can prevent an accident before it can occur. Much like our efforts in fire prevention, 'The best fire we can respond to is the one that we prevent.'

There are some critical clues to identifying links in the error chain. They are divided into:

- Operational factors
- Human behavior factors

The presence of any one factor (or more) does not mean that an accident will occur. Rather, it indicates rising risk levels in field operations and that firefighters and fire officers must maintain control through effective management of both risk and resources, in order to eliminate unsafe acts, unsafe conditions and unsafe behaviors.

Mission critical operational factors

1. Failure to meet competency
2. Failure to meet tactical objectives
3. Use of an undocumented/unauthorized procedure
4. Departure from Standard Operating Procedures (SOPs)
5. Violating limitations
6. Inadequate leadership
7. 'Tunnel vision'
8. Inadequate or inappropriate communication
9. Ambiguity/unresolved discrepancies
10. Confusion or empty feeling
11. Belief of invulnerability

1. Failure to meet competency
A firefighter must meet established benchmarks in order for him/her to attain qualification as a firefighter. The training process is ongoing and performance and competence standards need to be clearly defined, but, more importantly, fit for the purpose. There is no point in training firefighters how to fight high-rise fires or other emergencies using simple computer simulations alone. Operational performance and effective service delivery is directly linked to effective and realistic training and practical experience.

2. Failure to meet tactical objectives
In terms of tactical objectives, the incident plan must set realistic and achievable targets. If your firefighters are failing to make headway against a fire then take a look at their flow-rate. If the fire is vented it will most likely burn hotter and the heat-release may exceed the capability at the nozzle. Either anti-ventilate or increase the flow-rate.

Another example of failing to meet tactical objectives might exist where an attack has been underway for at least ten minutes with no discernible reduction in fire volume or change in tactics. Many firefighter lives have been lost in this way! Analyze just why they are failing to make progress and evaluate their position in line with the structure's integrity. Just as risking firefighters' lives to search for 'possible' life risk, never is the risk versus benefit conundrum so prominent as when firefighters are inside a burning structure for some time without gaining some clear advantage.

3. Use of an undocumented/unauthorized procedure
The use of a procedure, or procedures, that is/are not prescribed in approved training manuals, or operational safe practices, to deal with abnormal or infrequent conditions.

Example: Some larger departments operate with informal procedures accepted on a localized basis. These 'local' procedures can become confusing if they are not documented and approved. The author experienced this during the 1990s whilst attending a large number of incidents in the London Underground railway (tube) system. Shortly after the 1987 King's Cross fire, where thirty-one people including the initial incident commander were killed, a 'local' procedure had been suggested, agreed and implemented on a trial basis between three West End fire stations. This entailed all personnel entering the tube network from street level taking SCBA with them (against written SOPs at the time). Several SCBA-equipped firefighters would then be stationed at various sub-levels to set up radio communication links. The procedure was written up (by the author) and presented for consideration as a departmental change to the SOP for underground railway incidents. However, confusion arose after several months of unofficially using this procedure, where various company commanders from different fire stations began to interpret this trial procedure in their own way, resulting in conflicting approaches. The original (local) procedure was eventually written into the established brigade-wide SOP and any confusion was therefore removed.

4. Departure from Standard Operating Procedures (SOPs)

In some situations this may be defined as 'freelancing'. Intentional or inadvertent departure from prescribed Standard Operating Procedures is often the first link in the accident chain. Well-defined SOPs are the result of a synergistic approach to problem solving with the influence of time removed. As a result, in different situations, Standard Operating Procedures represent an effective means of problem resolution without the sacrifice of time, which is often not available. We are not suggesting that SOPs will resolve all problems. However, following established procedures will typically facilitate safe and effective operations. Failure to follow SOPs constitutes a link in the error chain and is a significant indicator of rising risk. If your organization has SOPs, train on them. If you train on SOPs, use them. If you vary from established SOPs then the person responsible for any such deviation from procedure must be held accountable. This may be acceptable where sound reasoning is presented but each situation should be investigated, as the potential for a SOP update may exist.

Example: Failing to have an uninterrupted, dependable water supply on the initial response to a structural fire violates most (hopefully all) SOPs. Further, ensure the self-discipline needed to avoid complacent responses or nonchalant approaches to routine calls is embedded in your firefighters' psyche and regularly monitored at all times.

5. Violating limitations

Violation of defined operating limitations or specifications either intentionally or inadvertently – as prescribed by manufacturers, regulations, manuals, or specifications – opens the door wide for an accident. This 'link' includes equipment specifications, operation limitations, and local, state and federal regulations relating to the safe operation and use of all equipment.

6. Inadequate leadership

A failure to establish or assert command, or inadequate performance in leadership, is questionably the leading cause of firefighter LODDs. The ability to take control of a situation, to establish authority, to formulate a viable and achievable strategy, and to communicate effectively in both dispatch and receipt of messages, are critical functions of an incident commander. Furthermore, besides having an in-depth understanding of operating procedures and the technical aspects of fire-ground management, building construction, fire behavior, hazards and safety, an effective incident commander possesses the ability to immediately recognize a situation that is placing his/her firefighters in a dangerous position, and will implement instant actions on the fire-ground to ensure their safety.

7. 'Tunnel vision'

It is often very easy to lose sight of changing conditions. This may occur where an IC is overloaded with tasks, or where he/she has established an incident plan but misses vital information that might cause this plan to be altered. It is essential to search out this information – which might be as critical as occupants who have escaped to the street whilst firefighters are being sent in to search for them, initial reports of a fire on the seventeenth floor being changed to the sixteenth floor, or cracks appearing in a wall of the structure. It is essential that on-scene intelligence is gathered and any amendments to the fire-ground plan are made immediately as practical. Take time to step back and see the overall picture and gain as much early information as possible within a minimal time-scale.

8. Inadequate or inappropriate communication

Failure to communicate effectively, or failure to communicate at all, are common problems at fires that may lead to a break in the error chain. Sometimes this will be down to technology failure but in most cases is due to typical human error!

- Establish what is critical information
- Establish who needs this information
- If you are unsure at least tell somebody!
- Communicate this information in order of priority
- Communicate the information precisely but clearly
- Ensure the communication was received, or re-send it until certain

Example: If one firefighter withholds observations or knowledge of existing hazards from another fire crewmember, or the incident, sector or company commander, a link in the error chain exists. Complete and effective communications are a must if we are to eliminate firefighter death and injury.

9. Ambiguity/unresolved discrepancies

Ambiguity exists any time two or more independent sources of information do not agree. This can include observations, radio reports, people, training manuals, SOPs, senses or expectations that do not correspond with existing conditions. This situation is often overlooked and reappears only after an accident occurs. Failure to resolve conflicts of opinion, information, or changes in conditions, or not raising issues that need to be brought to the attention of command or sector officers, generally has very negative consequences.

10. Confusion or empty feeling

A sense of uncertainty, anxiety, or bafflement (feeling clueless) about a particular situation. It may be the result of mentally falling behind the pace of operations, a lack of knowledge or experience. Perhaps it is caused by being pushed to the limits of one's training or operational capability or such physiological symptoms and effects as a throbbing temple, headache, stomach discomfort, 'gut feeling', or nervous habits.

Human factors researchers suggest that these signals are symptomatic of un-easiness and should be treated as indicators that all might not be right, leading to a potential accident.

Don't be afraid to ask for help.

11. Belief of invulnerability

Perhaps the most dangerous of human factors is the one of feeling that, 'I won't get hurt, that only happens to other people.' Whether willed into complacency by 'years of experience' of running into burning buildings or driven by the psychological effects of the adrenaline rush, this is a 'killer' feeling. This factor is oftentimes the foundation and precursor to additional error chain links which increase the risk and likelihood of a serious accident occurring. Individuals predisposed to this link are often prone to engaging in other high-risk off-duty activities, such as sky-diving and racing (vehicles, boats) among many others, which fit the widely held 'macho' perception and image of the firefighter.

1.7 THE SIXTEEN FIREFIGHTER LIFE SAFETY INITIATIVES[21]

An unprecedented gathering of the leadership of the American Fire Service occurred on 10–11 March 2004 when more than 200 individuals assembled in Tampa, Florida to focus on the troubling question of how to prevent line-of-duty deaths. Every year approximately 100 firefighters lose their lives in the line of duty in the United States; about one every eighty hours. The first ever National Firefighter Life Safety Summit was convened to bring the leadership of the fire service together for two days to focus all of their attention on this one critical concern. Every identifiable segment of the fire service was represented and participated in the process.

The National Fallen Firefighters Foundation hosted the summit as the first step in a major campaign. In cooperation with the United States Fire Administration, the foundation has established the objectives of reducing the fatality rate by 25% within five years and by 50% within ten years. The purpose of the summit was to produce an agenda of initiatives that must be addressed to reach those milestones and to gain the commitment of the fire service leadership to support and work toward their accomplishment.

The summit marks a significant milestone, because it is the first time that a major gathering has been organized to unite all segments of the fire service behind the common goal of reducing firefighter deaths. It provided an opportunity for all of the participants to focus on the problems, jointly identify the most important issues, agree upon a set of key initiatives, and develop the commitments and coalitions that are essential to move forward with their implementation.

The Sixteen Initiatives

- Define and advocate the need for a cultural change within the fire service relating to safety – incorporating leadership, management, supervision, accountability and personal responsibility;
- Enhance the personal and organizational accountability for health and safety throughout the fire service;
- Focus greater attention on the integration of risk management with incident management at all levels, including strategic, tactical and planning responsibilities;
- All firefighters must be empowered to stop unsafe practices;

21. National Fallen Firefighters Foundation USA

- Develop and implement national standards for training, qualifications, and certification (including regular re-certification) that are equally applicable to all firefighters based on the duties they are expected to perform;
- Develop and implement national medical and physical fitness standards that are equally applicable to all firefighters, based on the duties they are expected to perform;
- Create a national research agenda and data collection system that relates to the Sixteen Firefighter Life Safety Initiatives;
- Utilize available technology wherever it can produce higher levels of health and safety;
- Thoroughly investigate all firefighter fatalities, injuries, and near misses;
- Grant programs should support the implementation of safe practices and procedures and/or mandate safe practices as an eligibility requirement;
- National standards for emergency response policies and procedures should be developed and championed;
- National protocols for response to violent incidents should be developed and championed;
- Firefighters and their families must have access to counseling and psychological support;
- Public education must receive more resources and be championed as a critical fire and life safety program;
- Advocacy must be strengthened for the enforcement of codes and the installation of home fire sprinklers;
- Safety must be a primary consideration in the design of apparatus and equipment.

On 3–4 March 2007, a broad section of US Fire Service leadership gathered for the 2007 National Firefighter Life Safety Summit[22] in Novato, California, to continue to develop solutions to the ongoing problem of firefighter line-of-duty deaths, and by extension, firefighter line-of-duty injuries.

Here are some of most important recommendations affecting fire-ground operations:

- Define and advocate the need for a **cultural change** within the fire service relating to safety – incorporating leadership, management, supervision, accountability and personal responsibility;
- Enhance personal and organizational accountability for health and safety throughout the fire service;
- **Focus greater attention on the integration of risk management with incident management at all levels, including strategic, tactical, and planning responsibilities;**
- All firefighters must be empowered to stop unsafe practices;
- Develop and implement national standards for training, qualifications, and certification (including regular recertification) that are equally applicable to all firefighters based on the duties they are expected to perform;
- Develop and implement national medical and physical fitness standards that are equally applicable to all firefighters, based on the duties they are expected to perform;

22. National Fallen Firefighters Foundation, (2007), National Firefighter Life Safety Summit

- Create a national research agenda and data collection system that relates to the initiatives;
- Utilize available technology wherever it can produce higher levels of health and safety;
- Grant programs should support the implementation of safe practices and/or mandate safe practices as an eligibility requirement;
- Thoroughly investigate all firefighter fatalities, injuries, and near misses;
- National standards for emergency response policies and procedures should be developed and championed.

Chapter 2

Venting structures –
The reality

2.1 INTRODUCTION

1971 – One Fire – How far can you go?

0308 hours As we climbed the 50 ft wooden escape ladder the fire flashed over from a window below us. Flames came belching out onto the ladder, which had just transported us to the third floor level. My hands felt slightly 'crisp' but we had arrived safely. I watched out the window as the ladder was suddenly removed and wheeled away for urgent use elsewhere. The ladder itself was still on fire!

0311 hours There we all were, five of us huddled together on the filthy floor of a derelict abandoned structure. There was no glass in the windows, parts of the floors were missing; most of the doors were gone, but here we were in our little corner of London, all five firefighters huddled together on the floor at the third level, having been forced here by the fire. I can't even remember how these two crews came to be together without any water, but we were trapped and the fire was roaring up the stairway outside the room we occupied.

I remember shouting from the window for the urgent return of the 50 ft escape ladder that had originally brought us to this undesirable part of the world. The last time I had seen it, the burning wooden ladder was being moved around the side of the building to attempt a rescue of other trapped occupants. Hell they were welcome to it! We were trapped too and the fire was threatening us, but we were OK. In fact, as we sat facing each other, leaning against the walls of the room on this cool summer's eve, I felt we could have all had a quick hand of cards or something! There was no panic. We just sat and waited, taking turns at the window to shout for the ladder. Everyone in this room had been in this building before. It was our fourth fire here in a week, only this time the whole building was going up. We even knew the group of young squatters who occupied this rat-hole as they regularly bombarded us with buckets of urine each time we insisted on extinguishing the open fire they had constructed in the middle of a room downstairs.

I had joined this seasoned band of firefighters at Paddington (the busiest of London's 114 stations) on qualifying from London Fire Brigade's Southwark training school less than a year before this night's events. I had owed my place 'on the job' to one of life's great characters, my squad instructor Tom Stanton. He had nurtured me from a very young rookie who was terrified of heights, right through twelve weeks of gruelling training and hook ladder development. The hook (scaling) ladder was literally a short wooden ladder with a steel hook on one end, used by firefighters to reach points of difficult access in a structure. At the training school we would climb daily to heights in excess of 100 ft using a single ladder. There is no feeling in the world like this, hanging by nothing more than your fingers, placing your entire trust in the steel hook to hold you, 100 ft above ground!

I remember my arrival in Paddington, as a young eighteen year old, back in the winter of 1971. I remember striding up the steps of the Praed Street tube station (London having an amazing underground railway network known as 'the Tube') and out into the brisk winter morning air. It was my first day on the job and I was eager to get on that engine! As I walked towards the fire station I took a good look at the impressive buildings that surrounded me. They were unlike anything I had ever seen before and yet I had lived in south-east London all my life. The construction was deceptive for although the architectural façades from Queen Victoria's era were extremely well preserved, the interiors were often remnants from a period of repeated renovation and reconstruction. The once very grand buildings now housed a vast range of occupancies from tiny apartments to offices to hotels. The white or cream frontages were often decorated with dramatic pillars and other architectural effects, and there were sometimes interconnecting balconies at the lower levels. The rear of these buildings were never quite so grand in their appearance, and I could instantly see why the hook ladder was renowned for its use in some incredible rescues in this part of London (with narrow alleys and difficult structural recesses to access). But the area was rapidly changing to the north and west of the station's response area and this is where we were on this very night; Chippenham Road W2 to be exact!

In this sadly abandoned structure I could feel the radiated heat from the stairway on the side of my face. It really was beginning to 'roar' up the stairs and there were some anxious looks around the room. Finally the ladder came back to our position in what were probably just a few brief moments, but it seemed a lot longer. We were all grateful to abandon this situation and escape the clutches of a fire that was now threatening the stability of the entire structure. However, as we reached the ground a call came from the rear of the building for a hook ladder!

0315 hours I was straight off the escape ladder and up onto the roof of the engine to get the short steel-hooked ladder. All crews were committed to searching the last remaining uninvolved sections of the building for one of the 'squatters' who was still missing. We were at the back of the building within seconds and I went up with Dave Woodward. The absence of window glass made this an easy climb with the hook ladder. We had an escape line with us and were checking all the rooms at all levels as we arrived. The fire was pretty lively by now and was spreading into the rooms themselves from the stairway. We reached the top floor and then as we did so the room turned orange. Flames came belching out of the window and we made a hasty retreat back down to ground.

The squatter turned up later and thankfully had evacuated himself from the fire. Hey if my training instructor had seen all this he would have been proud! Tom, wherever you are – thanks for shedding my fears!

Words of wisdom from Those Gone Before Us

In 1992 I revived some interesting first-hand testimony[1] – from archaic dusty texts I had located in the British Museum – of 19th century (and later) fire chiefs who had led London's firefighters to understand the importance of maintaining control of the 'draft' (air-track) in a structure fire. These famous quotes were to become fashionable and were widely used by fire officers around the world to explain some of the lost art of firefighting. I was happy to have revealed these great men's words in such a way and I do so again throughout this text.

An American fire chief told me that he rubbed in the principle of ventilating by making his recruits extinguish a fire in a 'drill' building with all ventilation shut-off; they had a gruelling time of it. Then he gave them a similar fire with the building vented. They never forgot the lesson. **Ventilating must be done at the right time; air must not be encouraged to flow into a building until lines of hose are laid out and sufficient water is available.**

Chief Aylmer Firebrace CBE
London Fire Brigade 1938

1. Grimwood, P., (1992), *Fog Attack*, FMJ/DMG International Publications Ltd, Redhill, Surrey, UK

Firefighters in North America have, for decades, traditionally resorted to venting actions to open up structures in an attempt to release dangerous combustion products, smoke and heat from the interior. This tactical approach relies on well-trained crews operating under strict operational guidelines or protocols (SOPs), aligned with clear objectives. Such an approach generally results in a more intense fire as air also enters to feed the flames. Therefore, high-flow attack lines are needed in advance of venting actions if the firefighters are to remain in control of the situation. Success of venting actions relies heavily on sound precision, coordination and communication.

Somewhat in comparison, European firefighters (and many other nations) have formulated their strategy around lower-flow attack hose-lines operating into generally more solid construction. For example, 90% of UK building fires remain confined to the compartment or room of origin, whereas only 70% of US building fires remain within the original compartment before suppression is achieved, as firefighters continue to ventilate structures 'early and often'.

The author first noticed these differences in 1975, whilst working on detachment with the FDNY. The author later proposed that there was room for some middle ground; these two strategic approaches were so rigid in their implementation that they each failed to recognize situations where venting, or as an alternative – 'anti-venting' – presented the optimum approach to gaining some tactical advantage at fires.

In 1987 the author developed a unified strategy termed 'tactical ventilation', which was adopted universally in the UK and in many other parts of the world during the 1990s. It is based around a simple set of protocols and guidelines that are formulated on a risk-based approach. The tactical ventilation strategy reflects how changes in the *ventilation profile* – within a fire-involved compartment (or building) – are likely to influence fire development and affect interior conditions such as thermal balance, visibility, and convected and radiated heat levels. In simple terms, this 'middle-ground' approach is based around three areas of ventilation tactics, upon which we will broaden the discussion throughout the various chapters of this book:

- Incorrect location of vent opening
- Mistimed vent opening
- Inappropriate vent opening

2.2 US FIRE VENTILATION TACTICS

One of the major reasons that fires get out of control is the lack of proper and adequate ventilation . . . If you want to move in on a smoky fire, you must ventilate or you will be driven out. Yes, you can and should use masks to hold difficult positions. But most jobs [fires] will be readily controlled by good, fast ventilation and a crew determined to move in.

Deputy Chief Emanuel Fried
Fire Department of the City of New York
Fireground Tactics (1972)

Having worked at close quarters with South Bronx firefighters in the FDNY during an eighteen month detachment in the mid 1970s, I have seen how the US fire service operates. The South Bronx area during the late 1960s to mid 1970s was literally 'ablaze' with several working fires in almost every street, every night! The smell of smoke filled the air and a depressing foggy haze constantly hung over the southern part of the borough. The firefighters of this era had plenty of fire experience upon which to base their strategic approaches and it became clear that opening up structures by breaking windows and cutting holes in roofs was a daily and routine event. This action was taken to relieve interior smoke and heat conditions and assist firefighters in advancing inside the structure to rescue trapped occupants and suppress the fire.

It was certain that the high volumes of flaming combustion that emerged from exterior openings were something I had rarely seen during my five years of inner-city firefighting in London, preceding this assignment. The sight of such large structures alight on all floors reminded me of pictures and movies I had seen of bombings from the Second World War. Fire would come rolling out of multiple windows on all four faces of very large brick-built structures and sometimes take the roof as well, prior to successful suppressive efforts being achieved.

Furthermore, as a volunteer firefighter on Long Island, New York, I was trained and regularly deployed to open up and ventilate buildings by breaking windows and cutting holes in roofs at almost every working structure fire we attended. The training manual stated *'vent early and vent often'* and this was the creed by which we worked. I have to say, unlike the FDNY, the volunteers lacked the fire experience of their inner-city brothers and most efforts to ventilate structures seemed uncoordinated, imprecise and inappropriate.

I questioned the sense in opening up structures in this way; it appeared the buildings often suffered badly through the sheer extent of fire spread, as air flowed in freely to enrich the flaming combustion. However, I was soon to learn that New York City construction differed internally when compared to structures common in London, with small attic spaces termed 'cocklofts' and structural voids frequently located within buildings. Such structural features allowed fire to travel upwards to the roof with great speed, before mushrooming across to take the entire roof, then moving back downwards as the fire began to devour floor by floor. The large open staircases that were so common in tenements also exemplified how large numbers of people were often trapped by smoke and heat mushrooming into upper floors. A simple venting action at the head of the stairs, where roof teams removed or opened skylights, quickly relieved interior conditions and enabled the majority of occupants to escape unaided. I also witnessed some great roof operations where roof cuts of varying types and purpose undoubtedly saved structures from more severe fire damage, confining the fire to specific wings or parts of a structure.

On a wider note, my overall experiences of the ventilation strategy – whilst working on a series of lengthy detachments to fire departments across the USA – demonstrated that a large number of firefighters would break out windows blindly, with no apparent intent, direction or purpose. Even so, I acknowledged that the general concept of opening up buildings under specific circumstances would reap great rewards, providing the strategy was applied with a clear purpose, or intent and that the actions of firefighters were organized, disciplined and controlled.

2.3 EUROPEAN FIRE ZONING TACTICS

On the first discovery of fire, it is of the utmost consequence to shut, and keep shut all doors, windows, or other openings. It may often be observed, after a house has been on fire, that one floor is comparatively untouched, while those above and below are nearly burned out, this arises from a door on that particular floor having been shut, and the draught [air-track] directed elsewhere.

Superintendent James Braidwood
London Fire Engine Establishment
Fire Prevention and Extinction (1866)

In comparison to the US firefighters' strategy of opening up structures, European firefighters will practice limited, but highly controlled, ventilation tactics. The starting point will normally see fire 'closed down' or 'bottled up' in an effort to maintain control over any potential for rapid fire spread. There are major benefits in this strategy where the construction type is solid and where available flow-rates may be restricted. This strategy is termed *'anti-ventilation'* and has been practiced for many decades by London firefighters (for example) with some great successes.

There are specific 'zoning' tactics used in some areas to 'zone down' structures into manageable compartments. By initially closing doors, keeping windows closed, and restricting entry points to a minimum, the fire is located and managed; fire-free zones and approach routes are cleared of accumulated smoke. For example, if the fire is in one room on the first (ground floor) *with occupants reported trapped on upper levels*, the fire can be contained by closing the fire compartment down and utilizing Positive Pressure Ventilation (PPV)[2] to clear approach routes, stair-shafts and upper floor areas. This tactical approach normally works well because the fire subsides to a manageable stage of development as it becomes under-ventilated. The strategy relies on the stability of the compartment to hold the fire in-check whilst other parts of the accommodation are searched.

It is most certainly a strategy to be used by a crew attempting any kind of 'snatch rescue' where they are working ahead of, or, on occasions, above, the primary hose-line. Particularly in situations where it is known that occupants are within reach and where firefighters are committed to search before the primary hose-line is laid in and flowed, every attempt should be made to confine the fire to its room of origin.

It is common throughout US firefighter training texts to see statements such as, **'vent early and vent often'**, or **'ventilation saves lives'**. These texts go on to describe many distinct advantages that may be derived from breaking windows and cutting holes in roofs, early on during operations.

What these texts generally fail to emphasize are the clear disadvantages that may be experienced where venting actions are inappropriately or incorrectly applied. In fact most of the advantages (below) may also serve as reasons **not** to vent, as quite the opposite outcome may occur!

2. Note: At the time of writing, PPV has never been a strategy used by London firefighters. It has been used for some years, however, in several other major European cities, e.g. Paris, Manchester, Newcastle, Liverpool and Birmingham, to aid firefighting operations.

Ventilation contributes directly to accomplishment of basic firefighting objectives by:

- Reducing (**increasing**) danger to trapped occupants and extending time for rescue operations;
- Increasing (**reducing**) visibility thereby decreasing (**increasing**) danger inherent in other fire-ground operations and increasing (**reducing**) efficiency;
- Permitting quicker (**slower**) and easier (**more difficult**) access to allow either search operations or to advance lines;
- Minimizing (**increasing**) time required to locate seat of fire;
- Minimizing (**increasing**) time required to locate areas into which the fire has spread;
- Reducing (**increasing**) or preventing spread of fire through mushrooming or thermal radiation;
- Reducing (**increasing**) the chances of flashover or backdraft.

For example, how many textbooks on ventilation by firefighters also discuss the hazards of **thermal runaway** leading to ventilation-induced flashover, or describe the dangers of 'auto-ignition' which might lead to an ignition of super-heated fire gases as they mix with incoming air? Additionally, what exactly does the statement, 'Ventilation must be coordinated with fire attack', mean in practical terms? How is that principle applied on the fire-ground? There is reliable data that demonstrates at least 25% of structure fires deteriorate (the area of fire-involvement increases) after the fire service arrive on-scene! Why is that? Is it because we open doors to enter the building and locate the fire, then leave them open for some seconds/minutes prior to getting water on the fire, allowing air to enter and enrich the atmosphere? It may be that we are initially involved elsewhere in search and rescue actions or exposure protection, unable to immediately deal with the main area of fire-involvement; or perhaps we are searching out a water supply. Ideally we should be responding with the operational objectives of ensuring that once we arrive at a fire, we will take immediate actions that prevent a situation worsening.

This culture of 'vent early and often' is certainly one to be discouraged without first applying some strict tactical decision making based on a thought process that is guided by clear fire-ground protocols. Ideally these protocols should exist in a documented tactical ventilation SOP.

Having said all that . . .! The author recalls during his own experience as an inner-city firefighter in London, that many building fires were desperately in need of some form of venting to enable an advance in on a stubborn fire, or to assist in locating a hidden fire. There were several instances of severe heat conditions building up in stair-shafts and rooms above the fire floor. Wherever this build-up of heat was contained, conditions were punishing and restrictive on any advance up the stairs to reach and search upper floors, even after the fire was under control. The author can attest to victims being discovered incapacitated by smoke on upper floors, fifteen to twenty minutes following arrival on-scene. These victims might have been viable rescues had they been reached earlier, and there is no doubt that well-timed venting actions at the roof would have assisted a more rapid advance to upper levels.

At another fire it took several wears of SCBA cylinders to locate a smoldering fire in pipe lagging in a large volume structure. In this situation thermal image cameras were unsuccessful in assisting firefighters to locate the fire and the only avenue open was to ventilate the structure. This was undertaken whilst firefighters remained outside, for fear of causing a backdraft or smoke explosion. Subsequently, the fire was located after much of the smoke had cleared from the structure.

An effective venting strategy demands that a fire department is adequately equipped, well staffed, well organized and properly trained to operate under strictly documented protocols. In London (as with most European fire brigades) it has never been the case that equipment, organization, training – nor an operational documented procedure – have existed to enable effective or viable venting operations to be carried out safely or effectively. The culture prevents an organized assault on a fire building where gaining access to key areas at a very early stage, following fire service response and arrival on-scene, is critical to any success in gaining a tactical advantage. Therefore, the anti-ventilation process still remains the dominant strategic approach to fires in Europe.

What is required is some middle ground to both approaches that recognizes situations where a building is best left 'closed', as well as other situations where the creation of vent outlets will greatly assist the overall firefighting and rescue operation.

2.4 ANTI-VENTILATION

The men of the fire brigade were taught to prevent, as much as possible, the access of air to the burning materials. What the open door of the ash-pit is to the furnace of a steam-boiler, the open street door is to the house on fire. In both cases the door gives vital air to the flames.

James Braidwood
Master of Fire Engines, Edinburgh Fire Engine Establishment
On the Construction of Fire-engines and Apparatus, the Training of Firemen and the Methods of Proceeding in Cases of Fire (1830)

Anti-ventilation is the confinement, or isolation, of the fire compartment (room or space) from other areas that may be occupied. We do this by zoning off the fire room simply by closing the door. Such actions prevent air flowing in to feed the fire but, perhaps more importantly, will greatly reduce the amounts of combustion products, smoke, heat and flame transporting throughout the structure. This may also serve to reduce the dangers associated with rapid fire escalation, flashover, backdraft and smoke explosions.

Anti-ventilation may be the optimum strategy where:

- A clear objective or reason to create an opening has yet to be identified;
- A fire is demonstrating 'under-ventilated' conditions;
- A charged primary hose-line is not yet in position to attack the fire;
- The location of vent openings may spread the fire into roof spaces;
- A ventilation-controlled fire might advance towards flashover; and
- The flow-rate at the nozzle is unlikely to deal with such escalation;
- A snatch rescue (interior search without attack line in position) is in progress;
- Wind is entering the A side of the structure (for example the entry doorway) but we need to vent the B, C, or D sides for Vent-Enter-Search (VES) – **Close the entry door as much as possible until all VES operations are completed.** *Remember to close doors or control their opening widths where they may be feeding air in to escalate a fire – fire isolating or containment actions may serve as life-saving tactics on their own!*

2.5 TACTICAL VENTILATION

It is not necessary that every fireman should be profoundly versed in the study of the atmosphere known as pneumatics; but as he has to constantly deal with such substances [air] it is absolutely indispensable that he should thoroughly understand certain principles by which he is able to control their use.

Chief Sir Eyre Massey Shaw (1876)
London Metropolitan Fire Brigade

If you make as many vent openings in a fire building as possible, is this likely to relieve conditions and stabilize the fire, or will it make things worse? Venting is the tactical approach adopted by many. However, if you shut every building up tightly and consider ventilation as an after-thought, once the fire is under control, are you doing any better?

Establishing a middle-ground approach to fire venting strategy and tactics requires acknowledgment that there are both advantages and disadvantages to be gained by 'opening up' structures in some situations and leaving them 'closed' in others.

The author's original definition of his 1991 strategy states:

*Tactical ventilation is the venting **or** containment actions by on-scene firefighters, used to take control from the outset of a fire's burning regime, in an effort to gain tactical advantage during interior structural firefighting operations.*

Such an approach demands a cultural change – traditional firefighting tactics might need some updating. However, the changes may be minimal, since any venting strategy should have a starting point from which to base the tactical approach and this should be **anti-ventilation** – at least until a risk-assessed 'size-up' has been made. That means we are able to maintain control of the tactical approach from the outset under a risk-based approach, ensuring certain Risk Control Measures have been implemented or addressed, prior to making vent openings. This is not a time-consuming or resource-reliant process and is merely down to careful assessment and effective tactical decision making by the incident commander.

Prior to creating **any** openings in a fire-involved structure, an incident commander, or firefighter, must consider the following points, upon which a clear set of protocols can be established:

- There must be a **primary purpose (objective)** in creating the vent.
- Under *who's directive* is this vent opening being made?
- Does it conform to their plan (strategy) as communicated?
- Which direction is the wind blowing and what likely influence will it have?
- Where is the fire located and what conditions are presenting?
- Where are the occupants (if any) most likely located?
- Where is the primary attack line located?
- Where are other known locations of firefighters on the interior?

The first four points are **primary** to any decision to ventilate and the second four points may be equally as **critical**. Without the answers you cannot safely ventilate and *without a primary objective in mind, you cannot justify any sound reason to vent.*

Tactical ventilation – Protocols

(a) We must begin all operations from an anti-ventilation stance
The incident commander must ensure that all personnel approach each fire from the outset with an anti-ventilation mindset. If openings in the structure are pre-existing, then consider closing them, but certainly do not consider creating additional openings until there is:

1. A clear purpose or an objective;
2. An order or pre-assignment to vent;
3. Confirmed coordination with interior operations and fire attack.

(b) There must be a primary purpose (objective) in making openings
This is a primary concern of all incident commanders, for without an objective, purpose or intention to create an opening, there can be no justified reason to do so in the first place.
Primary purposes (objectives) are:

- **Vent for LIFE**
- **Vent for FIRE**
- **Vent for SAFETY**

Therefore, if there are no occupants reported trapped or believed possibly involved, you cannot justify 'venting for life' as a primary. If the fire has yet to be located it would be difficult to base any venting action on the 'vent for fire' scenario. You might though, in this kind of situation, 'vent for safety'.

(c) Under who's directive is this vent opening being made?
Who has ordered a vent opening to be made? It's as simple as that! Without a directive to ventilate or make openings in the structure, any firefighter should refrain from doing so. However, this is not calling for micro-management of operations where every action has to be sanctioned first! The fire-ground is an extremely dynamic environment and effective firefighting operations rely to some extent on experienced people knowing what needs to be done and doing it. However, this philosophy should not extend into areas where a building's ventilation profile is likely to take on major changes simply because individual firefighters take it upon themselves to break out some windows for the sake of it. In effect, this is 'freelancing' at its very best!

There are some tactical assignments that are pre-written in documented SOPs, and the FDNY *Ladders 3* and *Ladders 4* documents are typical of a venting strategy where responsible directives to 'open up' are given to certain firefighters, operating under assignments, even before they leave the firehouse. This process of pre-assigned directives allows firefighters and company officers more freedom to ventilate in accordance with a pre-plan for specific types of premises. It relies on the fact that those assigned to enact pre-written directives are experienced and well versed in reading fire conditions. It further relies on a pre-plan that generally places the primary, secondary and support hose-lines in pre-determined positions, in accordance with procedural objectives in specific scenarios and structure types. The overall application of such a strategy remains coordinated through effective fire-ground communication.

(d) Does it conform to their plan (strategy) as communicated?

A typical scenario might go something like this:

Ladder 4 transmit on the fire-ground radio, 'You have ventilation.' But where do they have ventilation, and why do they have ventilation (to assist what goal or objective)? Is this a pre-assigned task-based response action? Was it coordinated? Maybe there was no directive or pre-assignment and this was purely a 'freelance' action, based on what the ladder company thought was needed?

How about this – arrive on-scene and members of the ladder company proceed to take out every window they can access from the exterior of the fire building. Does this sound familiar? Is this an effective strategy? Is a hose-line in place? Has this action been risk-assessed and have Risk Control Measures been put in place?

- If the chief calls for ventilation during a *pre-assigned task-based response* – where SOPs assign specific roles and duties to first and second responders – he knows where these assignments should be located in any particular situation: normally roof and rear or side of structure. He can call for a venting action to assist a specific task, or company assignment. In some instances, the responsibility will be with the individual who will coordinate the venting action directly with the company undertaking fire attack, search or rescue.
- If the directive to ventilate is not pre-assigned, but is rather a reactive response to fire conditions, the chief (incident commander) giving the directive should **communicate clearly** and **coordinate any venting action** knowing that:
  1. *It has been requested by an interior crew;*
  2. *All interior crews are aware it is going to occur;*
  3. *There is sufficient water on the fire; or the fire is/has been isolated;*
  4. *There is a clear purpose in creating the opening(s).*

The chief's communication should follow something like: 'We need ventilation on the D side[3] at second floor level.' Where firefighters operate without clear directives to undertake a venting action, they may be placing interior crews and remaining occupants in great danger.

2.6 VENTILATION PROFILE

The term 'ventilation profile' may be defined as the amount of air available within a compartment. A 'ventilation-controlled' fire occurs when there is not enough air to burn all the materials and contents therein. A 'fuel-controlled' fire is where there are adequate amounts of air to ensure almost complete combustion of the fuel occurs. In compartment fires a time may come when the air needed for the fire to continue to burn efficiently is not sufficient and the level of fire intensity may reduce, until additional air is provided.

For the same fire load and fuel availability, a fire's potential to grow is entirely dependent on how much air is available. The ventilation profile is therefore dependent on the number of openings that are created and the size of those openings. The more openings we create (or which exist), and the larger they become, the wider the ventilation profile will reach.

---

3. From an incident command point of view, various authorities will/may use a range of communicative designators to assign specific sides, areas or floors in a structure for the purposes of assigning tasks and command functions. One method is A, B, C and D sides – A being the front, and then B, C, D clockwise when facing the front of the structure.

Ventilation profiling can be seen through scientific calculation and computer models where it is shown that doors and windows that are only partially open will slow fire development, as opposed to the same openings becoming larger, where fire development will progress more rapidly. This is only true of room and building fires in a ventilation-controlled or under-ventilated state.

Most fires that progress beyond the incipient stage are ventilation controlled at the point where the fire department arrives. If the ventilation profile changes to increase ventilation, the fire can rapidly increase in intensity.

- *Appropriate ventilation can significantly improve conditions inside the building.*
- *Inappropriate or unplanned ventilation can adversely impact conditions and speed fire development.*
- *Anticipate the effect of changes to the ventilation profile.*

2.7 PRE-EXISTING VENTILATION

On arrival at a structure fire there may well be pre-existing ventilation openings. The front door to the structure (or interior doors) may be wide open where occupants have escaped, or there may be windows that have broken through heat. It is important to make an immediate decision in this situation. The incident commander must decide if and how such openings should be closed, with the intention of isolating the fire and slowing fire spread.

The simple closing of the exterior door(s) may be enough to hold a fire in check until an attack hose-line is ready to go or a water supply is established. Perhaps waiting a few short seconds for additional crewing to arrive on-scene is practical, particularly where the structure is almost certainly unoccupied. In these situations, the closing of a door might be critical in saving the building as well as surrounding exposures.

Where interior doors are open and air is feeding into the fire, consideration might be given to closing these doors in certain situations (where possible and safe to do so), in order to protect escape routes from upper floors or to initiate a defensive venting action of such routes prior to taking the fire.

As in all cases, try to locate the direction of the 'air-track' within the structure and ascertain where the air is feeding into the building and where combustion products are leaving. Then assess the potential for altering or restricting the air-track to your tactical advantage.

2.8 UNPLANNED VENTILATION

The term 'unplanned ventilation' refers to situations where windows break through heat whilst occupants or firefighters are inside the building, or where fire burns through the roof. This unplanned ventilation may have devastating effects on the speed of fire development, the level of fire intensity and the direction of the air-track. A wind particularly may play havoc with the fire and cause firefighters and occupants to be in greater peril. Where unplanned ventilation occurs, it is sometimes possible to reverse the air-track where it has developed into a dangerous state. For example, if the failing of an exterior window causes fire to head in the direction of advancing firefighters, it may be possible to close doors to reverse the air-track away from their point of advance *(see Figs 2.1 and 2.2)*.

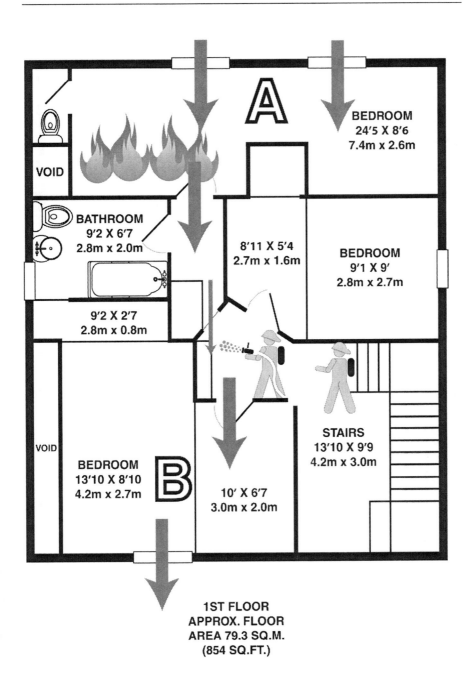

VOID

BATHROOM
9'2 X 6'7
2.8m x 2.0m

9'2 X 2'7
2.8m x 0.8m

VOID

BEDROOM
13'10 X 8'10
4.2m x 2.7m

B

A

8'11 X 5'4
2.7m x 1.6m

10' X 6'7
3.0m x 2.0m

BEDROOM
24'5 X 8'6
7.4m x 2.6m

BEDROOM
9'1 X 9'
2.8m x 2.7m

STAIRS
13'10 X 9'9
4.2m x 3.0m

1ST FLOOR
APPROX. FLOOR
AREA 79.3 SQ.M.
(854 SQ.FT.)

Fig. 2.1 – If the window at point B fails through heat in the overhead and becomes the 'outlet' vent, the air-track will travel from the air inlet at point A with fire and heat heading directly at the advancing firefighters.

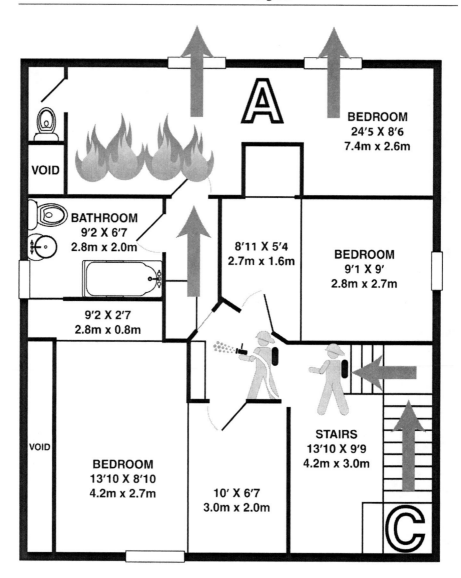

Fig. 2.2 – If the interior door to point B is closed and the entry door at point C remains open, point A most likely becomes the new outlet vent with point C providing the inflow to form the air-track.

2.9 OBJECTIVES OF VENTING

There has to be a clear purpose, or objective, for creating a vent opening. Before any such opening is made, consideration should be give to this objective, the likely effects of the opening in affecting the air-track or fire development, and what Risk Control Measures exist.

Tactical reasons or objectives for creating a vent opening:

- **Vent for LIFE**
- **Vent for FIRE**
- **Vent for SAFETY**

2.10 CONSIDERATIONS OF VENTING

An incident commander will need to consider:

- Should this fire be closed (anti-ventilated) at this stage?
- Do I know the fire's location?
- Are the stairways clear of smoke in mid-rise buildings?
- What is local strategy in this situation (SOP)?
- Do I have communication with all key personnel on-scene?
- Where are interior crews working?
- Where are likely/known occupant locations?
- Do I have an objective (purpose) to vent?
- Do I have a location to vent in mind?
- Do I have the staffing to vent in order of prioritizing tasks?
- Do I have the right equipment on-scene for the task?
- Where is the wind heading and at what velocity?
- How will the air-track be affected?
- Is my intended venting location behind or aside of the crews?
- How is the status of the entry doorway affecting the fire conditions?
- Where is this fire likely to be in fifteen minutes if we don't vent?
- Where is this fire likely to be in fifteen minutes if we do vent?
- Is there water on the fire at sufficient flow-rate?
- Is a back-up line being laid/crewed?
- Are all faces of the structure's exterior laddered?
- Remember B-SAHF (see Chapter Nine).

2.11 CREATING SAFE VENT OPENINGS

There are simple things to do before an opening is created. Sometimes, it's simply a shout up the stairs; other times it might need a Risk Control Measure put in place – such as a 'cover' hose-line – where any potential for exterior fire and radiant heat may cause problems.

- Don't ventilate where firefighters are on ladders above a window, unless a covering hose-line is in place and staffed below.
- Don't ventilate (open a door) onto a stairway, where occupants or firefighters may be located and vulnerable above – always clear the stairs first.
- Don't ventilate where an exposure problem may be created, unless a covering hose-line is in place.
- Vent with wind direction and velocity always in mind!

2.12 AIR-TRACK MANAGEMENT

The air-track is the 'point to point' route that is taken by air flowing into a structure and combustion products leaving the structure. Sometimes, a fully involved room on fire will have an open or failed window and yet there will be little or no smoke or flaming issuing. This may be because the window is serving as the air inlet point and flaming combustion, or smoke, is issuing at another point. In fluid dynamics, the term '**gravity** current' (or density current) is primarily a horizontal flow in a gravitational field that is driven by a density difference. Such flows may occur in air, water, snow, volcano lava, or in many other ways. A typical gravity current air-track in a fire structure sees the movements of cool air flowing in ('under-pressure') and hot air (smoke or flames) moving out in the 'over-pressure' area. Both flows are in opposition to each other but it is only at the interface of the two flows that they meet. In effect, we have air flowing in, and smoke flowing out, of the same opening.

It is useful to ascertain on arrival where the air-track (if in existence) is entering and leaving the structure. As information is relayed to the incident commander it then becomes apparent how the air-track might affect tactical objectives. The choices are to:

- Leave the air-track as it is;
- Alter the direction or velocity of the air-track;
- Use the air-track to assist application of water or compressed foam; or
- Close down the air-track.

Potentially, the first action may be to close down and control the air-track – this may have some stabilizing effect over fire conditions.

Positive effects of an air-track	Negative effects of an air-track
A moving smoke layer may develop in the overhead which will draw more air in on the 'under-pressure' and create better visibility at lower level.	The fire may develop rapidly, beyond the control of the flow-rate available at the nozzle.
Much needed air/oxygen may flow into areas occupied by trapped occupants.	The fire may advance into areas occupied by trapped victims or searching firefighters, or into structural voids.
Under-ventilated conditions may be reversed although the fire will most likely remain ventilation controlled.	Auto-ignition, thermal runaway (ventilation-induced flashover), or even backdraft may occur.

Fig. 2.3 – Both positive and negative effects may be derived from the existence of an air-track in a fire, and the fire commander must weigh up the benefits and disadvantages in each specific situation. Where no amount of water (or insufficient water) is flowing on the fire, it may be sensible to reduce the air-track or prevent it entirely.

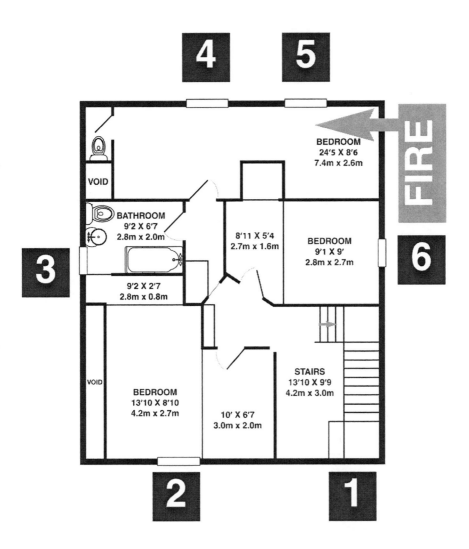

Fig. 2.4 – Air-track scenarios, with possible ventilation points at points 1-6.

Air-track profiling
Air-track profiling means assessing the various ways an air-track might form in a structure fire, demonstrating the 'point to point' pathway or route that the airflow might take, *from vent inlet to outlet.* It must be pointed out that not all fires will present an obvious air-track. In some cases, an air-track may not be in existence, even where there is sufficient air available from within the structure to allow a fire to develop and progress through the various growth stages.

- **Scenario A (1 to 1)** – Air flows in at the lower area of the entry doorway and heads towards the fire, whilst hot gases and smoke (or flame) head back towards the door, leaving the room or compartment at the upper area of the opening. In this scenario, smoke and heat will head directly towards the position occupied by advancing firefighters and subject them to varying amounts of radiant heat.
- **Scenario B (5 to 5)** – Air flows in at the lower area of an open (or broken) window located *near the fire* whilst smoke and heat (or flaming) leave the upper area of the same window. In this case, radiant heat from the air-track is mainly limited to the immediate area of the window and room of involvement.
- **Scenario C (3 to 5)** – Air flows in at the lower area of an open (or broken) window located *some way from the fire,* whilst hot gases and smoke (or flame) leave the upper area of the same window. In this case, a much larger area is exposed to radiant heat from the air-track, between the fire and the window.
- **Scenario D (1 to 4/5)** – Air flows in the entry doorway whilst smoke and heat (or flaming) leave the compartment/building via one or more windows. The greatest amount of radiant heat exists from and between the fire and the window(s).
- **Scenario E (5 to 2)** – Air flows in through an open or broken window whilst smoke and heat (or flaming) leave the compartment/building at another window – the two windows are some distance apart. In this case, the fire may be localized at a point somewhere between the two windows or it may involve the entire area. This may particularly occur where an exterior wind is directing the air-track, where a fire exists centrally between two points, or where the fire is on a lower level and an opening is created at a higher level.
- **Scenario F (one to avoid)** – This scenario is not related to the floor-plan in Fig. 2.4 but rather to a multi-level occupancy where the fire involves part of an occupancy on a lower floor (for example the fire is on the third floor of a twelve-story building). The air-track in this situation may occur in several ways but if the door(s) and pathway from the stair-shaft to the occupancy remain open, air will flow up and into the occupancy, feeding the fire. There are several possibilities here:
  1. The air-track is from stairs to fire to window outlet
  2. The air-track is from stairs to fire, back to stairs (windows intact)
  3. The air-track is from a window to the fire, back to a window

Any actions we might take that may alter this particular air-track, thus causing a negative pressure to occur behind advancing firefighters, may create a clear tactical disadvantage for the advancing fiefighters.

An example of this might be where we create an open path from stairs to occupancy with all doors open en route. If we add to this an open door in the stair-shaft at ground level and then create an opening at the head of the stairs on the roof (perhaps an automatic venting system will do this for us on detecting smoke in the stairway), we may see a sudden and massive air movement in the direction of the stair-shaft. It is this negative pressure, created by the updraft as air travels out of the roof vent, that can actually 'draw' or 'pull' fire out of the occupancy into access routes, corridors, hallways and adjacent areas.

At a fire in a Houston residential high-rise, one fire officer described this effect as follows:

They exited the apartment and headed down the hall, but a nasty thing happened when they opened the stairwell door, sources say. The stairwell acted like a ferocious maw, sucking heat and smoke down from the burning apartment. For Jahnke and Green the effect was overwhelming. The smoke grew thick as a blindfold; a torrent of hot air whirred past. The captains reportedly tried to beat a retreat by following their hose out of the apartment and down the hallway, a task made brutally complicated by the coiled, irregular pathway of their lifeline. The violent shift in the air current created high confusion by sucking the heat away from the fire. To Jahnke it seemed as if they were headed toward the fire, not away from it, as they followed the path of the hose, Hauck says.

This fire was a tragedy and Captain Jahnke lost his life. Another incident involved two London firefighters who also lost their lives while fighting a basement fire. Whilst firefighters were making their advance down into the fire, a stair-shaft was vented at the roof. This action created a reverse in the air-track that caused a sudden and intense development of the fire.

There are countless situations where stair-shaft venting actions have saved lives. There are also many instances where such actions have caused sudden reversal of the air-track, pulling fire out of an occupancy, and lives have been lost.

Therefore it is critical that we consider the following points:

- As much as possible, try to keep doors closed between the stair-shaft and fire-involved occupancy.
- Vent with an objective – as always!
- When venting the head of a stair-shaft in a building where fire-protecting lobbies are not constructed at each floor level, ensure that an adequate flow-rate is working on the fire and coordinate with the attack team.
- Close the roof vent if any sudden or unexpected reversal of the air-track occurs.
- The effect of stair-shaft ventilation requires openings to be made at both the top and bottom of the stairway.
- Where auto-smoke vents are fitted in a stair-shaft, good pre-planning will ensure that firefighters are aware of this arrangement. Be aware of any overriding facility that may exist to take control of sudden air-track reversal.

These are just a few examples of how air-tracks might form. The important points concerning air-tracks are:

- The point to point air-track is from **inlet** to **fire** to **outlet.**
- The air inlet may also serve as the outlet (may be the same window).
- The inlet and outlet may be the entry door.
- There may be radiant heat from the overhead, **between fire and outlet.**

- There may be more than one inlet/outlet.
- As further openings occur, the direction of the air-track may change.
- Any such change may be to the advantage or disadvantage of occupants or firefighters working the interior.
- We can sometimes take actions that will reverse the direction of the air-track to our advantage.
- We may sometimes take actions that will alter the direction of an existing air-track to our disadvantage!
- Air-tracks are greatly influenced by exterior wind and interior building pressures such as stair-shaft stack effects in tall buildings.
- The potential for an 'auto-ignition' of super-heated fire gases within a compartment is far greater in locations adjacent to vent inlets/outlets.

Take for example the air-track that exists in Scenario 1 where the inlet doorway serves as both inlet and outlet, with all other windows at this point intact. In this situation, if a ventilation-controlled fire is developing, the radiant heat is in the overhead along the approach route to the fire. By purposely creating an opening at Point 5, and possibly Point 4 (providing wind direction supports this), and by closing down the entry doorway, the direction of the air-track can be reversed, along with the radiant heat in the overhead.

This is how the principles of Positive Pressure Attack (PPA) work, by reversing or creating air-tracks to the tactical advantage of interior crews. The big learning point here is that any opening we might make should first be considered for its likely effect on any existing or potential air-track.

Momentum and inertia forces

As a confined fire develops to create a highly pressurized environment within a structure, there may be heavy amounts of hot fire gases and combustion products in smoke that are simply waiting to be released. There is a distinct possibility that where openings are made (particularly at high levels) to relieve highly pressurized smoke and heat, any such release of this pressure from a structure may initiate a **sudden** and **massive** movement in the air-track. If this occurs, the effect is one of great momentum and changes in inertia as the air-track rushes at the fire, possibly causing a sudden escalation leading to a forced flashover or backdraft.

On a windy day at home, where you leave a front or back door open, with another opening elsewhere, there may be a sudden loud closing (slamming) of an interior door as this massive air movement occurs. This effect may take some minutes, but as conditions are optimized for air entering and leaving the structure, the door will surely slam shut. This is similar to the massive air movements occurring in a structure fire. As there is a sudden air depression in a room where an opening is made, the momentum of the air-track leaving the structure may cause a greater depression at the air inlet point, which draws in even more air.

At a three-story house fire (at ground level) in Illinois, the firefighters working the interior (second floor) were calling for some ventilation. The street door to the structure was open, serving as the air inlet point. As crews in a tower ladder broke out the windows on the top floor there was a sudden and massive air movement up the stair-shaft. It was this sudden release of high-pressure smoke that caused a massive movement of air up the stairway, pulling fire up through the structure, and forcing several firefighters to jump from the upper floors.

2.13 SELECTING VENTILATION LOCATIONS

One of the many failings of tactical firefighting operations is the selection of inappropriate or incorrectly sited ventilation openings. The cause of this is often indiscriminate venting, or creating openings without an objective or plan. Thus, any opening made where an *objective* has been determined, but the location or the stage of development of the fire is unknown, should be discouraged. For example, interior crews may call for ventilation to reduce heat and lift the smoke layer in the area they are working. So where is the fire? And where should you open up? It is often the case that **fire and heat will head directly for any opening that is made** and if the fire's location is unknown, any vent made near to the firefighter's location may worsen conditions rather than improve them.

IFSTA (USA) recommends a number of factors that the officer in charge should consider when choosing a site for ventilation, including the following:

- The seat of the fire
- Roof construction type and condition
- Continuous observation of the roof
- Charged attack and protection lines ready
- Wind direction and intensity
- Coordination with attack companies operating inside
- Keeping track of elapsed time into the incident

As this book will convey over and over again, a common error that may lead to firefighter injuries, or even fatalities, is the failure to account for **wind direction and velocity** when selecting an opening point. At one fire, used as a case history in this book, firefighters were certain that the point of entry for the fire would be as follows – in this order and whatever the circumstances!

1. The front door (main entrance)
2. The non-fire side of the structure

In this case, their entry point suited both of the above requirements. However, they were entering into the leeward side of the structure. This meant that if any opening was created on the windward side, either through planned or unplanned ventilation, then the interior of the structure was likely to become untenable. Does it truly make sense to attempt to gain ground against a headwind in this way? In reality, the fire could have been more effectively approached from the 'fire side' via the 'rear' entrance! Even though there is a fear that the advancing hose-line and wind may drive the fire throughout the structure, **here is a definite opportunity to control the air-track** by closing the entry (rear) door.

Note: If wind is entering the A side of the building (entry doorway) and we need to **VES** the B, C or D sides, then we need to control/close the entry door as much as possible whilst this is occurring.

If wind is entering the A side (entry doorway) and openings are non-existent elsewhere in the structure, we either need to create one as near to the fire as possible, or maintain the entry doorway, closing it as far as possible to prevent wind influencing fire development. Would you Positive Pressure Ventilate (PPV) a fire building without having first created a vent outlet?

2.14 TIMING VENTILATION OPENINGS

Another serious failing when conducting ventilation operations at fires is known to be mistimed openings. How many times do we read that ventilation must be **coordinated** with the attack team? Yet, at the same time, in how many instances do we read LODD reports where this was not the case?

The critical factors again come back to having a **directive**, an objective, knowing the location of the fire, and knowing the direction and velocity of the wind; but, above all in this case, **communication** is the key. Fire commanders must assert control and make sure nobody creates a vent opening without having been given a clear directive or the responsibility to do such a thing. They must also await the request from an interior attack team before directing such an opening be made.

Let's consider a tactical example here: The primary attack line is advancing on the fire at ground floor level when a call for ventilation comes from the search team operating on the floor above them. All teams have entered on the A side with the entry street door remaining open; the post-flashover fire is located on the C to D corner on the ground floor level; the wind is heading B to D side. In this situation can the IC give a directive for a vent opening? If so where should this occur in the structure? Where is the likely air-track?[4]

Another example: The same situation as above but the primary attack hose-line is delayed. There is a search being undertaken on the floor above the ground floor fire on the C to D corner by two firefighters who call for a venting action. The wind is heading C to A. In this situation can the IC give a directive for a vent opening? If so where should this occur in the structure? Where is the likely air-track?[5]

Documented protocols (SOPs) are of course, also critical in this respect.

2.15 VENTING FOR LIFE (INCLUDING VES)

Openings made under this category are to clear escape routes of combustion products, to provide much needed air for trapped occupants, to enter under the VES concept, to create an outlet for PPV, and as an attempt to raise the smoke layer from the floor to assist firefighters searching etc.

One of the most effective openings that can be made in this respect is at the **head of a stairway** serving multiple occupancies on several floors. Wherever the fire is located in a structure, if combustion products, smoke, heat, gases or fire get into the stairway, they will travel up and then mushroom back down, cutting off the escape route for occupants and creating an extremely hot route for firefighters to ascend. The author can attest to several hotel and multi-occupancy fires in London where this occurred during the 1970s and 1980s. With so much heat in the stairway, it became almost impossible to ascend above the second or third floor without relieving the conditions, through some form of venting action. Quite often, this was

---

4. The coordination of search team and fire attack team are confirmed; the air-track is in existence from A side to D side; a venting action of the floor above the fire D side, followed by the B side, should be effective in clearing combustion products, fire gases and smoke from most of this area.

5. The coordination of search team and fire attack team are not confirmed; the air-track is in existence from C side to A side; it might be dangerous to create openings on the upper floor for fear of pulling fire up the stairs, trapping the firefighters. An immediate effort to get the primary line in place, protecting their means of escape, should be undertaken prior to any venting actions occurring. In this situation the wind direction is a major factor and even committing firefighters into such a situation is extremely hazardous.

easily achieved by the simple removal, or opening, of a roof access hatch over the stairway. However, this required some reactive tactical decision making by the incident commander that often came late in the incident, despite immediate access being available via aerial ladders, or other adjacent roofs. The author was concerned about the lack of thought being applied to such tactics and began a campaign in the 1980s to reverse this tactical failing that appeared common throughout the UK.

The concept of venting stairways from the roof was widely practiced by FDNY firefighters who have this written into their SOPs and the author learned this valuable lesson during his 1970s detachment from London to the New York Fire Department.

Vent-Enter-Search

Another valuable search tactic is that of Vent-Enter-Search (**VES**) which requires an outside vent firefighter to position, according to documented pre-assigned tasks (SOPs), either on the face of a building using a fixed metal fire escape, or the side or rear of a structure. This assignment's role was to provide ventilation (for fire) or to assess where there were access points (windows) that might lead to rooms near, above, or adjacent to the fire room where a quick entry might be made, and a search completed, before returning back to the relative safety of the access point.

On occasions, VES would be used in the fire room itself. Many times a ladder has been placed, or an outside vent man (OV) has worked from an exterior fire escape, to locate and enter a window serving the fire compartment itself. A quick venting action, followed by a quick entry through the window and a rapid search of areas near to the window, have enabled some dramatic rescues to be completed. Often, a baby lying in a cot has been pulled from the clutches of fire in this way.

Being successful at using the VES concept relies on the following considerations:

- The concept of VES must be written into SOPs.
- Firefighters should train for VES.
- It may be a pre-assigned task or a reactive decision.
- It must be communicated to all interior crews that this venting action is occurring and at which specific location (D side second floor etc.).
- If wind is entering the A side of the building (entry doorway) and we need to VES the B, C, or D sides, then we need to close the entry door as much as possible whilst this is occurring.
- Ideally, VES is undertaken by a minimum of two firefighters in full PPE and SCBA, with only one entering the room and the other remaining at the head of the ladder or outside the window being used.
- Close the door to this room (see below).
- The firefighter entering will make a quick sweep search of the room.
- On completing the search, return and exit via the entry point.
- Do NOT proceed further inside the structure to search other areas.
- Report back that the room has been searched and look for other potential VES points to repeat the process.

Some important points concerning VES:

- Look to spend a maximum of thirty seconds in the room, depending on fire conditions and occupant status as reported.

- Be certain to consider whether the creation of such an opening will fit in with the incident commander's plan for venting.
- On leaving the room, if the door was closed (it should be whilst you search, if you can get to it) do you then open it again? There is much debate on this. The author would argue against it. There is a counter-argument that this vent opening may be needed to assist firefighting operations and if the door has been closed, this will negate that opportunity. However, it is more likely that one, or a number of, VES opening(s) may serve to destabilize the structural fire conditions and allow for rapid and uncontrolled development and spread of fire. Therefore close the door (where possible) to protect your occupancy of the room and leave it closed on exit.

Note: PPV (PPA) in pre-attack mode and VES are not normally a viable combination of tactics. The location of a PPV outlet may not be the ideal location of a VES point, and vice versa. Where they are combined, strict monitoring, control and communication are critical.

In all venting for life situations, there must first be 'known' occupants or their presence must be considered most likely. In discussing 'vacant' and 'abandoned' structures, interior search operations for 'unlikely' occupants should be discouraged in a 'risk versus gain' balance. However, there might be a useful debate concerning the viability of VES in smaller occupancies of this nature, for where this approach is used safely and correctly, two-person teams may safely and effectively search 80% of the structure. This is done simply by taking rooms from the exterior and by not placing oneself in any unwarranted danger.

2.16 VENTING FOR FIRE

If you corner a rat it will most likely attack you! Where an attack hose-line crew is advancing against a confined fire it is almost certain that the fire, heat and water to vapour expansion will head right back in their direction. This will inevitably cause some discomfort and might even force them to withdraw the hose-line off the fire floor. In this situation it is logical to create a vent on the other side of the fire to allow an escape route for all the heat as they 'push in' on the fire.

Another 'vent for fire' situation might be that of a **trench** cut in a roof. In effect a strip of roof is removed or a one meter (3 ft minimum) trench is cut right across the roof, avoiding roof supports. This is done to prevent fire in a common roof void, attic or cockloft from spreading to involve several other properties. In New York, firefighters will often use this strategy to limit fire spread and protect sections of large structures. This is also a common approach to fires in row-frame (terraced) houses.

London's water-fogging and venting combination tactics

In the 1980s, during the London Fire Brigade's pilot research into Swedish firefighting tactics, the author developed the notion of combining 'venting' tactics with 'fogging' tactics. Termed **combination tactics**, the water fogging of hot gas layers was used to cool and 'inert' smoke prior to it being vented to the exterior. This 'fogging' action was very effective and prevented auto-ignitions of fire gases as they were vented. The tactical approach also reduced the chances of interior auto-ignitions occurring near the venting points.

Where an entry is made to a fire compartment using correct CFBT door entry techniques, several short 'pulses', or a couple of brief 'bursts' of 35–40 degree (cone) water-fog are often enough to cool the overhead and inert the gas layers, whilst maintaining thermal balance in the room. An exterior venting action of the window then takes place and the water vapour escapes within the smoke. This allows firefighters to advance into a safer environment to fully extinguish the fire.

2.17 VENTING FOR SAFETY

This approach is one that may offer several other options. Therefore it is one that may be used by some but not by others. Where a fire is tightly closed and harboring backdraft-like, or under-ventilated conditions, it may be a viable approach to release these hazardous conditions from the structure prior to committing firefighters. In commercial occupancies out of hours, or other situations where there is no 'known' or 'suspected' occupancy, the venting of such a building may cause the fire to show itself.

2.18 VENTING LARGE FLOOR SPACES

There is much practical experience as well as scientific research of venting large floor spaces. One thing is certain, fires are likely to develop very quickly in large areas where a heavy smoke layer forms at the ceiling. If the fire load is plentiful a fire will burn in a fuel-controlled state for far longer than where confined in small rooms. This means that a heavy build-up of combustion products will accumulate as a vast amount of radiant heat is transmitted to surrounding objects.

The routine tactical approach may often draw firefighters deep into a structure fire of this nature without them realizing the dangers. Flaming combustion may exist high above their heads, hidden in the smoke layers. Sometimes this fire may exist behind a large volume false ceiling. Due to the large floor expanse, modern lightweight construction may see some steel or wood trusses holding the roof up. Such trusses are likely to fail fairly quickly, within a few short minutes, once they become involved in flaming combustion.

Once they become well involved in fire, few structures of this nature will be savable. Unless the construction is solid, expect to lose the structure but save your firefighters.

Real fire experience has demonstrated over and again that pre-installed roof vents, based on fire test analyses and building codes, are rarely able to deal effectively with large accumulations of hot smoke and combustion products created by a heavy fire load. In some cases, a roof opening may cause such super-heated rich fire gases to auto-ignite either within the structure, and/or to the exterior, often burning freely with some ferocity.

The use of PPV has also been researched in large volume structures[6] where it was suggested that such use against a working fire was likely to jeopardize the safety and working conditions of firefighters by dropping a normally stable smoke layer to the floor.

6. Svensson, S., (2002), Report 1025, Lund University Sweden

Cross-ventilation tactics in large structures are generally only effective for very small fires where vast amounts of smoke are generated, for example in pipe lagging or similar. Again, in larger fires, the creation of horizontal ventilation openings may be counter-productive.

2.19 HORIZONTAL VENTILATION – THE GLASS RULES

Horizontal, or cross-ventilation is used as a means of removing dangerous combustion products including heat, smoke and fire gases from a fire building. It entails a firefighter, or crew, removing windows from the outer walls of a fire building either from an exterior, or sometimes interior, position.

What are the 'Glass Rules' that some fire departments lay down for their firefighters? These are protocols for cross-venting, SOGs or even SOPs, or sometimes they are unwritten rules. The Glass Rules relate to the horizontal openings we make in structures, usually by breaking out glass windows. They are basically a list of 'dos and don'ts'.

Some basic Glass Rules:

- Don't break glass until you are **directed** to do so*.
- If you are assigned a venting task, confirm the **location** and the **timing.**
- If the **wind** is strong at your back, check with the IC before you vent.
- Take out the entire window and clear all jagged edges.
- Make sure you clear the curtains, blinds, or any interior obstructions.
- Make sure you have full PPE and SCBA where necessary.
- If wind is entering the A side of the building (entry doorway) and we need to VES the B, C, or D sides, then we need to control/close the entry door as much as possible whilst this is occurring.

* *This may be a pre-assigned directive via SOP.*

2.20 VERTICAL VENTILATION – RESOURCE DEPENDENT

Vertical ventilation entails opening, or cutting into, a roof to release rising smoke and gases at the structure's highest point. This operation is often very successful in clearing interior escape stairways of smoke and heat, preventing mushrooming fire spread, reducing backdraft potential, preventing fire spread through common roof oids and accessing difficult attic and cockloft fires. The strategy is, however, fraught with hazards and many firefighters have been fatally wounded whilst undertaking such operations.

As time goes on, firefighters are expected to deliver an ever-increasing list of tasks or roles using the same, or even a reduction in, already limited resources. Many inner-city fire stations are closing and unit staffing is constantly being cut. It is certain that as additional responsibility to fulfil various needs or tasks on the fireground becomes necessary, where resources are limited these tasks sometimes move down in the order of prioritization and become a secondary response assignment rather than a primary response function. Things still get done, but now they sometimes get done a bit later in the operation.

One such task is that of vertical ventilation where several firefighters, and sometimes entire companies, are needed to undertake safe and effective cutting operations on roofs. This task has now moved down the line of prioritization in some areas as it has been affected by Rapid Intervention Team (RIT) duties or FAST truck (Firefighter Assist Search Team) assignments, for example, which will often be a primary response function.

2.21 POSITIVE PRESSURE VENTILATION (PPV)

The use of PPV to create a *forced draft 'point to point' air-track*, in order to clear a structure of smoke, is now a commonly used strategy in **post-fire** situations where the fire has been declared under control. In some (but not all) situations the fire may not be fully extinguished but a major knock-down of the fire has been achieved.

There are a wide range of PPV ventilators on the market with differing designs that may produce slightly different effects. The objective is to get a high amount of air forced into the structure, moving at a good velocity.

Two types of ventilator:

- Conventional air-stream
- Turbo air-stream

Whilst the larger fan-blade configuration of a 'conventional' air-stream produces a wider-spaced cone of air that appears to create a 'seal' around the entry point (door), the configuration of a greater amount of short, stubby blades of a 'turbo' air-stream will form a faster moving narrow cone of air that appears to draw additional air into the stream (and opening), due to its high velocity. Test house research shows that both designs of ventilator produce excellent performance. The turbo units are generally smaller but still produce the same high airflow **through a structure** as some of the larger units and this may be seen as an advantage where stowage space is at a premium. However, the airflow from the larger conventional units may be more stable as they progressively cover a larger surface area at the entry point, although much of the airflow they produce admittedly fails to even enter the structure due to this fact, striking exterior walls and door surrounds.

Post-fire ventilation of smoke, using PPV ventilators, is normally considered a 'safe' operation but this may depend at what stage in the fire this occurs. In the UK a national **three-phased approach** was used over a ten-year period to introduce PPV in manageable stages. This ensured that firefighters were effectively trained to apply the various tactical concepts associated with post-fire and pre-fire attack PPV.

GRA 3.6 (UK) Risk Assessment – PPV
Key control measures:
When applying PPV during firefighting operations, there are a number of key Risk Control Measures that will need consideration, including:

- Pre-planning
- The training of crews
- Command and control
- Fire-ground communications
- Application techniques
- A phased approach to introduction

Three-phased approach:

- **Phase One** – Post-fire use for smoke clearance only (fire completely extinguished).
- **Phase Two** – At a stage where the fire was declared 'under control' but remained burning to some extent – smoke clearance.
- **Phase Three** – Pre-fire attack (pre-entry) for clearing a path of heat and smoke to enable a rapid entry and advance.

Despite the belief that Phase One and Two PPV operations were hazard-free, there were several instances where the fire was re-instated to a point where structures burned out of control, having already been suppressed to a stage of damp-down, turn over and overhaul.

This was caused by small amounts of hidden fire remaining in voids and attics that fed on the forced draft to develop and burn with some greater ferocity. There was also the effect of re-instating the pyrolysis process. This occurred as hot wall linings and surface fuels, which had been mostly extinguished, started to produce flammable gases from a state of smolder to a stage where these gases may actually ignite from the sparks being driven out of the surfaces, in the forced draft created by PPV. This effect has led to flashovers (thermal runaway) occurring even after the fire had been controlled or almost extinguished.

Despite these drawbacks, the concepts of PPV were being advanced (at the time of writing) in a wide number of UK fire brigades, including six of the seven large metropolitan fire authorities in England and Scotland (not London Fire Brigade), as well as many other parts of Europe.

When purchasing PPV equipment fire brigades will need to consider the following:

- The suitability of the selected fan
- Fan performance
- The necessary stowage and maintenance arrangements
- The necessary mobilizing and call-out arrangements
- The training of personnel
- The manual handling implications (weight and portability)
- The levels of noise

PPV should not be introduced as part of fire-ground operations until firefighters have a clear understanding of the use of tactical ventilation and its **effect on fire behavior.**

2.22 POSITIVE PRESSURE ATTACK (PPA)

The introduction of vast amounts of forced air into a fire-involved structure is intended to remove smoke and combustion products, cool the atmosphere, provide much needed air to any remaining occupants within, and to provide a smoke free path to the fire, for firefighters to gain rapid entry into the building.

There are simple rules that should be written into SOPs and followed where PPV is used as part of the fire attack strategy:

- The fire's approximate position in a structure must first be located
- An outlet must then be made as close to the fire as possible

- The air inlet point must be geometrically suited to the air outlet
- The outlet opening must be at least 50% in area of the air inlet point
- Firefighters must not block the airflow at the inlet point
- No PPV where conditions present warning signs of backdraft
- No PPV in large compartments where the fire is ventilation controlled
- No PPV unless the IC has clear communication with the interior crews
- The control of the fan must be an **assignment** and must be **staffed**
- The placement of the fan is critical – **not too close!**[7]
- Known voided properties or balloon-frame structures may not be suited to this strategic approach.
- Thermal image cameras (TICs) may assist in locating such fire spread.
- Consider the effect of automatic venting systems, where installed.
- Where VES is practiced, PPA may not be a viable tactic unless **carefully coordinated** with a **single room entry** (vent point).
- Risk Control Measures should include cover hose-lines at points where intense exterior flaming may cause exposure problems.
- The PPA air-flow should never be applied after entry has been made.
- A period of at least 30 seconds should occur between PPA airflow being initiated and entry being made, to allow for some stabilization of the smoke mixing and the creation of a directional forced draft (NIST suggest up to 120 seconds before stabilization occurs).
- If, at any stage, the fire conditions appear to worsen *inside the structure*, recall the interior team to evacuate and direct the airflow away from the inlet opening **as they exit**, but *where any such fire development is threatening their escape route, direct the airflow away from the inlet point immediately.*[8]

Tactical awareness

There have been some suggestions that the narrow air-cone of the turbo units may allow the potential for some blow-back of flaming at the entry door. This may be the case if the air is flowing directly into the room involved and the vent outlet has not been created, or is not large enough to handle the air exhaust rate.

There is also some potential for a very large PPV ventilator to be too powerful for PPA in a small area or compartment. In this instance, a very large vent outlet is needed, or the speed of the fan must be reduced to decrease the amount of air flowing

---

7. Positioning of the PPV ventilator in pre-attack (PPA) is critical because if the unit is placed too close, the potential exists for some 'blow-back' from the fire gases as they roll out of the entry inlet (doorway) and ignite, rather than being directed through and out of the exit outlet (window). The potential for flashover inside the structure also exists where the path to the exit outlet is restricted in any way. This can occur where an interior door is closed, where firefighters overcrowd and block the route, or where the exit outlet is not created prior to fan placement, or is not large enough. The fire conditions must be closely monitored in order to assess what effect the forced draft from the ventilator is having on fire development.

8. This point is worthy of debate amongst students – if fire conditions are deteriorating and the fan's air-flow is directed away from the inlet opening (doorway), both visibility and interior heat conditions may rapidly deteriorate and greatly hinder the interior crew's escape from the structure. At the same time, it is natural to remove the believed cause of the fire's sudden deterioration by turning the fan away. This is a **critical decision** to be made by the fan assignment (as staffed) and the air-flow should be maintained into the structure, where occupied by firefighters, for as long as possible. Many firefighters have been able to escape flashover conditions where the airflow has been maintained.

in. In this case, a high air-flow may lead to **thermal runaway** and **flashover** as the combustion products are unable to escape from the fire compartment fast enough.

The decision to initiate PPV should only be made by the incident commander following a dynamic risk assessment, which should include the availability of sufficient resources. Ideally the unit should be deployed in readiness, but should only be activated on the instructions of the incident commander (and not as an automatic function) who will consider various factors such as:

- The size of the compartment to be ventilated;
- The location and stage/extent of fire development;
- If known occupants are trapped, establish their location;
- Check for signs of rapid fire development;
- Wind direction;
- The location for the outlet vent;
- The location of the SCBA Entry Control Board (air management) may need to be away from fans due to their operating noise;
- Hose-lines to cover outlet vent exposure risks (water is NOT to be directed **into** the outlet vent under any circumstances).

Size of the outlet opening

There are varying recommendations concerning the optimum size of the outlet vent. Some say it should always be smaller than the inlet opening whilst others suggest it may be up to twice as large and still be effective. This all depends on fan size (performance) in relation to the area and configuration of the compartments being ventilated. What is most important is that the fan's performance does not overpower the ability of the combustion products to leave the opening, as discussed above.

Sequential ventilation

Where multiple rooms or floors require ventilation, the process of sequential ventilation will achieve the best results. This entails providing a maximum volume of pressurized air to vent each area in turn and will minimize overall ventilation time. The doors to all rooms should be closed initially, then, starting with the room nearest the fan, open the door and window to maximize the positive pressure available. Once cleared, this room can be isolated and others tackled sequentially in the same manner. The same principle is used for multiple floors starting at the lowest affected area. For large volume buildings it may be possible to use sequential ventilation if the area can be divided into smaller compartments. This will dramatically improve the effect of PPV.

Zone control tactics (safety zoning)

Taking a similar approach to sequential ventilation, the fire compartment itself is tactically isolated in this case, (or is pre-isolated) by closing the door to the room. What follows is a smoke and 'combustion product' clearance by PPV from all surrounding areas, or areas adjacent to the fire compartment, prior to taking the fire itself.

In effect, what this does is remove or reduce the hazardous nature of smoke and fire gas accumulations within the structure, prior to opening up the fire room.

This approach may also be used where, for example, a mattress is alight underneath and within, a foam sofa has smoldered for some time, or where a pile of plastic bags has smoldered away inside a cupboard. In these scenarios the compartment itself may have accumulated a heavy layer of combustion products, smoke and flammable (even cold) fire gases within. Prior to lifting the mattress or the plastic bags, or cutting into the sofa to reveal the fire, PPV (or hydraulic or natural venting) may be used to remove the dangerous combustion products from the immediate zone. This simple act may prevent a 'smoke explosion' and save lives!

Advantages of safety zoning:

- The rooms and areas adjacent to and above a fire compartment will be made 'safe' from subsequent smoke explosions and rapid fire progress.
- The fire compartment itself may be made safe where a simple 'small' but potentially deadly fire exists.
- Visibility is greatly improved.

Disadvantages of safety zoning:

- There will be a delay in entering the fire compartment.
- Such a delay may allow the fire to breach compartment boundaries, lead to some structural involvement/collapse, or delay rescue of occupants who may still occupy the fire compartment itself.

Overcoming wind pressure

A UK fire research project[9] demonstrated the effects of creating PPV airflow against a headwind in a four-bedroom house. When there was no wind blowing, or a negligible wind, the trials showed that use of a PPV fan could improve ventilation, reducing both smoke logging and air temperatures near the inlet opening. In this situation, the inlet opening should be selected so that any slight breeze assists the fan if possible but, if this is not possible, the fan should be able to reverse a slight breeze. The report states that in the latter case, a large inlet/outlet area ratio should be used. Reducing the outlet dimensions will reduce the amount of air flowing in.

Where the natural wind opposed the fan, it was possible for the fan to overcome the opposing component of the wind, provided that the wind was not too strong and the inlet/outlet area ratio was arranged to be in the fan's favour (large inlet, small outlet). However, in this situation it is possible for the effect of the fan to cancel out the effect of the natural wind, and impede ventilation.

The trial's results suggested that, even if an inlet/outlet area ratio of 2:1 can be achieved (a single doorway to a single window), there would be no point in attempting to reverse the air-flow caused by an opposing wind component of about 2.5 meters/second or more (6 mph).

The report went on to show that in laboratory measurements, an inlet/outlet area ratio of about 1:1 gives somewhat **higher volumetric flow-rates** than a ratio of about 2:1. However, in practice it was concluded that an inlet/outlet ratio of about 2:1 would be a good ratio to aim for, and gives a PPV fan a good chance of improving the ventilation of a building. It would be advantageous to ensure, at least, that the inlet opening is larger than the outlet opening. This is in order to try to ensure that the airflow setup in the building will be, and will remain, in the required

9. Fire Research & Development Group (UK Home Office), (1996), Report 17/96

direction, should the strength, and/or direction of the wind, change during the ventilation process.

Burning rate
A room fire will develop towards flashover, providing it has adequate amounts of fuel and air/oxygen. In a large room with high ceilings and items of stock or furniture spread widely apart, any progressive development towards flashover may be hindered, as any fire spread from a single burning item (unless very large) through convection, conduction or radiation is unlikely to occur. However, in smaller rooms, convected heat will reach the ceiling and radiated heat may well reach surrounding fuels where closely spaced. If sufficient air is available then the fire will develop to flashover. The heat output of the fire is dependent on these facts, along with the potential fire load in the room. A burning fire load can only burn to around 50% efficiency where air is supplied through normal-sized windows and doors. However, where air is forced into the compartment, space or room by an exterior wind, or PPV air-flow, it is just like blowing on barbecue coals; they will glow and burn more efficiently and fiercely. The energy is released more rapidly from the fuel (fire load) where this occurs and NIST research showed that the burning rate of a room fire might be increased by up to 60%[10].

This raises a question about firefighting flow-rate. If we are to accept that a compartment fire is likely to achieve an increased rate of burn (up to 60% greater) where PPV is used over natural ventilation, perhaps we should also be considering how effective the available flow-rate at the nozzle is likely to be. Another question addresses the potential for an increased rate of burn causing compartmental boundaries to become breached by fire at an earlier stage. Such an effect might then lead to earlier structural collapse. Whilst the rate of burn (heat release) may increase in this way, fire compartment temperatures, on the other hand, may not increase, as the incoming air from the PPV air-flow serves to cool the environment. This is an effective way to demonstrate to students the differences between **heat** and **temperature**.

However, a further series of test burns[11] in a three-story fire training building were scientifically monitored by NIST and provided a range of typical results. It was suggested that floor temperatures in the fire compartment **were** likely to increase in situations where PPV caused a room fire to burn with greater intensity, despite the cooler airflow from the PPV:

> *The [NIST] data indicated that, with both natural and Positive Pressure Ventilation techniques, using correct ventilation scenarios resulted in lower temperatures within the structure at the 0.61 m (2 ft) height, where victims may have been located, and at the 1.22 m (4 ft) height, where firefighters may have been operating. There were only limited ventilation configurations where the temperatures in rooms other than the fire room exceeded the victim or firefighter threshold temperatures with either ventilation technique.*
>
> *The use of Positive Pressure Ventilation resulted in visibility improving more rapidly and, in many cases, cooled rooms surrounding the fire room. However, the use of Positive Pressure Ventilation also caused the fire to grow more quickly, **and in some***

10. Kerber, S. & Walton, W., (2005), NIST Report NISTIR 7213, Building & Fire Research Laboratory
11. Kerber, S. & Walton, W., (2006), NIST Report NISTIR 7342, Building & Fire Research Laboratory

cases, created higher temperatures at the lower elevations within the structure. Overall, this limited series of experiments suggests that PPV can assist in making the environment in the structure more conducive for firefighting operations.

Each test in this series had a fire load that consisted of six pallets and 7.5 kg (16.5 lb) of field-cut dry hay. The fire load was selected to achieve flashover or near flashover conditions in the fire room with up to a 2.5 MW rate of heat release. The research proposed that vent outlets for PPA were ideally located where the vent from the fire room opened directly to the outside of the structure and did not cause the fire to be vented via paths leading through uninvolved rooms.

Temperatures at the floor when using PPV

In the second series of NIST tests the maximum fire room temperature for the naturally ventilated test was 550 °C (1020 °F) and the maximum temperature for the PPV ventilated test was 780 °C (1440 °F). In the room adjacent to the fire compartment, the temperature with PPV was nearly 50 °C (90 °F) higher than the naturally ventilated test.

At the 0.61 m (2 ft) height, where victims may have been located, the maximum temperature in the fire room was 180 °C (356 °F) for the naturally ventilated test and 370 °C (698 °F) for the PPV ventilated test. At the 1.22 m (4 ft) height, where firefighters may be located, the fire room temperatures were also higher in the PPV ventilated test.

The temperature in the PPV ventilated test was 190 °C (374 °F) greater than in the naturally ventilated test, which was most likely due to the mixing created by the fan. This is a significant increase, although the researchers pointed out that victims in the fire room would have been subjected to the 100 °C (212 °F) incapacitation threshold for either of the ventilation tactics.

Where there were rooms between the fire and the vent, the use of PPV increased the floor temperatures substantially in all rooms, but, again, in all cases either with or without PPV, victims in all of these rooms would have been subjected to the 100 °C (212 °F) incapacitation threshold for either of the ventilation tactics.

NIST researchers demonstrated that there was, in general, a rapid increase in temperature after ventilation. In the naturally ventilated fire, the temperature increased at a rate of 3.35 °C/s (6.03 °F/s) reaching a maximum temperature of almost 700 °C (1290 °F). In the PPV ventilated test, the temperature increased at a rate of 4.43 °C/s (7.97 °F/s).

It is worth noting that in one of the NIST tests in this series (configuration twelve), the use of PPV to ventilate the fire compartment, using a window in a room adjacent to the fire room, may have caused an ignition of fire gases in the adjacent room being used as a path for ventilating the fire. In practical terms, such an event is quite possible where firefighters locate two windows, serving different rooms: one with fire and one without, but neither demonstrating anything but a closed window with dark smoke seeping out. In this situation where the wrong window is selected for the outlet, temperatures in the adjacent room will soar where any rapid fire progress occurs and remaining occupants will suffer badly.

Contrast the above NIST data with previous research undertaken by Chiltern Fire (with Tyne and Wear Fire and Rescue Service) in the UK[12], and the University

---

12. Grimwood, P., Hartin, E., McDonough, J. & Raffel, S., (2005), 3D Firefighting, Oklahoma State University, p177

of Texas[13], which generally concluded that temperatures at floor level were improved or only slightly affected by PPV where occupants remained on the floor. Both researchers in these cases commented to the author as follows[14]:

Texas University USA (Dr. O A Ezekoye): *'In the first study we noted evidence that suggested that PPV with downstream venting might not be completely harmless. While the temperature increases in the lower layers of the downstream-vented room were not sufficiently large to absolutely imply that injury [to occupants at floor level] was definite, a risk seemed to be exposed. The first tests were not quite as well characterized as the second tests, and in these tests we found the magnitude of the heating in the lower layers did not pose a hazard.'*

Chiltern Fire UK (Mostyn Bullock): *'It is not my intention to give the impression that I would support the idea that heat flux at the casualty location is always reduced by PPV. Indeed our data [regarding Test 3] indicated that the reverse was true in that heat flux levels reached 33 kW/m² at the casualty location as a result of the offensive use of PPV accelerating a flashover of the fire. I would support a view that offensive PPV needs very careful deployment, especially where occupants may be trapped downstream of the fire.'*

Oxygen at the floor

The NIST research demonstrated that the oxygen concentration in the fire room dropped as low as 5% at the lower level of the fire room as the 2.5 MW fire became ventilation limited, but increased to 15% at the lower level at the time of natural ventilation. For the PPV scenario, the oxygen concentration returned to the ambient value of 21% much faster than in the naturally ventilated fire, especially at the lower level.

NIST researchers' conclusions (extracts)

A number of the fire experiments were designed to compare correct and incorrect ventilation scenarios with a fire located in a given room within the structure. A scenario is defined as correct when the ventilation opening occurs near the seat of the fire and localizes the fire. Scenarios were considered incorrect when the flow from the fire had to pass through other rooms before reaching the vent.

During actual firefighting operations, the selection of a ventilation procedure will depend on additional factors such as access to the structure and the location of victims or firefighters operating within the structure. In addition, firefighters may not know the exact location of the fire prior to entering the structure.

The use of PPV caused the fire to grow more quickly and in some cases created higher temperatures at the lower elevations within the structure. The use of PPV ventilation resulted in visibility improving more rapidly and in many cases cooled rooms surrounding the fire room. Overall, this limited series of experiments suggests that PPV can assist in making the environment in the structure more conducive for firefighting operations.

13. Grimwood, P., Hartin, E., McDonough, J. & Raffel, S., (2005), 3D Firefighting, Oklahoma State University, p179
14. Grimwood, P., Hartin, E., McDonough, J. & Raffel, S., (2005), 3D Firefighting, Oklahoma State University, p178/182

PPV in high-rise
Between 1985 and 2002 there were approximately 385,000 fires in US high-rise buildings greater than seven stories. These fires resulted in 1,600 civilian deaths and more than 20,000 civilian injuries[15] and between 1977 and 2005,
20 firefighters died from traumatic injuries suffered in high-rise fires in the USA[16].
Note: These figures do not include the World Trade Center losses of 11 September 2001.
Firefighters often rely upon built-in fire protection systems to help control a high-rise fire and protect building occupants. In many cases the buildings do not have the necessary systems or the systems fail to operate properly.
In a later series of tests undertaken by NIST researchers[17], 160 experiments were conducted in a thirty-story vacant office building in Toledo, Ohio. The aim was to evaluate the ability of fire department PPV fans to pressurize a stairwell in a high-rise structure in accordance with established performance metrics for fixed stairwell pressurization systems. Variables such as fan size, fan angle, setback distance, number of fans, orientation of fans, number of doors open and location of vents open, were varied to examine capability and optimization of each. Fan size varied from 0.4 m (16 in) to 1.2 m (46 in). Fan angle ranged from 90 degrees to 80 degrees. The setback distance went from 0.6 m (2 ft) to 3.6 m (12 ft). One fan up to as many as nine fans were used, which were located at three different exterior locations and three different interior locations. Fans were oriented both in series and in parallel configurations. Doors throughout the building were opened and closed to evaluate the effects. Finally a door to the roof and a roof hatch were used as vent points. The measurements taken during the experiments included differential pressure, air temperature, carbon monoxide, meteorological data and sound levels.

The NIST conclusions from this research:
PPV fans utilized correctly can increase the effectiveness of firefighters and survivability of occupants in high-rise buildings. In a high-rise building it is possible to increase the pressure of a stairwell to prevent the infiltration of smoke if fire crews configure the fans properly. When configured properly PPV fans can meet or exceed previously established performance metrics for fixed smoke control systems. Proper configuration requires the user to consider a range of variables including fan size, set back and angle, fan position inside or outside of the building, and number and alignment of multiple fans.
The data collected during this limited set of full-scale experiments in a thirty-story office building demonstrated that in order to maximize the capability of PPV fans, the following guidelines should be followed:

- Regardless of size, portable PPV fans should be placed 1.2 m (4 ft) to 1.8 m (6 ft) set back from the doorway and angled back at least 5 degrees. This maximizes the flow through the fan shroud and air entrainment around the fan shroud as it reaches the doorway.

15. Hall, J.R., Jr, (2005), *High-rise Building Fires*, NFPA, Quincy, Massachusetts
16. NFPA Database, *Traumatic Firefighter Fatalities in High-rise Office Buildings in the United States*
17. Kerber, S. & Walton, W., (2007), NIST Report NISTIR 7412, Building & Fire Research Laboratory

- Placing fans in a V-shape is more effective than placing them in series (this was also noted in a European research project the author was associated with in France in 1999–2000).
- When attempting to pressurize a tall stairwell, portable fans at the base of the stairwell or at a ground floor entrance alone will not be effective.
- Placing portable fans inside the building below the fire floor is a way to generate pressure differentials that exceed the *NFPA 92A** minimum requirements. For example, if the fire is on the twentieth floor, placing at least one fan at the base of the stairwell and at least one near the eighteenth floor blowing air into the stairwell could meet the *NFPA 92A* minimum requirements.
- Placing a large trailer mounted type fan at the base of the stairwell is another means of generating pressure differentials that exceed the *NFPA 92A* minimum requirements.
- Fans used inside the building should be set back and angled just as if they were positioned at an outside doorway.

* NFPA 92a – Recommended Practice for Smoke Control Systems (NFPA Standards)

Carbon monoxide and PPV
A fire has the potential to produce a very large amount of carbon monoxide (CO). This amount could be in the order of 50,000 parts per million (ppm) in an under-ventilated fire[18]. Tenability limits for incapacitation and death for a five minute exposure are 6,000 ppm (0.6%) to 8,000 ppm (0.8%) and 12,000 ppm (1.2%) to 16,000 ppm (1.6%) respectively. CO is the major toxic gas in approximately 67% of fatalities in structure fires. Using PPV fans to keep the CO produced by the fire, along with the other harmful combustion products, out of the stairwells, greatly increases the chances of safe evacuation.

The CO produced by the PPV fans was at least one order of magnitude less than that created by a fire. As long as the PPV fans were not placed in the stairwell with the door shut, the NIOSH ceiling exposure (200 ppm) was not exceeded. However, the NIST report advised that CO readings **less** than 50 ppm are unlikely with a gasoline/petrol powered PPV fan and the author can confirm such readings in excess of 50 ppm on several occasions. It is important to use gas-monitoring devices in conjunction with PPV, as well as full PPE and SCBA in high exposure areas (refer to local regulations).

Always be sure to check CO levels following ventilation of a structure and prior to allowing occupants to return inside!

Note: The National Institute for Occupational Safety and Health (NIOSH) ceiling limit for CO exposure is 200 ppm, which should not be exceeded at any time. The UK Health and Safety Executive ceiling is 50 ppm for a maximum of thirty minutes. The American Conference of Governmental Industrial Hygienists (ACGIH) excursion limit for CO is 125 ppm (or five times the threshold limit value time-weighted average [TLV-TWA]), which should not be exceeded under any circumstances. The Environmental Protection Agency National Ambient Air Quality Standard for one hour CO exposure is 35 ppm.

18. Purser, D., (2002), 'Toxicity Assessment of Combustion Products', in *The SFPE Handbook of Fire Protection Engineering Third Edition*, National Fire Protection Association, Quincy, Massachusetts

Concentration of CO in the air	Time of intake before illness
87 ppm	15 minutes
52 ppm	30 minutes
26 ppm	1 hour
9 ppm	8 hours

Fig. 2.6 – The maximum level of carbon monoxide and exposure time that cannot be exceeded without causing illness. Source: World Health Organization.

PPM CO	Exposure	Symptoms
35 ppm	8 hours	Maximum exposure allowed by OSHA in the workplace over an 8-hour period.
200 ppm	2–3 hours	Mild headache, fatigue, nausea and dizziness.
400 ppm	1–2 hours	Serious headache, other symptoms intensify. Life threatening after 3 hours.
800 ppm	45 minutes	Dizziness, nausea and convulsions. Unconscious within 2 hours. Death within 2–3 hours.
1,600 ppm	20 minutes	Headache, dizziness and nausea. Death within 1 hour.
3,200 ppm	5–10 minutes	Headache, dizziness and nausea. Death within 1 hour.
6,400 ppm	1–2 minutes	Headache, dizziness and nausea. Death within 25–30 minutes.
12,800 ppm	1–3 minutes	Death.

Fig. 2.7 – Exposure to carbon monoxide: symptoms and effects.
Source: www.carbonmonoxidekills.com

PPV ventilator noise levels
Another concern with the use of PPV ventilators is the noise they create. In the NIST high-rise research, noise levels were monitored in certain locations throughout the experimental series to estimate the level of impact on the fire crews and command officers. Ambient noise measurements were 60 to 65 decibels (dB). This value rose to 80 dB when traffic went past the building. Measurements next to the compartment size fans were approximately 100 dB to 110 dB depending on the size of the fan.

Source of Sound/Noise	Sound Level (dB)
Threshold of hearing	0
Quiet bedroom at night	30
Conversational speech	60
Curb-side of busy roadway	80
Heavy Truck	90
PPV	**100–110**
Jackhammer	100
Chainsaw	110
Threshold of pain	130
Instant perforation of eardrum	160

Fig. 2.8 – Comparisons of sound levels (PPV as recorded during NIST research).

2.23 LIMITED STAFFING ISSUES

It is certain that primary response tactics are dictated by the weight of attack and depth of our resources. It is also certain that, **whatever politics are involved,** structural fire response in many parts of the world is restricted to a single engine with three to four firefighters. In some situations, this single engine crew is going to remain alone in their response for 15–30 minutes or more!

Where firefighters are forced to operate without support, hose-line back-up, safety officers and other means of controlling risks and ensuring their security, then they must carefully adapt their approaches and prioritize the critical fire-ground tasks. With this in mind the author has developed a range of SOPs[19] for limited staffed crews (see Chapter Five). What is important here is that never is it more crucial to isolate fire spread, if possible, and vent the building effectively, than where staffing is restricted in such a way.

The use of Positive Pressure Attack (PPA) offers an ideal tactical solution in this respect. Whilst any actions taken by limited staffed crews must not place them into situations where they face increased levels of risk compared to a full primary response of firefighters (minimum base standards depending on structure size and number of floors), there are certain approaches that may be made within reasonably safe parameters. A typical example is where:

- The building is not large;
- The fire involves one small room on the outside wall;
- The fire has self-vented to the exterior;
- There are possible occupants remaining therein;
- There are no major exposures of a critical nature.

19. Grimwood, P., (2006), *Standard Operating Guidelines 4242 for Limited Staffed Crews*, Firetactics.com

In this situation, the use of PPA for clearing a smoke and heat-free path to the fire is something that a limited-staffed crew may undertake within reasonably safe parameters. The vent outlet is already in existence; one firefighter can place the ventilator; two firefighters can enter with a hose-line after a period of 60–120 seconds of PPV, allowing the entry path to be created. They may then advance to the fire room and continue with full suppression of the fire or isolate the fire (close the door) and search all other areas of the structure prior to returning outside and taking the fire from the exterior window.

2.24 FDNY *LADDERS 3* – OCCUPIED NON-FIREPROOF TENEMENTS[20]

The basis for ventilating fire-involved structures in New York City is a strategy enshrined in tradition developed by, and based upon, the experiences of many hundreds of battalion chiefs and firefighters who have served before us. The tactical approach is very proactive and is aimed at handling a typical range of scenarios in specific structure types common to various boroughs of the city.

Brownstones
Brownstones were built in the late 1800s as private dwellings. They are typically three to four stories with a basement on the first floor – and a cellar beneath. They are usually 20–25 ft wide and up to 60 ft deep. They can be built with party wall construction. The roof is normally flat and has a small parapet wall to the front and usually no parapet wall on the rear. Access to the roof is from the top floor via a scuttle. Although three to four stories on the front, Brownstones may have four to five stories on the rear.

Rowframes
Rowframes vary from two to five stories and are 20–30 ft wide, and 40–60 ft deep. They are of balloon-frame construction and can be set in a row of up to as many as twenty buildings. Walls separating the buildings may or may not be firewalls. There may be a common cockloft (attic).

Taxpayers
This term applies to a one or two-story commercial building, with exterior masonry walls, and wooden interior construction. Size can vary from 20 ft wide and 50 ft deep to as much as an entire block. Taxpayers may be sprinklered – but usually only the cellar is. May have a common cockloft and many void spaces.

Old law tenements
These were built before 1901 and can range from four to seven stories. They are 20 to 25 ft wide, and 50 to 85 ft deep. Non-fireproof, with brick walls and wood beams and floors, there can be two to four apartments per floor. Old law tenements have an internal stairway and a fire escape and may have a fire escape front and rear.

20. FDNY, (2000), *Standard Operating Procedures – Ladders 3 (Tenements)*

Original new law tenements
Built between 1901 and 1916, these range from six to seven stories, and are 30–50 ft wide and 85 ft deep. Interior stairs are enclosed and 'fireproof'. Walls and partitions are fire-stopped at each floor.

Newer new law tenements
Built from 1916 to 1929, these may have a large floor area: 150×200 ft is not uncommon. Floor areas are partitioned into units of less than 2,500 sq ft. (Fireproof construction is required if 2,500 sq ft is exceeded.) Partition walls only extend floor to ceiling. Large cockloft – building may have an elevator.

'H' type
Masonry bearing walls, wood beams, steel beams, and girders.

Stairway types vary: may be wing type (located in the wing), or transverse (stairwells located in each wing and connected by a hallway). Although 'H' is the most common, there are other types: 'E', 'O', 'U', and 'Double H'. The narrow area that connects the wings is referred to as the throat.

The Bronx is still loaded with six-story H-types, as far as the eye can see. Many of these are vacant, but many are still thriving after renovation. The author is told the far south of the borough probably has the most lingering damage from the 'war years' in the late 1960s and early 1970s.

The FDNY venting strategy is pre-planned and documented across several SOPs. Two of the most detailed of these documents are termed *Ladders 3* (Tenements) and *Ladders 4* (Private Dwellings). These two documents are fairly precise in assigning the 'task-based primary response' to roles, whilst the FDNY command, control and fire-ground radio procedures ensure that operations are effectively coordinated.

> ***Saving life is the primary function of ladder companies.*** *Any immediate, limited ventilation is justified if it is coordinated between the inside team and the outside team and it will help facilitate an interior search for occupants. Bear in mind that ventilation for search purposes will generally intensify the fire and could endanger other occupants of the building.*

FDNY Ladders 3 (p12).

There are three key 'venting' roles (termed positions) pre-assigned on the ladder company and these are:

- The roof-position
- The outside vent position (OV)
- The chauffeur[21]

These three firefighters comprise the 'outside' team on a ladder company in New York and the *Ladders* SOPs will pre-assign their tasks. They know pretty much what their role is before they arrive on-scene. That is not to say the assignments are so

21. Note: The chauffeur (driver) is also responsible for operating the aerial ladder. However, the additional ventilation duties are assigned on the basis that two ladders respond on the primary response to a structure fire. The ladder chauffeur should remain on the turntable when members have entered the building by aerial ladder and are in precarious positions such as: a floor over a heavy fire, the roof of a building with a heavy fire condition etc. The chauffeur should keep alert as to the Who? When? and Where? of members using the aerial ladder.

rigid that they cannot be varied or redeployed more effectively, but they remain very proactive in anticipation of any potential for common roles that need fulfilling. For example, the chauffeur and OV will commonly and automatically take up positions on the front fire escape (where in existence) to cross ventilate the structure. They will do this in support of the interior attack teams, or for the purposes of VES (Vent-Enter-Search) to locate victims in immediate danger in, or adjacent to, the fire-involved room(s).

The OV position:

Except for assisting the chauffeur in front of the fire building when aerial or portable ladders are needed for rescue or removal, assignment is to ventilate the fire area from the exterior providing lateral ventilation. This is generally done from the fire escape landing of the fire apartments. Access is via the front or rear fire escape. Some buildings have one or two apartments per floor with one fire escape. In this case the OV's choice is eliminated and he/she uses that fire escape.

Other buildings have three or four apartments per floor (or more) and the building will have both front and rear fire escapes. In this case he/she must choose the correct one to attain a position on the exterior of the fire floor. If the location of the fire apartment is not obvious from the exterior of the building the OV should communicate with his/her officer. Once the location is verified the OV can then reach the correct fire escape via a window from a lower or adjacent apartment or from a drop ladder/ portable ladder at ground level.

FDNY Ladders 3 (p12)

There are occasions when the OV position is varied:

- *Store fire: Ventilate the rear of the store from the exterior. If this would expose people above on a fire escape, ventilate immediately after they are out of danger. If a delay in ventilation is encountered and/or anticipated, notification should be made to your company officer.*
- *Top floor fire: Proceed to roof with saw and Halligan tool. If possible, descend fire escape and provide ventilation. Entry and search will be completed if he/she teams up with the second OV (or another available member). If unable to descend the fire escape notify your company officer, attempt to vent fire apartment from roof level, and then assist the roof firefighter with roof vent.*

In both situations, they will affect the removal of any occupants but still must consider fire severity or extinguishing operations, which may endanger him/her. This task may prove difficult due to window bars and or gates.

When the OV must assist the chauffeur in a removal operation, or the OV is unable to descend the fire escape from the roof, the officer may dispatch a member of the forcible entry team to perform outside ventilation after they have forced the door to the fire apartment. Entry and search will be completed if he/she teams up with another available member.

FDNY Ladders 3 (p13)

Where a tower ladder (with bucket/basket) responds as opposed to an aerial ladder, the tactics alter slightly. The OV will operate from the basket and the chauffeur will remain on the pedestal to take overall control of the basket's operation.

The roof position assignment is perhaps even more critical in the task-based response plan.

The roof firefighters' access to the roof is achieved via:

- Adjoining building
- Aerial ladder
- Rear fire escape
- NEVER by the internal stairs

The *Ladders 3* SOP continues:

> *The duties of a roof firefighter demand an experienced, observant and determined firefighter capable of decisive action. The responsibility of this position covers three broad areas: life, communication, and ventilation. Roof ventilation is critical for search, rescue and extinguishment of the fire. NOTHING SHALL DETER the member assigned the roof position from carrying out the assigned duties. The roof firefighter should always confirm his/her way off the roof as soon as he/she reaches the roof. The roof firefighter is responsible for the following:*

> - *Opening the bulkhead door and skylight, or scuttle and roof level skylight over interior stairs;*
> - *Probing bulkhead landing for victims;*
> - *Probing for roof level skylight draft stop;*
> - *A perimeter search of the building for persons trapped and those who may have jumped or fallen. This search shall include the sides, rear and shafts of the building;*
> - *Locating the fire and making a visual check for extension across shafts or by auto exposure;*
> - *Transmitting vital information to the incident commander, either directly or through the company officer, on conditions observed from that vantage point;*
> - *When necessary, team up with OV to VES fire floor and, if not needed for search on that floor, proceed to VES the floors above the fire;*
> - *When necessary, team up with second roof firefighter to VES all floors above the fire;*
> - *At top floor fires, venting top floor windows from roof level. He/she is also RESPONSIBLE FOR UTILIZATION OF THE SAW to vent the cockloft and top floor when necessary AFTER COMPLETING INITIAL DUTIES;*
> - *Conveying information to second ladder company. Informing them of the extent of the search completed, so that all floors above the fire may receive a thorough search. Also informing the second ladder company when proper examination of exposed interior stairs and public hall has not been made due to other duties. The second ladder company shall complete the above-mentioned examinations;*
> - *Reporting back to their company officer (generally located on the fire floor) when assignment is completed or when relieved by second ladder company and apprising them of all pertinent information.*

An analysis of the FDNY approach to venting assignments sees primary responsibility devolved to the individuals who must locate themselves speedily and effectively in positions from where they will operate. Their task assignments are numerous but are based in order of prioritization and needs as determined. They must however, according to the directives in *Ladders 3*, communicate and coordinate their actions with each other.

In getting into such key positions early in the firefighting operation, there is much opportunity to gather and relay vital information to others on the fire-ground and early opportunist rescues can be made.

FIRST FDNY LADDER COMPANY TO ARRIVE:
1. Ladder company operations on fire floor.
2. Determine life hazard and rescue as required.
3. Roof ventilation and a visual check of rear and sides from this level.
4. Laddering as needed.
5. If second ladder company will not arrive within a reasonable time, make interior search and removal of endangered occupants above the fire.

SECOND FDNY LADDER COMPANY TO ARRIVE:
1. *All floors above the fire floor* for search, removal, ventilation, and to check for fire extension.
2. Confirm roof ventilation (assist first unit).
3. Check rear and sides of buildings.
4. Reinforce laddering and removal operations when necessary.

2.25 FDNY *LADDERS 4* – PRIVATE DWELLINGS[22]

Originally built for one or two-family occupancy, these structures are usually one to three-stories in height. They may be attached to adjoining buildings, semi-attached or detached. The interior of split-level homes however, may have as many as five levels within a three-story building. In the UK these structures may be termed Houses in Multiple Occupation (HMOs).

Due to the relatively small height and area of the private dwellings, as compared to the multiple dwelling discussed in *Ladders 3*, the first to arrive ladder company is responsible for forcible entry, ventilation, and search of both the fire floor and floors above. The second ladder to arrive will primarily be used to augment the search for life and then to assist as needed.

FIRST FDNY LADDER COMPANY TO ARRIVE:
1. Rapid and comprehensive exterior size-up of fire situation. Determine life hazard and rescue as required.
2. VES of all occupied areas of the dwelling either via the interior, or by a combination of an interior/exterior approach.
3. Nothing must delay primary search, but an examination of entire building must be made as soon as possible.
4. After arrival of the second ladder company, the first ladder is generally responsible for the fire floor and floors below.

SECOND FDNY LADDER COMPANY TO ARRIVE:
Report to the officer in command and prepare to:
1. Augment or supplement laddering operations of first ladder, where required.
2. Search areas not yet covered by first ladder company.
3. As soon as possible, assume responsibility for operations above the main body of fire, to include opening of the roof if necessary.

FDNY Ladders 4 (p4)

22. FDNY, (1997), *Standard Operating Procedures – Ladders 4 (Private Dwellings)*

It can be seen here that the FDNY tactical approach remains pre-assigned, but to a much lesser degree. The second ladder company is, for example, directed to report to the IC for its assignments. VES is normally the tactic of choice by FDNY in these situations where the COMPLETE removal of glass, window sash, curtains, blinds, etc., from the window selected for entry/search/rescue is directed. This is accomplished in preference to rapid, incomplete ventilation of all available windows, with the sole intent of facilitating the inside operation.

Roof operations are generally not feasible during initial fire operations at fires in private dwellings with peaked roofs*. Therefore, the roof firefighter can be used to advantage in the VES operation. The roof firefighter will normally take the front of the building and the OV person the rear or side for VES, although these positions are interchangeable. Any other venting action will normally be in support of the advance of the primary attack hose-line and therefore, such venting will not normally take place until the engine company are advancing their charged line in.

For a fire on upper level, ventilation must be accomplished via ladder. In addition to ventilation of the fire room, ventilation must be provided to facilitate movement of the engine company up the interior stairs. There is often a window right at the head of this stair. In other buildings, a bathroom located at the top of the stair may be vented to improve the interior situation.

2.26 RISK MANAGEMENT – VENTING STRUCTURES

Besides the obvious risk management principles associated with operating machinery and ventilators – as well as cutting saws and forcible entry tools – what risk management considerations should be applied on the fire-ground when venting structures? The base protocols are:

- Establish what the risks are
- Select a safe system of work
- Implement Risk Control Measures
- Monitor the dynamic processes on the fire-ground
- Are the risks proportional to the benefits or gains?

Tactics – Venting structures (tactical ventilation)

- Venting **assignments** may be pre-assigned or deployed on-scene
  1. Pre-assigned through SOP (as FDNY)
  2. Deployed on-scene based on fire conditions and tactical priorities (as London)
- Venting **actions** may be automatic or awaiting a directive (request)
  1. Automatic venting as in VES, or in venting stair-shafts (FDNY)
  2. Venting action pending a request or directive from the interior crew(s), fire floor commander or incident commander

These issues concerning '**assignments**' and '**actions**' may dictate, to some extent, the training needs. For example, if a firefighter is pre-assigned into position with

---

*A particular style of larger private dwelling in NYC, of somewhat older construction (the Queen Anne), is however vented at the roof as soon as possible, where needed. The particular roof construction sees large voids hidden in attics, valleys, ridges, dormers, and around hips. In this particular type of structure, one window that shall not be entered for VES is the one immediately over the side entrance door. This window is generally at the top of an interior stair.

venting as a primary task, the training need may simply be *how to create a vent opening*. The 'why' is arguably countered by the need to wait for a directive, or request. There is no decision made on the part of the person creating the opening other than where to locate, as the decision to open up comes from another source. Having this, a trained and experienced firefighter may relay vital information back to the interior team asking for ventilation in situations where 'rapid fire' indicators are in existence.

If, however, the firefighter has final responsibility to decide the **if, when and where** of ventilation, then a greater need for more in-depth training clearly exists. A high level of experience, awareness and understanding is needed in relation to fire behavior and fire dynamics. How is a fire likely to develop? How is a fire likely to be affected by air-flow? How are the dynamics of building stack effect, wind and other interior air movements likely to influence a fire's growth and direction of spread?

The risks associated with venting structures

- Venting may increase the fire's burning rate;
- Firefighters or occupants may be caught by rapid fire progress;
- Fire development may overpower the available flow at the nozzle(s);
- May increase smoke production inside the structure;
- May allow wind to enter and 'blow-torch' the fire.

The risks associated with NOT venting structures

- Heavy smoke-logging;
- High heat build-up, particularly at the floor;
- Under-ventilated fire;
- Poor visibility;
- Potential for rapid fire progress;
- CO levels rising with low O_2 levels where occupants may be trapped;
- Potential for 'unplanned' ventilation whilst firefighters occupy the structure.

Venting structures – A safe system of work
A safe system of work for ventilating structures would follow protocols that assign responsibility for **who** can vent **what, how** and **when,** calling for precision (location), communication (timing), coordination (timing) and:

- There must be a **primary purpose (objective)** in creating the vent.
- There must be a **directive** for any vent opening being made.
- It must be clear who is responsible for making openings.
- Wind direction and force must be a primary concern.
- Where is the fire located and what conditions are presenting?
- Where are the occupants (if any) most likely located?
- Where is the primary attack line located?
- Where are other known locations of firefighters on the interior?
- Where is the optimum position to create an opening in each situation?
- Have interior crews requested or confirmed their desire for this opening?

Venting structures – Risk Control Measures

- Pre-planning.
- Provide a SOP with clear instructions and well defined protocols.

- Provide adequate staffing for the structure on the primary response.
- Provide adequate resources and equipment on the primary response.
- The effective **training** of crews in ventilation procedure.
- The effective **training** of crews in fire dynamics and fire behavior.
- Effective command and control structure.
- Fire-ground communications and procedures (confirmation of receipt).
- **Begin** all operations from an **anti-ventilation** stance.
- Where openings exist, consider closing them. Don't ventilate where fire-fighters are on ladders above a window, unless a covering hose-line is in place and staffed below.
- Don't ventilate (open a door) onto a stairway, where occupants or firefighters may be located and vulnerable above – always clear the stairs first.
- Don't ventilate where an exposure problem may be created, unless a covering hose-line is in place.
- Vent with wind direction and velocity always in mind!
- Don't vent in situations that place interior crews between the fire and the vent opening.

Monitoring the ventilation process
The following roles should be assigned to personnel who are effectively trained to read compartment fire conditions (B-SAHF), recognize changing conditions, understand what these mean, and act upon (and communicate) key fire behavior hazard indicators.

- 'Door control' assignment
- Interior crews
- Fire floor commander
- Incident commander

Of these, the most critical role is perhaps that of the door assignment, who may be the first person to notice visual signs that are indicative of changing circumstances or hazardous conditions.

Are the 'risks' proportional to the potential 'gains'?
Are the risks of venting (**or not venting**) proportional to the potential benefits or gains?

- Reducing mushrooming heat and smoke;
- Reducing chances of rapid fire progress;
- Lifting the smoke layer, improving visibility;
- Providing much needed air to trapped occupants;
- 'Lighting-up' the fire so that crews may locate it quickly;
- Directing fire in the overhead away from advancing firefighters;
- Removing heat, flammable gases and combustion products from the structure.

If you have addressed all the above bullet points in your risk assessment and size-up and you possess a clear understanding of fire behavior indicators, then you should now be in a position to answer this question.

2.27 VENTING AND RAPID FIRE PROGRESS

The extent of fire development in an enclosed area, or compartment, will have great influence over our decisions to create openings and control the door openings. The fire conditions may present in a number of ways, as follows:

- Fuel-controlled fire
- Ventilation-controlled fire
- Under-ventilated (cool conditions)
- Under-ventilated (hot conditions)

In the fuel-controlled scenarios, a fire will be in the incipient or minor growth stages of development. There may well be light to heavy smoke production but the conditions will generally be 'cool' throughout the compartment. In this situation, is there any possibility of rapid fire development? If so, how might this occur? The fire gases will generally be below their Lower Flammable Limits (LFLs). The only phenomenon relevant here is that of flashover, as the fire finds sufficient amounts of fuel, in an abundance of air, and progresses towards this sudden growth stage that culminates in full room involvement with flames issuing from windows.

In the 'ventilation-controlled' scenarios, the fire gases will generally exist within a wider range of limits, either side of the LFLs.

Under-ventilated fires will lead to large accumulations of fire gases existing above the Upper Flammable Limit (UFL) and before they can take part in the combustion process they must be mixed with air/oxygen. This may occur where air enters the fire gases, or where the gases themselves transport on the convection currents to other parts of the structure, or to the exterior. If this mixing occurs then the fire may develop rapidly (even explosively) if an ignition source is present. Alternatively, the combustion process may redevelop in the form of a ventilation-controlled fire heading towards flashover (ventilation-induced) termed 'thermal runaway'. Where fire gases are very hot they may auto-ignite without the need for an ignition source. This may occur both inside and outside the compartment. A backdraft may also result in a very intense fireball, possibly with explosive force.

It is this side of the flammability limits (above the UFL) that we must be especially concerned with when creating ventilation openings as the admission of air may lead to:

- Ventilation-induced flashover (thermal runaway)
- Backdraft
- Auto-ignition

Where the fire is under-ventilated, the creation of a vent opening (including that of the **entry doorway**), should be approached with caution using Compartment Fire Behavior Training (CFBT) door-entry techniques and careful selection of vent openings. If for example, a window is showing signs of high heat within, (maybe it is starting to crack or craze, or is going very black and stained) then maybe we should confirm if this window should be vented under these conditions with either the interior crews, the fire floor commander or the IC.

Another possibility is that the air-track created by an opening into an under-ventilated fire might cause the stirring up of the base fire, sending a flaming ember up on convection into fire gases existing between the LFL and UFL at the ceiling.

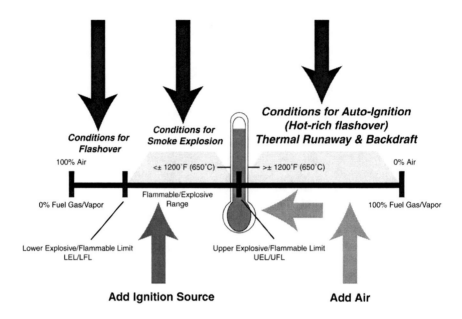

Fig. 2.9 – Limits of flammability and conditions for flashover, smoke explosion, auto-ignition, ventilation-induced flashover (thermal runaway) and backdraft. Note: Above the UFL, air is a possible 'trigger' for an event, whereas between the LFL and UFL, any source of ignition might serve as the 'trigger' for a smoke explosion or flash-fire.

The result: a 'smoke explosion' or 'flash-fire'. This is even more likely where PPV is used.

Burning regime or state of fire	Most likely phenomena
Fuel-controlled regime	Progression to flashover
Ventilation-controlled regime	Heat induced flashover Ventilation induced flashover
Under-ventilated (cool conditions)	Smoke explosion Flash-fire
Under-ventilated (hot conditions)	Smoke explosion Flash-fire Backdraft Auto-ignition

Fig. 2.10 – The 'most likely' phenomena associated with rapid fire progress for different regimes of burning or state of enclosed fire conditions.

2.28 COMBINING US-EURO TACTICS

Throughout the 1980s the author presented several controversial papers and articles, based mainly upon his own operational research and experiences as a firefighter both in the UK and the USA, that closely examined structural ventilation practices as carried out by firefighters around the world. His proposed concept of 'tactical ventilation' (a term he originally introduced and defined in 1989 through his book *Fog Attack* and several earlier articles in the UK *Fire Magazine*) was to encourage an increased awareness of 'tac-vent ops' and PPV, and present a safer and more effective tactical process for the ventilation of fire-involved structures by on-scene firefighters, paying particular attention to the influences of air dynamics and fire gas formations. Following work with Warrington Fire Research Consultants (*FRDG* 6/94) his terminology and concepts were adopted officially by the UK Fire Service and are now referred to throughout revised Home Office training manuals (1996–97).

In 1984 (9/84 *Fire Magazine*) he posed the question of whether US-style roof venting methods should be utilized at an earlier stage in the fire attack and discussed some previous UK incidents where venting may have helped. His five-page article in 1985 (10/85 *Fire Magazine*) described the tactical implications of using roof cuts to vent fire gases and discussed a wide range of tactical options used to create safer working conditions for firefighters and trapped occupants through the creation of openings in the structure. It was here, in 1985, that he first introduced and discussed the benefits of Positive Pressure Ventilation. In 1987 (5/87 *Fire Magazine*) he called for a Home Office review of UK strategy and prompted some research into tactical venting methods, and by 1988 (12/88 *Fire Magazine*) he was describing how such tactics might have been used to save several large structures that had recently incurred major financial losses where it was thought a lack of ventilation had contributed to such loss. He wrote: 'Over the past four years I have attempted to educate and prompt discussion on the topic of tactical ventilation by firefighters in fire-involved structures,' and acknowledged that the recent interest by a Chief Fire Officer (John Craig of Wiltshire) in the theory and practice of 'tac-vent ops' was a major step towards national acceptance. He was personally requested by CFO Craig and the Wiltshire Fire Brigade to assist in writing the first UK SOP document (Operational Note) on 'Tactical and Positive Pressure Ventilation' in 1989.

At this time, the UK Fire Service was guided in venting operations by a single thirty-five-word paragraph in Book Twelve of the *Manual of Firemanship* which stated that rooftop venting operations should only be undertaken 'as a last resort'.

The tactical ventilation strategy was founded upon a combination of US venting operations with UK anti-ventilation tactics. All operations started from an anti-ventilation stance where it was equally important to zone-off compartments by closing doors once rooms had been searched. If the fire compartment itself were located, this too would be closed off, unless an immediate attack hose-line was advancing in. The basic core principles of the author's original tactical venting strategy supported limited roof operations (particularly on inner-city mid-rise flat-roofed buildings) and VES tactics, with PPV (PPA) a viable alternative in smaller structures.

The combined US-Euro **tactical ventilation** protocols:

- Start all operations from an anti-ventilation stance;
- Locate the fire;
- Establish the stage of fire development and area of involvement;
- Establish any existing air-track and its influence on the fire;
- Read all fire and building conditions – B-SAHF;
- Ventilate the stair-shaft in mid-rise at the earliest opportunity;
- Select viable cross-ventilation points if needed;
- Establish a viable purpose to create an opening (for FIRE or LIFE);
- Ventilate only under the directive of interior crews;
- Consider VES operations in a carefully controlled manner;
- Utilize combination fogging/venting tactics where viable;
- PPV (PPA) in compartments of limited volume or size (may include high-rise but not large volume buildings);
- Consider defensive PPA where viable (zoning off the fire compartment), to clear adjacent compartments (zone-control) prior to taking the fire.

In order to effectively apply these protocols, firefighters must be well versed and trained in fire dynamics and the fundamental principles of tactical application, including CFBT.

Chapter 3

Venting structures – International round table discussion

3.1 INTRODUCTION

Contributors
The author wishes to point out that the views and opinions expressed by contributors in the following round table discussion are personal views of the contributors, and not necessarily representative of an official view held by their fire authority.

- **Deputy Assistant Commissioner Terry Adams (London Fire Brigade)**
- **Battalion Chief Ed Hartin (Gresham Fire and Rescue, Oregon USA)**
- **Chief Jan Südmersen (City of Osnabrück Fire Service, Germany)**
- **Major Stephane Morizot (Versailles, Paris, France)**
- **Firefighter Nate DeMarse (City of New York Fire Department)**

- **Training Officer Tony Engdahl (City of Gothenburg Fire Service, Sweden)**
- **Lieutenant Daniel McMaster (Alexandria, Virginia, Fire Department)**
- **Captain Juan Carlos Campaña (City of Madrid Fire Brigade, Spain)**
- **Captain Jose Gomez Antonio Milara (City of Madrid Fire Brigade, Spain)**
- **Firefighter Matt Beatty (Rescue One, City of New York Fire Department)**
- **Paul Grimwood (London Fire Brigade, retired) (Author)**

3.2 ATTRIBUTES OF A PRE-ASSIGNED TASK-BASED VENTING STRATEGY

> **What do you see as the advantages or disadvantages of a pre-assigned (proactive) task-based response to venting structures, where fire-fighters deploy automatically, according to written directives in their SOPs?**

Adams (London) – London Fire Brigade's Deputy Assistant Commissioner Terry Adams is a thirty-three-year veteran of London Fire Brigade. He still remains very much a 'hands on' firefighter who, like the author, worked his early days in the heart of the London's busy West End district when fire action was ever present and rescues were almost a nightly occurrence across the nine fire stations serving the district. He believes that there are clear differences in the tactical approaches used by the two big city fire departments (London and New York) and there are good reasons for this.

Whilst he acknowledges that planning and resourcing an incident, from a command perspective, may be made far easier where following a pre-assigned task-based primary response system, similar to that used by the FDNY, Mr Adams believes that such an approach may be too inflexible and this may hinder the sudden transitions in deployment that are often needed in structural firefighting. Quite often, what you expect to happen doesn't, and you need to be able to react quickly to control those changing events.

DAC Adams further believes that whilst specialist trained crews may be at individual advantage when deployed as teams to undertake specific tasks, there is greater strategic advantage in having cross-trained crews who, he states, may be more effectively deployed into rapidly evolving situations.

Hartin (Gresham) – Ed Hartin has thirty three years of service and is the Training, Safety, and EMS Division Chief in Gresham, near Portland, Oregon. Chief Hartin lectures on an international level in fire behavior and tactical ventilation and he considers these two topics to be very closely aligned.

I do not see pre-defined ventilation assignments as 'proactive', but as a reaction to prior experience (not all bad, but not necessarily proactive). Pre-defined assignments provide a simple algorithm base approach (if, then) to fire-ground tactics. When many similar incidents are encountered, this type of assignment provides a consistent response that potentially works much of the time.

Building factors are a major consideration in tactical ventilation, but not the only one. Pre-defined assignments without any consideration of burning regime (fuel or ventilation controlled), stage of fire development, and the possibility of fire spread, have the potential to result in undesirable fire behavior. This potential is increased if firefighters simply learn the 'plays' and do not understand why they are performing specific tactics.

Südmersen (Osnabrück) – Mr Südmersen is a very experienced Operational Chief and Training Officer in the city of Osnabrück in Germany. He has advanced firefighting strategy and tactics across Germany through his many articles and lectures and is a big supporter of PPV concepts. He believes that effective teamwork and standardization of training are critical to success in any venting strategy. He supports the reactive approach and believes firefighters are more adaptable to changing circumstances in this system, but only where communication is effective.

Morizot (Versailles, western Paris suburbs) – Stéphane Morizot is a thirty-one-year veteran of the French Fire Service. In 1976, he joined his home-town volunteer fire brigade where he served for three years. Then, during his one-year military service, he was assigned to a fire service unit. He became a professional firefighter in 1980 at Yvelines county Fire and Rescue Service in the western suburbs of Paris. He has been assigned in several of the busiest fire stations in the county. He was trained to CFBT in the UK and Sweden, and imported the concept to France, with Yvelines Fire and Rescue being the first French fire brigade to use a CFBT container. He is now assigned at the fire academy, and in charge of structural firefighting, CFBT and ventilation, even though he keeps an operational duty several days a month. He is a regular technical contributor to *Soldats du Feu* Magazine.

In this case, the positive aspects are that ventilation is always part of the operational tactical approach. Especially in a case such as FDNY where they have a full response (two engines, plus two trucks, plus one battalion chief and the according manpower) and when SOPs include an OV and roof man in each ladder company. In this case, ventilation isn't an option, it's mandatory, at least to take it into consideration.

DeMarse (FDNY) – Nate DeMarse has served in the US Fire Service for thirteen years. Prior to joining the FDNY in 2003, he previously served nine years in the Midwest, working in three suburban departments. He is currently assigned to a FDNY Engine Company in the Bronx. Nate is the photo editor of *Fire Engineering* Magazine and has been a Hands On Training (HOT) instructor at *Fire Engineering*'s Fire Department Instructor's Conference (FDIC) in Indianapolis, Indiana since 2006. He believes there are several advantages of pre-assigned tactical ventilation response systems and three of the most important of these are:

1. By coordinating horizontal ventilation with the attack line's advance, it will provide an opening for the intense heat, steam and the products of combustion to be expelled from the building. This is crucial as the attack line advances to extinguish the seat of the fire.
2. By coordinating horizontal ventilation with the members operating inside the fire apartment, the smoke will lift and the fire will 'light up'. In many cases this will allow members to pinpoint the exact location of the seat of the

fire very quickly. By placing your head on the floor and looking under the smoke, victims may be seen and room and furniture layouts can be observed.

3. The member providing pre-assigned ventilation also acts in a dual role. The member will search for victims that are trapped behind the fire. This procedure, also called VES (Vent-Enter-Search) has resulted in many successful rescues of civilians who would have otherwise perished as the attack line moves through the fire area.

One disadvantage of pre-assigned horizontal ventilation may take place when an inexperienced member is responsible for making a decision to ventilate a window or not. If a charged hose-line is not moving towards the seat of the fire, the member must resist the urge to vent the window until the line has been charged and ready to move in. If the window is broken prematurely, members searching the fire area for victims, and to locate the seat of the fire, can be overtaken by 'rapid fire' progress, as fresh air is drawn into the fire area.

The same restraint may be needed if a heavy wind condition is present. The member that is responsible for horizontal ventilation must recognize wind conditions and the adverse effect that they could have if blowing directly into the fire apartment. In this case, the window may not be vented until after the fire is darkened down.

Engdahl (Gothenburg) – Tony Engdhal is a Training Officer with the City of Gothenburg Fire Service in Sweden. He states: *'Ventilation is one of the many tools that we have to assist extinguishing a fire. Use it, but use it with other tools as well (water). What we often forget is how to protect the adjacent areas to the fire, just using a fan. We must also talk about ventilation with fan and without. In Sweden we are very good using Positive Pressure Ventilation in apartments and small houses, but not in big volume (industry).'*

McMaster (Washington DC) – Daniel McMaster is a twenty-year veteran of the US Fire Service and has previously served Ladder and Engine assignments in New York City (FDNY). He is currently a Lieutenant on a ladder company in Alexandria, Virginia (on the southern border of Washington DC). He is also a renowned proponent and avid supporter of aggressive ventilation tactics, which he favors over the more passive approaches used by some departments.

Lt. McMaster believes that, *'The biggest advantage of pre-assigning ventilation positions is that members will automatically be in place to perform needed tasks, from the outset, no matter what the ultimate strategy will be. Members can reach assigned positions, perform individual size-ups, relay key information, and then turn to actual vent operations as indicated. If ventilation from that position is not indicated, members can then turn to other jobs that take advantage of their position in or around the building, such as a targeted search of a room or area; if ventilation is indicated it can begin relatively quickly, as no orders or direction are needed to start the process. At worst, accessing these assigned positions allows for a clearer picture of the fire problem from the outset of operations, with an extra eye toward victims and hazards that may have not been readily apparent on arrival.'*

He continues: *'I do not feel that pre-assigning positions has any inherent disadvantages or flaws, provided that manpower levels allow for the assigned positions and tasks to be focused and manageable. If a single truck company carries six members, an effective inside and outside team can be assigned with a narrow window of responsibility; if staffing of a company is four or less, the assigned areas may be too large for efficient coverage, or the list of assigned tasks may be too long to allow for good results. Pro-active approaches are always good because they put guys in the right spots, in a quick, efficient manner; if staffing does not allow for total coverage from one company, then additional units should be assigned to fill-out the coverage of roles and positions, without calling for resources later.'*

Campaña/Milara (Madrid) – Juan Carlos Campaña and his colleague Jose Gomez Milara are both twenty-year veteran Sergeants (Captains) in central Madrid, Spain, who have driven the concepts of CFBT training in their city for the past five years. They also strongly believe that venting tactics should form part of their fire brigade's primary response strategy but at this time this is not the case:

There are clear situations in which it can be very useful to have one or two specialist trained firefighters outside the structure with clear pre-assigned tasks to venting. But I think that not in all structural fires is this needed or even convenient when venting the structure, and the final decision must be taken by the company officer. We think that fixed and pre-assigned ventilation could possibly be dangerous to the interior teams, to the spread of the fire towards these teams and towards unaffected areas, and to the safety of the occupants, unless carefully coordinated.

Beatty (FDNY) – Matt Beatty is a very experienced firefighter serving with Rescue Company One in downtown Manhattan, New York City. Mr Beatty has been a NYC firefighter for twelve years, serving three Ladder Company assignments and two Engine Company assignments before his transfer to Rescue One. He is currently working towards a bachelors degree in fire service administration.

There are basically two positions that do the bulk of venting from the outside. The roof position and the outside vent. Although the positions are pre-assigned, the tactics are not necessarily automatic. Horizontal ventilation assigned to either position is done after consulting with the truck company officer on the inside. When he requests horizontal ventilation, it is then carried out by the pre-assigned roof or OV position. The only automatic ventilation that occurs is that of the roof position. This is the venting of the bulkhead door, skylights or scuttle. The purpose of this is to relieve heat and smoke from the interior stairs to facilitate the interior firefighter's ability to 'get above' the fire for searches, and increase civilian survival on the interior stairs and floors above. The only time any cutting of the roof is done is when fire has entered the cockloft (space between the top floor ceiling and the roof). The FDNY rarely vents peaked roofs of private dwellings, as horizontal ventilation is generally sufficient.

So to answer, I believe the pre-assigned position is critical to a well-run department. It assures the position is covered. It assures the member covering this position knows exactly where he is going as he gets off the rig, and that he has the proper tools with him. It allows the officers to simply call the assigned position ('Ladder 103 to Ladder 103 Roof'), when needed. It avoids time consuming instructions, as the member assigned already knows what their duties are when they arrive at the fire.

I do not see any disadvantages. The venting of the interior stairs (in non-fireproof buildings)[1] is always an advantage. Any horizontal venting is undertaken only after consultation with the interior firefighters, so it shouldn't be an issue.

1. The term 'non-fireproof' buildings refers to older premises without protected stairways or interior fire compartmentation. The two primary considerations in 'fireproof' construction are design and materials. Fire-resisting walls, floors, and partitions to limit the spread of fire should subdivide a building. Elevator and stair-shafts, walls, light wells, and other vertical structures must be isolated for the same reason.

3.3 ATTRIBUTES OF A REACTIVE CONDITIONS-BASED VENTING STRATEGY

> **What do you consider may be the advantages or disadvantages of a reactive stance to the needs of venting structures based on fire-ground situations as they evolve?**

Adams (London) – '*Reacting to fire conditions, as they develop, offers a far more flexible approach that is better able to cope with unpredictable events as they evolve on the fire-ground.*'

Hartin (Gresham) – '*As in question one, I do not view this approach as necessarily "reactive". If ventilation tactics are selected based on assessment of conditions and **anticipation** of future fire development and spread, this is a "proactive" approach.*

The primary advantage of ventilation based on current and anticipated fire conditions is the ability to positively influence fire behavior and conditions within the structure. While not a "disadvantage" from my view (selection and implementation of ventilation tactics on the basis of conditions), this approach requires thinking fire officers and firefighters with an understanding of building construction, fire dynamics, and the influence of tactical operations.'

Morizot (Versailles, western Paris suburbs) – '*A proactive situation may bring some disadvantages if personnel aren't well trained. As we all know, there are some situations where it's better not to vent (considering the direction and strength of outside natural wind for example). So, in this case, it's absolutely necessary that firefighters are properly trained, are in a sufficient number and have the good communication equipment to analyze the situation before they act.*'

McMaster (Washington DC) – '*I think a reactive approach to venting allows for a complete size-up of conditions before establishing incident priorities; therefore, men and equipment can be more efficiently directed to areas of need, rather than deployed to positions where they may not be needed. In areas where staffing is low, it may be impractical to assign roles in advance, as the first arriving members will have to prioritize tasks based on the initial size-up, and then direct additional companies to cover other, secondary tasks. Although this approach may seem to delay some venting functions, it seems that it could prevent incorrect venting in many cases, as members would only vent in response to a direct order or assignment, whereas pre-assigned members may be more likely to vent without permission or order.*

It would seem to me that, in some cases, longer reflex times associated with a reactive stance would defeat potential benefits of rapid venting. Obviously some crews and battalions will be more experienced and efficient, but I do see a potential for delays and confusion if responsibility is not clear on arrival. Poor communication and understanding of priorities can further delay venting operations, which may end up defeating the initial purpose. Vertical ventilation will be particularly susceptible to these types of delays, as those operations usually take more time to accomplish than horizontal vent operations, and also become increasingly more hazardous to perform as time passes.'

Campaña/Milara (Madrid) – '*We consider this stance is not reactive. We believe that we have to work in a structural fire according to the needs, according to the conditions, and according to the safety of the firefighters and victims. For us the main advantage of this "reactive" stance is a more controllable environment by the company officer, who – according to his visual information (exterior) and the information from the interior teams about what they see and what their requirements are in order to carry their assignment (search, rescue, extinguish . . .) – decides how, where and when to vent.*

The decision to ventilate can't only be based on the experience of one person. The person who has the high responsibility to take the decision to ventilate must be very highly trained in fire dynamics, construction, ventilation techniques, and of course, be a very experienced officer. Of course, the rest of his team also must be well trained in order to advise their officer. Otherwise, this person can make a wrong decision and put the situation at risk.

In Madrid there is neither proactive nor reactive approach to venting. We have not the training, and even our chiefs aren't aware of the advantages and disadvantages of ventilation.

In our experience, there were a lot of situations of structural fires in which particularly a lack of ventilation had put at risk the interior teams and the result of the entire operation, simply because the ICs have not contemplated the possibility of ventilation as a tactical option.'

Beatty (FDNY) – '*A reactive approach really depends on the type of buildings in the area. The FDNY's buildings can be broken down into several different types, which is why we are able to have clear procedures on the buildings. An advantage to a reactive approach is that it would place total control over ventilation with the officer in charge. This would help to avoid indiscriminate ventilation. However, I do think overall this is a disadvantage, because the officer in charge now has to specifically instruct the firefighters to ventilate; find firefighters who are not already engaged in other operations to do it; get them in position, and carry it out. In a pre-assigned approach, the firefighter is completely focused on his ventilation duties, even while en route to the fire. As the FDNY books state "**Nothing** shall deter the roof-firefighter from carrying out this assignment".*'

3.4 TACTICAL ERRORS WHEN USING EITHER STRATEGY

> **In your experience has either of the above strategic approaches to venting resulted in a tactical error?**

Adams (London) – '*Yes! Increasingly incident commanders are not "time served" and consequently lack experience to make the right call all of the time.*

The venting plan can become uncoordinated which can put people at risk if rapid fire spread occurs as a result of poor tactical venting, particularly in very windy conditions. Unless hand-lines are available, or even well placed monitors (ground or aerial) positioned, when an area is vented the fire intensity will initially increase. You must have adequately positioned jets to control this. I have also seen poor tactical venting of the floor above present an easy route for flames to loop back to that floor from below.'

Hartin (Gresham) – '*I believe that neither pre-defined assignments nor those based on conditions are inherently prone to error. However, unthinking application of pre-defined assignments can result in poor or hazardous outcomes.*

For example, our department previously had a standard practice of Positive Pressure Ventilation by the first arriving company as the default option. In a number of cases, this resulted in poor outcomes due to a lack of foundational knowledge about the influence of this tactic on fire behavior. A shift to selection of ventilation tactics based on conditions and development of a sound knowledge base has significantly improved fire-ground effectiveness.'

Morizot (Versailles, western Paris suburbs) – *'Yes it does. As mentioned previously, the "bad" aspect about a proactive approach is that some firefighters may have an automatic action instead of a fully-thought and analyzed one – just act like robots (I do this and act like that because it's the way it was showed to me). I have seen several situations where firefighters have broken windows under the fire level just because they had been told to vent the building/house from smoke. They were doing so without considering they were feeding the fire with air, and at the end the fire really developed because of the wrong action of the fire service. This kind of situation was mainly observed at two or three-story house fires.*

I do remember a furniture store fire to which I responded. At that time I was a firefighter and on that day I was driving the pumper. It was on a Saturday evening and the store was closed, we were the first engine on-scene. Obviously, the building was burning and even though we didn't see any flames, there was quite a heavy smoke condition on the outside. The first thing decided by the IC was to force entry which was done quite quickly by breaking the glass doors giving access to the show room. After the doors were opened, several lines were stretched and ladders were raised. I remember, from where I was, seeing the situation worsen in a few minutes.

Smoke became blacker and deeper and the smoke layer was quickly banking down. Then suddenly flames could be seen in the smoke layer and soon erupted from the store. I was in a good location to observe and it was obvious that there was a link between the opening of the doors on the level of the fire and its quick and powerful development.

I still remember that fire but only understood several years later why and how this sudden development occurred. I remembered this experience even more after the tragedy which occurred in Charleston, killing nine brother firefighters in 2007.

In my experience, I would say that the percentage of success is about 40:60. I mean 40% positive and 60% negative because often people break glass to remove smoke and do forget to consider the air intake and its consequences.'

DeMarse (FDNY) – *'I have operated in buildings that were severely "under-ventilated", which have resulted in tactical errors. Those fires occurred when I was working in the Midwest, and all occurrences were in balloon-frame buildings where fire was running the voids. In my experience, if the building is not ventilated sufficiently, members operating inside will be unable to see to overhaul and expose hidden fire. All overhaul and extinguishment functions will have to be done with diminished visibility, which greatly slows down the strategic approach. Additionally, the stress load is increased on the firefighter. The end result is usually members being pulled out of the building for a defensive attack.'*

McMaster (Washington DC) – *'The majority of the errors we have seen with pre-assigned venting have been associated with incorrect timing of horizontal vents. Members have broken windows before the initial line was in a good position, allowing the fire to grow and spread and making the advancement and extinguishment more difficult. Pre-assigned vertical ventilation has worked well when members were clear on their assignments and had* been trained to perform those tasks correctly. Occasionally, a member detailed from another company was put in a position with which he was unfamiliar, resulting in various problems.

From a reactive standpoint, any "errors" we have experienced were related to poor size-ups and decision-making on the part of individual and command officers. Members who are making "where and when" decisions regarding ventilation must be able to assess key fire and building factors, and quickly assign members to address key tasks. Incorrect or poorly timed orders have resulted in ineffective venting, or have allowed interior members to endure difficult conditions that could have been avoided.'

Beatty (FDNY) – 'If firefighters are trained properly, there should be really no major issues. I've always taught the younger firefighters that all actions on the fire-ground should be taken for a specific reason, and with forethought. This forethought may have been thought out years ago, when the procedures were written, or may have to occur right then and there. But if ventilation is carried out, with reason and forethought, it should not be an issue. Now, that is not to say things can't go wrong. An example: a decision to ventilate by an officer, or a standing order even, could lead to an intensification of the fire that wasn't expected, such as a shift in wind direction.'

3.5 STAFFING REQUIREMENTS FOR PRIMARY RESPONSE VENTING TACTICS

> **In your opinion, what is the minimum number of firefighters, forming a primary response to a six-story multi-occupancy building, that would ensure ventilation could be undertaken by that first response?**

Adams (London) – '*London Fire Brigade protocols three engines (no ladder company) totalling twelve to fourteen personnel. However, I would personally consider a minimum response of sixteen firefighters would be required to ensure that venting operations were undertaken as part of the primary response tasks.*'

Hartin (Gresham) – '*I believe that this is dependent on the magnitude of the fire. However, as a baseline, 26 personnel (four engines, two trucks, and two chiefs) would provide a solid starting point.*'

Südmersen (Osnabrück) – '*In Germany, sixteen firefighters minimum are needed.*'

Morizot (Versailles, western Paris suburbs) – '*To me, the minimum response to reach this purpose is sixteen firefighters. In France, we work with groups composed of two firefighters and in an engine or pumper there are two groups (four), plus a sub officer, plus a driver (two) = six firefighters.*

So I consider you need two pumpers plus an aerial (crew of three) plus an IC (incident commander or officer in charge) = sixteen firefighters.

- One group = attack line
- One group = search/vent
- One group = water supply (hydrant)/vent

- One group = safety crew
- Ladder crew = rescue/attack/ventilation
- One IC = coordination

According to the situation (most of the time you don't need to get hydrant water supply, or the chauffeur manages to handle alone this mission), the water supply group becomes available for ventilation.

Considering the basic tactic = open up/close down – one group or ladder is in charge of opening up, one group puts a PPV fan in place and the two sub officers and the officer in charge, after coordination by radio, order PPV or just handle natural ventilation according to the situation.'

DeMarse (FDNY) – *'In my opinion, the minimum number of firefighters needed to cover the **initial** tasks at a structural fire is approximately 30–40. Please keep in mind that six-story buildings that we respond to and operate in do not have standpipe systems and are not of fireproof construction. Some of the buildings require up to sixteen 50 ft lengths of hose to reach the top floor. With that said, most of the departments in the USA do not run with this type of manpower. Many departments, some which cover urban areas, respond severely undermanned. Thus, they are unable to accomplish all of the critical fire-ground tasks simultaneously upon arrival at a structural fire. This greatly reduces the safety and efficiency of the operation. Obviously, they make it work, but it is not an ideal situation.'*

Engdahl (Gothenburg) – *'Five firefighters plus one officer.'*

McMaster (Washington DC) – *'I think that the absolute minimum number of truck (ladder) company members to allow for proper interior and exterior operations, is ten. These ten would include a two-man inside team on the fire floor, two men on the floor above, two members on the roof, and two members performing horizontal venting and searches above the fire from the front and the rear. The number of units initially assigned to the incident and the assignment of members to the various positions would depend on the staffing of the individual departments; some departments will assign units from one company to cover various assignments, while others will decide to combine members from different companies, as they arrive.'*

Campaña/Milara (Madrid) – *'Our primary fire response to a structural fire consists of one fire truck (pump), in which there are one company officer, one driver or pump operator and six firefighters; and one ladder, in which there are one driver and two or three firefighters (total eleven or twelve).*

We consider this first response enough if the conditions are normal, and the fire is manageable. Of course if there is fire spread and/or other circumstances, this response is increased. As said before, we don't usually use tactical ventilation, and although each firefighter has their own assignment, we think that with this first response it would be enough to do it as well. It would just be a matter of organization and some of the station officers are trying to introduce it.'

Beatty (FDNY) – *'At a minimum, strictly to ventilate the building (not to accomplish other tactics) would require four. Two firefighters to go to the roof and vertically vent the interior stairs, and two firefighters to ventilate the outside horizontally, the fire floor, **and** the floors above.'*

3.6 SITUATIONS WHEN NOT TO VENTILATE

> **In what specific circumstances might you NOT ventilate?**

Adams (London) –

- 'In buildings with computer controlled "air con" or venting systems where opening up will unbalance the system
- Sprinklered premises possibly – again a fire-engineered solution.'

Hartin (Gresham) – '*In offensive operations, the question is generally not "Would I ventilate?" but "**When** would I ventilate?".*'

Südmersen (Osnabrück) –
'*I would not ventilate in situations where there is:*

- Pressurized smoke;
- Under-ventilated fire with unknown seat of fire;
- Fires in large, unknown buildings.'

Morizot (Versailles, western Paris suburbs) –
'*I would not ventilate in situations where there is:*

- An obvious backdraft situation. No PPV, just create an exhaust in the upper part of the room for hot pressurized gases;
- When the room/building on fire is under the wind and the creation of an opening will blow the fire (even with a fan, there are situations where the wind strength is more powerful than a blower);
- In case of an isolated room without windows or opening. In this case I'd rather keep the room with rich gases as there is less "energy" to be absorbed, so I'd rather cool the gases and reduce the flammability range using water sprayed by a fog or combination nozzle;
- When I can't see the way the building is designed (when I can't identify precisely the inlet and the exhaust[s]);
- In high rise buildings (according to the level of the fire and the strength and direction of wind) considering that they are generally fitted with an internal ventilation system;
- No window horizontal venting behind the attack line group on their way to the room in fire – risk of rapid fire progress towards them;
- Avoid horizontal venting in under-ventilated fires, especially if crews are committed in (Osceola-Charleston);
- Basement fires: if door control isn't possible, beware of venting low level windows while an attack line group progress downstairs towards the fire (they are in the "chimney tube" and you open the air inlets).

In large open floor spaces, the main problem is to be able to put a large floor space in overpressure. The other thing is to consider being able to locate the fire so that it doesn't spread too quickly. Fire prevention rules in building construction recommend creating sectors/areas which keep the fire confined.

It's also even more difficult to vent them by traditional PPV methods because these buildings are generally fitted with sky domes. These allow smoke and hot products of combustion to get out, but limit the possibility to put the building in pressure.

The actions to undertake with a reasonable kind of success concern the control of openings just to avoid bringing in too much air.'

DeMarse (FDNY) – *'If I am operating on the interior of a building, there are a few circumstances in which I would not horizontally ventilate:*

1. High-rise fires

   Wind conditions must be carefully evaluated and communicated before window ventilation takes place. Premature horizontal ventilation in a high-rise building could cause members to be overrun by a wind-driven fire condition. The members that are sent to the floor above to search should evaluate the wind conditions on the fire-side of the building and report to the members operating below before horizontal ventilation takes place.

2. Fire will be pulled

   While searching and locating windows, those windows should not be broken if it will cause fire to be drawn to your location. This could cut off your means of egress, trapping you, or the search will have to be abandoned and the victims will perish.

3. Signs of backdraft

   A third instance may be the case of a commercial structure presenting with signs of a backdraft. In this case horizontal ventilation must be delayed until lines are stretched and vertical ventilation is attempted.

4. Operating on the floor above

   If I am operating on the floor above the fire and horizontal ventilation will allow fire to enter the floor that I am operating on via auto-exposure, I will not ventilate the window. If the window is open, I will close the window to deter auto-exposure.

If I am operating on the exterior of a building, there are a few instances where I might **delay horizontal ventilation:**

5. If members are present on a ladder or fire escape and are directly above and in the path of the fire and gases, I would delay window ventilation until they reach a safe location.

6. If ventilating the windows would cause civilians trapped above further harm and complicate rescue efforts, I would delay window ventilation until the civilians are removed.

7. I will also delay horizontal ventilation if the initial attack line has not been charged. In most cases, horizontal ventilation should be delayed until the attack line is charged and ready to advance to the seat of the fire for extinguishment. The only exception to this rule is to save a life (VES). For example, if a charged line is not in place but a victim is known or suspected trapped in the room serviced by the window you stand in front of,

ventilation may be performed to save that life. Your primary goal upon entering the window should be to close the interior door leading to the fire area in an effort to limit the fire from extending to your location.

If you are delaying window ventilation for any reason, especially in the case of a trapped civilian, you **must** *communicate to the members operating inside. The same is true if you are delayed getting into position. Communication is very important if your department expects you to perform horizontal ventilation (proactive) as the attack line moves in on the fire. Any delay should be communicated.'*

Engdahl (Gothenburg) – *'Where signs of backdraft are present.'*

McMaster (Washington DC) – *'As a general rule, if it can be vented, we will vent it. The compelling difference in our venting tactics comes from the timing and placement of the vent openings. Dangerous structural conditions will obviously preclude operations in certain areas, such as lightweight metal or wood roof assemblies that are involved in fire, or peaked roofs with dangerous pitches.*

In commercial, mercantile, or industrial buildings, high ceilings, large open floor spaces, and exceptional fire loading – without appropriate fire attack capabilities in place – will cause venting to be delayed, so as not to allow rapid fire spread to uninvolved areas.

Any building that has a significant exterior wind condition, such as a high-rise or waterfront structure, will often not be vented until all visible fire has been extinguished, if the officer in charge feels that the potential for violent fire spread is significant. Fireproof commercial high-rise buildings are often equipped with windows that do not open; even if windows can be opened, the lack of interior compartments and the high-wind potential often keep horizontal venting from taking place.

If no life hazard is present or suspected, our department will not ventilate until the fire can be controlled. If an attack line is in position to control the fire, or the fire can be held in place by interior doors or compartments, then selected venting can occur. If the construction of the building provides significant compartmentation, venting in areas distant from the fire seat can be performed as conditions indicate.

Vertical ventilation should not take place in lightweight roof assemblies that are exposed to fire on arrival, unless the members can be independently supported from a ladder or bucket. Tightly sealed buildings, or those with "smothered" fire conditions, should not get early horizontal venting at lower levels, but should be vented at the highest point before entering.

I would estimate that in my experience with private dwellings, non-fireproof apartments, and older row frames, the breaking of glass for ventilation has been generally safe and effective (70% good, 20% improper but manageable, 10% dangerous). There have been times in recent years where energy-efficient buildings have slowed fire growth to the point where routine glass-breaking has led to flashovers with members operating in the building. The tendency to respond to high temperatures encountered while searching, by breaking windows from the inside, has led to rapid fire growth in some cases. Older members, used to seeing fire on arrival and encountering relatively well-ventilated structures, have had to adjust to the perils of energy-efficient buildings and "pre-flashover" fire conditions on arrival.

The majority of our bad experiences have come from window failures in high-rise buildings, or smaller buildings exposed to high-winds. While members had previously waited for the line to be charged and "moving" before venting, this strategy has been troublesome when high-winds are involved. In high-rises, the glass will not be broken until the door is closed to the fire apartment, or water is being applied to the fire. In smaller buildings, the incident commander will give the order to break or not break the glass, based on the situation at hand.'

Campaña/Milara (Madrid) – *'We personally would not ventilate during the first moments of the response, until I would know all the circumstances of the fire (location and possible spread of the fire, victims etc.) and, of course, until all the necessary lines are charged and ready to protect all the exposures and to begin the advance over the fire.*

Of course we would not ventilate in the specific circumstance of a possible backdraft, if we could not do direct and vertical ventilation.'

Beatty (FDNY) – *'I would not ventilate immediately if the interior crews were still searching for the fire, unless they requested it. I would also not ventilate if there is a probability of a wind-driven fire being created, particularly at high-rise buildings, at least until water was being put on the fire. Also, I would hold off ventilating a bulkhead (roof door) on a high-rise building, until it can be determined what effect it will have on the fire floor. I would not cut a peaked roof private dwelling, unless fire was directly under the roof, as horizontal ventilation is almost always sufficient in these types of buildings, and personnel can be utilized elsewhere.'*

3.7 SITUATIONS WHERE VENTING SHOULD BE A PRIMARY ACTION

In what specific circumstances is ventilation a primary consideration?

Adams (London) – *'In my opinion, just about any time, the sooner the better, but you must have water with you. It is letting the heat and dangerous unburned gases out. Definitely in fire-protected stairwells, especially if they are not double lobby approach protected. Hose-lines will compromise the door seals from the fire floors and opening up also keeps the staircase cooler (wind effect) on crews who might be located there.'*

Hartin (Gresham) – *'I would generally perform tactical ventilation at all offensive operations at some point (before, during, or after fire control). However, these tactics may be combined with anti-ventilation (not opening up until hose-lines are in place, controlling air-flow to the fire to limit fire growth), depending on fire conditions and the building configuration. The monitoring of an air-track is critical to the tactical ventilation process. Tactical ventilation operations serve to influence the air-track by establishing a selected channel for both exhaust of smoke and heated gases and fresh air entering the structure.*

Reading the air-track can also provide an indication of the adequacy of ventilation openings (exhaust and inlet).'

Südmersen (Osnabrück) – *'I would recommend the use of PPV in just about any situation, except in relation to the hazardous circumstances described above. To me, it is simply a "GO" or "NO GO" decision with PPV. If you go to the interior of the structure for fire or rescue purposes, then you must ventilate. Maybe this sounds too simplified but it will work!'*

Morizot (Versailles, western Paris suburbs) – '*I would say mostly all the other situations.*

The condition to ventilate is to know why, where and how. When you consider that smoke and hot gases are fuel, then it's always a good help to be able to lower the quantity of fuel as well as heat as we know that heat produces fuel.'

DeMarse (FDNY) – '*If I am operating on the interior there are several circumstances where I would ventilate. I will elaborate:*

1. To alleviate conditions while searching for trapped victims or the fire location. As I stated above, this will allow the smoke condition to slightly lift off of the floor, increasing visibility and the chances of finding victims and the seat of the fire. If you predict that the fire will be pulled toward your location, do not ventilate the window until a charged attack line is ready to advance to the seat of the fire.
2. If the main body of fire is knocked down and members are opening walls and ceilings, checking for fire extension, I would remove windows in the immediate fire area. This will allow members to visually operate instead of operating by touch and feel only.
3. If I am working on the exterior, and horizontal ventilation is my assigned task, I will move into position opposite the attack line's advance. When the attack line is charged and ready to move in, the window should be taken.
4. Additionally, if I arrive in position prior to the attack line being charged and there is reason to believe that someone is trapped behind the fire, I will also ventilate the window. After the window has been removed and the initial "blow" of the heat and gases leaves the structure, I would enter to search for the victims (VES). My first action at this point would be to locate an interior door to close and limit the fire from extending in my direction. By closing the interior door, this will also allow the room that I am searching to lift for an easier search.

While I was working in the Midwest, we didn't have the luxury of opening stairway doors and skylights at the roof level since there were very few buildings with those features. Now I see what a positive impact those ventilation openings have on the outcome of a fire. I think it is absolutely crucial to the operation to ventilate the stairways and skylights of non-fireproof buildings as early in the operation as possible. It is so crucial that at least one member should be assigned to do this task immediately upon the fire department's arrival at the scene. Stairways, scuttles and skylights should be vented before any cutting operation is performed on the roof.

Opening stairwells and other vertical arteries prevents the smoke from banking down to lower floors. Mushrooming is described as the smoke rising to the top floor, and then banking down to lower floors due to the lack of vertical ventilation. Once the vertical arteries are opened, mushrooming is prevented. This makes stretching and operating attack lines, as well as searching and checking for fire extension, faster and more efficient. Venting the stairways will also allow civilians to evacuate normally instead of via fire escapes or fire department ladders.'

Engdahl (Gothenburg) – '*In all situations, PPV is a primary consideration and well-used tactic. We generally attempt to confine the fire and clear smoke and gases from areas* *adjacent to the fire room before fighting the fire. If this is not possible we will ventilate the fire room, as close to the fire as possible, and use the airflow from the fan to create a safer path for the firefighters.'*

McMaster (Washington DC) – *'Once members are in a position to control the fire, horizontal ventilation can begin in all areas, as indicated; once the fire is "knocked down", more complete ventilation and smoke removal can take place as needed. If a life hazard exists, or is suspected, selected venting of windows can begin before the fire is contained, with careful consideration of the likely effects. Usually this venting occurs at the fire room and/or the likely location of victims, depending on smoke and fire conditions and the interior layout of the building. In these cases members must take control of interior doors and remain in relatively safe positions, in anticipation of rapid fire growth and spread. As interior operations are taking place, a simultaneous effort to contain the fire in a room or area must be undertaken, until sufficient fire flow can be brought to bear.*

Vertical ventilation can occur immediately on arrival, provided that the building type and fire location indicate it. Stairwells in multiple dwellings, natural openings on one-story commercial buildings, and flat roofs on rowframes or private dwellings are all areas that should be accessed and ventilated as soon as possible, as their opening can have significant positive effects on occupants and the lateral spread of fire in these buildings. Attached similar buildings, or other structures where lateral fire spread to exposures is likely, must receive early, aggressive vertical ventilation in order to prevent mushrooming and to slow lateral spread sufficiently for interior attacks to be effective. Top-floor fires, attic/cockloft fires, and fires in balloon-framed structures all require early venting and control of the void spaces; fires in these buildings which have not reached the top floor or vertical voids may still benefit from vent openings over the interior stairwell, which can improve conditions on the top floor.

If a mid-rise building is non-fireproof or ordinary constructed with open stairwells, it is critical to open bulkhead doors or skylights in the stairwell to improve conditions in the common areas. If the building is fireproof, with rated stair enclosures, it is less critical initially, but may become important as time passes. When lines are advanced from stand-pipes, or victims and operating members open and close stairwell doors, the stairwell may become contaminated and difficult to pass for victims stuck above the fire. If victims cannot be "protected in place" on the upper floors, pressurization of the stairwells with rooftop openings must be made to clear poor conditions.'

Campaña/Milara (Madrid) – *'In our opinion this is one of the situations in which a rapid defensive ventilation action (vertical or PPV) can make a difference in the survival of occupants above the fire, and to stop the interior spread. I think that this is a critical priority, especially in the case of unprotected stairways.*

In our service the stairway is considered as a tactical priority, but as I said before, we don't usually ventilate it because of the lack of training and knowledge. We usually assign one or two teams for going to the stairway at the same time or after the attack team is in place. I think this must not be done before. We have found a lot of people in the areas above the fire, some dead and others in very bad condition.'

Beatty (FDNY) – *'I would always ventilate the interior stairs at non-fireproof multiple dwellings by removal of the bulkhead door and skylight. I would always ventilate as water is being applied to the fire, and to facilitate searching for life. Venting the stairway in* mid-rise buildings really depends on the construction. Is it fireproof or non-fireproof? Are the stairs enclosed (a fire door on each floor)? Or open? In a stairwell that is open, it should be a priority to vent the interior stairs. This will increase the survival chances of civilians on the upper floors, or in the stairwell, and it will allow firefighters better conditions to operate in on the upper floors and stairwell, as well as in the fire apartment. When the building is fireproof, and the stairs are enclosed, it becomes less critical, especially as the height of the building increases. Search of the stairways is still essential, but there is relative safety on the upper floors, since the hallways can be isolated from the stairs. In fact, in the FDNY, we rarely remove occupants from the floors above a fire in fireproof multiple dwellings. We will instruct them to remain in their apartment, and open the window if there is a light smoke condition. This is usually safer than walking them down the stairs, past the fire floor. A search still must be conducted in the stairwells for civilians who may have been overcome while attempting to flee the building. Eventually the stairwell door will have to be chocked open to relieve the smoke conditions. So to answer, generally if the stairwell is open, then stairwell ventilation is essential; if the stairwell is enclosed, it is not as critical early on.'

3.8 SIMPLIFYING THE TACTICAL APPROACH TO VENTING STRUCTURES

In what way can the tactical ventilation of fire-involved structures be simplified for less experienced firefighters and fire departments?

Hartin (Gresham) – 'I am not sure that ventilation strategies such as tactical ventilation or tactical anti-ventilation can (or should) be reduced to an algorithm. These decisions must be based on an understanding of fire behavior and the influence of tactical operations.

Consider the influence of increased ventilation when the fire is in the fuel-controlled or vent-controlled burning regime. If the fire is fuel controlled, increasing ventilation may slow or prevent transition to a fully developed fire. However, if the fire is vent controlled in the same compartment, increased ventilation may lead to a vent-induced flashover.

Similarly, a crew conducting search operations may take a window to improve conditions in the immediate area. This can have a negative impact on fire spread and place the crew at risk (by being between the fire and exhaust opening) if the door to the compartment being searched is not closed.

The problem is not simplifying ventilation enough for the firefighters, but finding methods to develop knowledge, skills, and expertise for firefighters to operate safely while in harm's way.'

Morizot (Versailles, western Paris suburbs) – 'When teaching ventilation, I often try to refer to a very logical and easy to understand example.

My reference is a chimney, I just remind everyone how a chimney works: there's an opening at the top, a tube and a fire at the base. When you want to "push" the fire, you open the air inlets at the lower part of the chimney. So, knowing that, when you act as a firefighter on a fire, just do the opposite. You don't want to accelerate the fire, you want to *slow it down and then extinguish it. So, limit as much as you can the air supply feeding the base of the fire. Then allow the smoke and hot gases to escape by the vent outlet and in doing so, you'll include ventilation in your tactics.*

A chimney is also a good example because you can see or imagine the process (fire, exhaust, inlet, tube) and you can easily duplicate this to a fire-scene. To make ventilation efficient, you need to know the "geography" of the building/room.

Concerning PPV, it's something different requiring another form of training which takes more time. There is also one thing I often say which is when you don't know, when you're not sure, leave the fan in the pumper. Only use PPV when you know how, where and what you're doing, and keep ready to stop the fan and close the inlet at any moment according to the evolution of the situation.'

DeMarse (FDNY) – *'I think creating a loose SOP/SOG on when and when not to ventilate could simplify the approach of ventilation. Obviously, a rule can't be written for every instance, but general operations can be stated. I say a "loose" SOP because it would have to be fluid and able to be changed. These ventilation rules should be discussed and drilled into new recruits and reiterated at company level drills to keep even the more experienced members up to date on the ventilation procedures. Whether your ventilation is proactive or reactive, some form of rule(s) could be created.'*

McMaster (Washington DC) – *'I think the best way to avoid mistakes in ventilation is to tie together all the different decisions and factors from the very earliest levels of training. Too often members are content to learn how to physically ventilate with no true comprehension of what they are doing to the overall fire condition and attack and search effort. Fire behavior, building construction, and all other pertinent areas of study should be repeatedly tied together, so that an overall understanding is gained, rather than simply a mindless reproduction of behaviors. Ventilation must be viewed as a tactical priority, which (in my opinion) is second only to water application in importance. If the gravity of these tactics are never impressed on a new member or officer, then no amount of regulation or direction will make them "get it".*

If it is not practical for all members to have such an understanding or decision-making ability, then company officers should maintain strict control over all ventilation tactics and should, in advance, make clear the guidelines for company operations. The man making the decision, whoever that man is, has to see the whole picture before he does something to affect it (positively or negatively) for all involved.'

Beatty (FDNY) – *'The first consideration should be educating the firefighters of why and when, or why not to ventilate. I think it is important for firefighters to have a thorough understanding of ventilation in the various types of buildings to be encountered in their response area, and to have SOPs based on the types of buildings they will encounter. They should understand exactly in which type of buildings vertical ventilation will occur at all times, and in which buildings it will not, or permission must be granted (and the **reason** why given).*

They should understand that horizontal ventilation should not be initiated until contact is made with the interior crews, and they request it. In a situation where a search will be made for life, and will be accomplished from the exterior, obviously ventilation will have to occur for entry to be made. In this situation, they should at least inform the officer on the interior that they are about to ventilate, and give them the opportunity to respond, prior to venting.'

3.9 BASIC GLASS RULE CONCEPTS

> **In your view, what should the basic Glass Rules state?**[2]

Adams (London) – *We have had hose-lines feeding dry risers punctured by falling glass, which compromises the water supply – not good. This places personnel at risk who may be operating in the street.*

Always try to open or remove windows before breaking glass, but if they must be broken, ensure the area is clear and that crews are expecting falling glass. Such an operation must be coordinated.'

Hartin (Gresham) – *'In much the same way as cutting a roof opening, once glass is broken you cannot change your mind (relative to location of the opening).*

Bad outcomes usually relate to inappropriate location or excessive openings (without regard to fire location).'

Südmersen (Osnabrück) – *'Because we have dual-pane windows, we normally (more than 90% of the time) open it up instead of breaking it.'*

Morizot (Versailles, western Paris suburbs) – *'To me, the rule depends on the situation, but, generally speaking, I prefer to operate in a "rich" (smoke) environment rather than in an over-ventilated compartment/structure, due to the risks of rapid fire development. The general idea would be to facilitate smoke and gas removal **in front** of the attack line and prevent air coming in from behind.'*

DeMarse (FDNY) – *'In my opinion, basic Glass Rules should be initially limited to opposite the attack hose-line's advance. After the fire is knocked down, then additional windows can be taken to assist ventilation. If there is a life-hazard, members will do whatever is needed to get to the victims. Whether your operation is VES, interior searches or both, windows will probably be broken to alleviate conditions on the interior.*

With that said, those same rules should deter "freelance ventilation". Freelance ventilation is unacceptable and could cause additional injuries to members. The practice of a member running around haphazardly smashing out windows is unacceptable and unprofessional.

*I have mixed experiences with **breaking glass** to horizontally ventilate. These mixed experiences did not necessarily cause the fire to progress, but I find it easier to simply open replacement energy efficient windows (or EEWs) than to break them. I have never opened a window with the intent to later close it.*

The way that I have been taught to remove EEWs is to unlock the window and raise it part way. This will break the integrity of the middle sash. Then strike one of the sashes outward with a tool. This will normally remove the entire half of the window. Repeat the steps for the other half of the window.

---

2. 'Glass Rules' are either documented (SOP) or local 'unwritten' rules that provide guidelines for firefighters of when windows should be vented; where they should be vented; why they should be vented and under whose directive such actions should occur. As with any venting operation, openings should serve a purpose; be closely coordinated with interior operations (crews); and be effectively located to achieve the objective to hand.

The easiest way to remove EEWs when time is less of a factor is simply to use the tabs located on the top of the sash. Open the window part way, operate the tab and pull towards you. Then twist the window out of the frame. This technique is very useful on food-on-the-stove or minor fire type runs where damage should be limited. Even in a fire situation, it is very quick if heat conditions allow you to stand up and operate the tabs.'

McMaster (Washington DC) – *'In high-rises, the glass will not be broken until the door is closed to the fire apartment, or water is being applied to the fire. In smaller buildings, the incident commander will give the order to break or not break the glass, based on the situation at hand.'*

Beatty (FDNY) – *'The decision ultimately to break glass should rest with the interior officer who knows the conditions inside the fire apartment. I find it essential to clear all large shards of glass from the window, and broken glass can be extremely slippery on fire escapes and on the ground. It is not only important for firefighters to drill and practice on breaking glass, but also on removing the sash and window frame/screens, as this is equally important to venting a window. As we say in NYC, "You are making the window into a door"!'*

3.10 AVOIDING THE ERROR CHAIN IN VENTING TACTICS

> **Effective venting demands communication; precision (where to vent); coordination (when to vent); knowing the fire location; and an instruction to do so (may be pre-assigned) – If this is all true, what are your experiences (if any) where a link in this error chain has failed in some way?**

Adams (London) – *London now have all operational staff with personal issue radios. People thought (myself included) that this might lead to general talking and consequent loss of focus – but it hasn't. What people seem to do is listen to what is being said between teams, and this has had real benefits to overall command and coordination at incidents. Radio discipline has remained good and people only talk when they need to.'*

Hartin (Gresham) – *'NIOSH often identifies the need to "closely coordinate ventilation and fire attack" in recommendations included in line-of-duty death (LODD) reports. Failure to coordinate these two tactical elements can result in increased potential for extreme fire behavior. This is particularly true when increases in ventilation are unplanned, exhaust openings are made in the incorrect location, or charged hose-lines are not in place and ready to commence fire attack when ventilation is performed (this includes opening the access point). One of the greatest problems in the application of tactical ventilation is inappropriate use of PPV. Many fire departments in the USA have an understanding of the general concept, but not how this tactic influences fire behavior. This leads to significant risk to firefighters and potential for extreme fire behavior.'*

Südmersen (Osnabrueck) – *'Experience demonstrates that this link normally fails, that is why I see, for example, "anti-ventilation" as being a very critical strategy for the German Fire Service.'*

Morizot (Versailles, western Paris suburbs) – '*I remember a demonstration that had been organized in the south of France by a PPV fan manufacturer a few years ago. The building used was a vacant warehouse for meat in which some pallets and mattresses had been set. The demo fire crew was in place backed up by another crew. The fire was started, then when heat was measured at 600 C, a door at the opposite end of the storage room was opened and afterwards, a PPV fan was started. Immediately the temperature started to diminish and a 70 mm line was put in action. Suddenly, a large quantity of black smoke began to issue from the inlet point and the fire developed very quickly with flames and smoke pushing out from both inlet and outlet openings and also through the roof.*

In fact, what happened was that the rear (exhaust) door had been closed by a reverse wind feeding in as nobody was controlling the back of the building. The IC quickly tried to understand what was going on, and when he realized, he had a group of firefighters open the back door. At the same time, an additional, larger PPV fan was started up in support, with a 45 mm line added to the 70 mm one. In less than ten minutes, this fire was brought under control, proving, if necessary, that PPV may have a major influence on fire behavior.

The lesson learned is that every link of the chain has to be connected by a means of communication and that the inlet as well as the outlet has to be under control at every moment to be able to react according to what may happen.'

DeMarse (FDNY) – '*In my opinion the system that we use works very well. Most members are not assigned to the outside vent position until they have gained experience as can, irons" and roof firefighters. The experience as the can and irons firefighter provides firefighters with a good base on when ventilation is needed, when it is not needed or when it should be delayed.*

With that said, the outside vent firefighter knows where he has to be, how to get there and when ventilation should or should not take place. Personally, I have never seen an error in that chain that wasn't immediately resolved. For example: the OV picks the wrong fire escape, but immediately realizes the error and repositions to the correct one.'

McMaster (Washington DC) – '*Our biggest issues have come from horizontal openings made before the line was actually in position to control the fire. Members have become over-eager, or misread the actions of the engine, sometimes resulting in a tougher fire to fight than was necessary. In some cases, where reactive orders were given, the reflex time associated with equipping, accessing and making the openings, has led to poor results. More often than not, these are training and motivation issues that could be easily addressed for the future.*

As far as poor location of the openings is concerned, these problems are usually found on the roof, where members cannot quickly determine the proper location from their personal size-up. Rather than asking for help in determining the location, some members simply "gave it their best shot", resulting in inefficient and poorly placed openings. However, there have also been some cases where too many "for life" openings were made in the wrong spots, resulting in excessive, uncontrolled fire growth.'

Campaña/Milara (Madrid) – '*Neither us or our fire service have enough experience about ventilation for answering your question, and for this reason we consider that our biggest mistake is not considering this "chain" at all.*'

Beatty (FDNY) – *'I have seen situations where venting was done too soon, and the fire increased in size, because the line was not moving in on the fire. I have also seen ventilation implemented too early, which caused unnecessary auto-exposure of the building and its occupants. I have further seen situations where insufficient ventilation has caused fire to spread further than it would have, if complete ventilation were done. The FDNY has had several incidents in which firefighters were killed (and many more incidents where nobody was seriously injured) where fire "blow torched" on firefighters because of ventilation caused by **failing windows** (not vented by firefighters).'*

3.11 CREATING AN OPENING – WHO IS RESPONSIBLE?

Who is immediately responsible to make a decision to break glass? (Window horizontal ventilation.)

Adams (London) – *'Whilst the incident commander has overall responsibility to initiate ventilation, the person actually undertaking it should be thinking what is below (danger of falling glass onto street); if venting the floor above the fire, can fire auto-expose into the openings created?; and how any vent opening might effect the intensity of the fire's development. Open it rather than break it – once it is broken you cannot close it!'*

Hartin (Gresham) – *'This would somewhat depend on your operational approach. Within our department, the incident commander is responsible for defining the ventilation plan, but specific task activity (like breaking the glass versus opening the window) is left to the company performing the task.*

Use of horizontal ventilation must not be indiscriminate, but integrated into the overall vent plan to control the fire and provide improved conditions on the interior. This does not preclude horizontal ventilation of an isolated compartment as part of search as long as the compartment remains isolated and this action does not negatively influence the overall ventilation effort.'

Südmersen (Osnabrück) – *'Battalion chief level.'*

Morizot (Versailles, Paris) – *'The IC is in charge to determine if glass has to be broken and especially where and when. The decision may also be the responsibility of the attack crew while progressing inside.'*

DeMarse (FDNY) – *'The individual firefighter is immediately responsible. From what I have noticed, most officers expect their members to take appropriate actions for ventilation in low or mid-rise apartment buildings and private dwellings. In high-rise multiple dwellings or office buildings, the officer dictates when and where window ventilation should take place.'*

McMaster (Washington DC) – *'We do not have a department-wide rule on the decision to break glass, but many battalions and companies follow their own informal policies. In companies with experienced officers and men, the decision is often left to the member actually making the opening – with clear guidelines for appropriate decisions. In some companies where junior members are common, the truck or engine officer will call for ventilation from the fire area, when they are ready for it.'*

Beatty (FDNY) – '*Personally, I strongly believe the decision to ventilate horizontally should be placed mainly with the officer on the fire floor and not the commander in the street. An example would be an interior team of firefighters who have not been able to locate the fire in the building. This might not be a good time to ventilate horizontally. Once the fire is located, and a hose-line is in position, the officer **inside** can call for windows to be taken from the exterior.*'

3.12 DOOR CONTROL AND AIR-TRACK MANAGEMENT

> **Are firefighters/company officers or crew commanders trained and likely to practice door control – in terms of opening/closing entry doors after firefighters have entered to control airflow feeding the fire?**

Adams (London) – '*In London, yes, this should be a consideration. However, there is always the need to protect means of egress (sometimes you need to get out in a hurry!!) and the two are not mutually supportive. Taking hand lines in tends to make this somewhat difficult in any event.*'

Hartin (Gresham) – '*We have placed increased emphasis on maintaining door control, but this is not a standard (do it always) practice. Closing the door after entry requires a specific decision on the part of the attack team and/or command.*'

Südmersen (Osnabrück) – '*Not yet – but I consider this an important tactical approach and one that is of use to the German Fire Service.*'

Morizot (Versailles, western Paris suburbs) – '*In my brigade, we introduced this particular kind of training some four years ago. This training is provided to recruits and to sub-officers recently promoted. Like previously mentioned, our main problem is that we don't have a facility adapted to provide this training, so we use the flashover FDS container and an abandoned building when we can.*'

DeMarse (FDNY) – '*From my experience, the door to the fire apartment or fire area is chocked open to allow the attack line to smoothly advance to the seat of the fire. In some cases, I have seen a ladder company officer leave a member at the entry door, but mainly for orientation purposes and not to control the air-flow to the fire area. Additionally, doors inside the fire apartment or building are also closed in an attempt to slow or contain fire extension.*'

McMaster (Washington DC) – '*All members of the department are trained in the importance of door control, particularly when searching ahead of, or above the attack line. If the first crew to reach the fire area is a truck, they are trained to search the room and then close the door upon exiting; if they are unable to enter the room, they are to close the door and search the areas away from the door, until the fire can be attacked. Most of these principles are followed quite well.*

Unfortunately, the complacency and bad habits developed by some members allow unsafe door management to take place in areas remote from the fire. Although they may understand its importance to safety, poor discipline often finds members searching remote rooms with the doors open, or hanging out on stairways, or standing directly in doorways; all of which could cause serious trouble if the fire behind, or below them, grows rapidly.

I will say that containment of the fire in its original compartment is a critical part of our initial attack operations, as the inside team from the first truck will proceed immediately to the fire area in order to search the fire room, and then close the door upon exiting. Members operating ahead of, or above the fire are also cautioned to close the doors that separate them from the spreading fire conditions. Members performing VES will immediately locate and close the door to the room they enter, before searching for victims; once the search is complete, the member will then evaluate the conditions in the rest of the building before deciding to reopen the door, or leave it closed.

The closest we come to employing the "air-track management" procedures is in a high-rise apartment building. Members will move as quickly as possible to the fire apartment and close the door to contain the fire; if conditions allow for entry into the apartment before the hose-line is in place, the door to the apartment will then be closed behind the operating members, even though they are essentially "closed in with the fire". This is done to protect the common hallway, and also to prevent the growth and spread of the fire within the apartment.

I suppose the most difficult part of making vent decisions regarding air-track management would be the trade off of poor visibility and high heat with closed doors and windows, for the relative improvement in conditions when openings are made, but the potential for rapid fire growth in return.'

Campaña/Milara (Madrid) – *'Since 2005, when we began our CFBT program, we have been running courses for our firefighters, but at this time it seems we have a long way to cover. The training will provide much needed knowledge of fire behavior, fire dynamics and ventilation factors. Currently not all firefighters or officers have attended the course, but we are beginning, very slowly, to taking into account the great weight of the tradition and the oldest tactical ideas in our chiefs' minds. We consider that we are doing a hard work, trying to change slowly a lot of things. Sometimes it is very frustrating.*

In our courses we teach them door control techniques and how to read the signs and conditions of the fire. We teach them to close the door behind the attack team to be sure the fire conditions remain stabilized.'

Beatty (FDNY) – *'In the FDNY a lot of emphasis is placed on door "control" and "management". At times an officer has the option of leaving one firefighter at the door to control it, and even keep it closed, but unlocked, so he can act as a beacon for the firefighters within the apartment. He can also keep the door closed if the wind conditions are aggravated by having the door open.'*

3.13 EXTERIOR WIND HAZARDS

> **In your experience, how much has wind direction and velocity caused problems in the fire attack where inappropriate glass has been taken out by firefighters? Is it correct to take glass out where wind may enter?**

Hartin (Gresham) – *'Wind is a major issue. However, we have had a greater problem with unplanned ventilation caused by fire effects on window glazing.*

The appropriateness of taking window glass (or leaving the door open) on the windward side is dependent on circumstances and the tactical approach being taken. Cross ventilation can be quite effective if it is applied in a planned and systematic manner.'

Südmersen (Osnabrueck) – *'I experienced a "blowtorch backdraft", where the windows failed during the attack. Wind velocity and direction should be considered – especially in high-rises.'*

Morizot (Versailles, western Paris suburbs) – *'I remember the situation of a room fire in a building that occurred on the morning of the massive thunderstorm which hit Europe in December 1999. The strength of the wind was so great that we couldn't handle doors of vehicles. We had two doors of the pumper completely ripped off and destroyed by wind as well as the hatchback of the IC car.*

At the fire, the window was broken by wind (the window frame and glass had been weakened by fire), then the wind blew into the room on fire. It looked like a blowtorch in the corridor.

I confirm that, generally speaking, it isn't correct to take glass out where wind may enter, especially if the wind is strong – it may "push" fire, smoke and hot gases into the building concerned by fire.

Venting is really an important aspect of tactical firefighting. It is a big step and many firefighters need to be properly trained and at least be able to understand how important air control is and how effective it is on fire development.'

DeMarse (FDNY) – *'In a low or mid-rise building or private dwelling, I have never seen wind direction and velocity come into play on the outcome of the fire. At fires where a high wind condition is at an angle that would blow into a fire apartment, horizontal ventilation is delayed until water is on the fire.*

I have observed window failure from fire conditions has a drastic outcome on firefighting efforts in a high-rise building. The fire was on an upper floor of a forty-one-story high-rise multiple dwelling. The windows failed as the "inside team" of the first due ladder company reached the apartment. Wind conditions (50 mph) whipped the fire into a blowtorch condition, which burned members of both the engine and ladder companies as they retreated to the hallway. Several attempts to close the door were made, but the door could not be fully closed. The hallway and attack stairway became untenable.

The original fire apartment was fully involved. Fire entered the apartment above via auto-exposure and that apartment was completely involved. The fire continued to extend to two floors above, when multiple 2.5 inch attack lines were able to advance and extinguish the fire. In order to advance on the fire, walls had to be breached from adjoining apartments. Additional lines were advanced across exterior balconies where the aluminum safety railings had melted away.

Once again, it is important to point out that these windows were not taken by firefighters, but failed due to fire conditions. I included this story to reiterate the impact that wind conditions had on this fire.'

Campaña/Milara (Madrid) – *'We had to fight some fire in which the spread of the fire was totally towards the inside, and with wind blowing into the building there were not any flames or smoke coming out from the street side. For this reason I consider that the wind is a very powerful factor to take into account when we want to open up the structure.*

We should not open up any building in the side in which a wind, with enough velocity, may enter. We don't ever know how this mass of air can affect the conditions, although sometimes it can work as a PPV. Of course, there will be situations in which it can work well (to our advantage).'

Beatty (FDNY) – *'It has happened many times in the FDNY where wind has killed and injured our members. For this reason, taking windows should be a thought-out, and planned tactic, especially in high-rise residences where the wind conditions can be severe, due to the layout of the building and the height. One of the ways the determination is made in the FDNY, is that the roof firefighters of the first and second due trucks go to the apartment directly above the fire apartment (remember . . . roof ventilation is not an initial consideration in this type of building). Their job is to chock open the apartment door, and open the windows in the apartment, to find out what will happen on the fire floor below if those windows are removed. He then notifies the officer in the fire apartment of any potential for wind problems. The officer in the fire apartment will then decide whether or not glass should be taken, and if the apartment door will remain open.'*

3.14 AUTHOR'S SUMMARY

Grimwood (Author) – A summary of the round table discussion

Firstly I want to thank all of the contributors to this round table debate on venting fire-involved structures, for sharing their views and experiences. It was both enlightening and interesting to bring these international views from highly experienced fire officers together.

It is clear that there are two distinctly different tactical approaches developing here between the two continents, and please remember, there are several urban approaches compared here. The US inner-city approach is based very much on the basic core principles of creating openings in structures very early in the operation, even during the **primary** response stage. In contrast, the European approach is generally more aligned to venting structures as a **secondary** response task and even then, only where fire conditions denote that a clear need exists. However, there is a clear trend in Europe to move towards the use of PPV for venting structures and increasingly, this is being seen as a fire attack tool, used to control and stabilize interior conditions prior to entry. This trend may also be shared by many fire departments throughout suburban USA and Europe.

One thing was quite clear, and all contributors agreed in their responses, that **inadequate staffing and/or a lack of effective training** were the two main factors that most likely led to a breakdown in venting operations where critical factors such as precision, coordination and communication may fail.

The author's experience would further confirm that to gain a true tactical advantage from any venting actions, it is absolutely critical that such openings be made at the earliest opportunity on the **primary response**. In order to do this a fire department would need:

1. Documented protocols for primary response (SOPs);
2. Directives for assigning levels of responsibility;

3. Training in protocols, operations, application and fire behavior;

4.Adequate staffing.

Note the clear need for written instructions (SOPs) and directives that assign responsibility for venting actions. *Ladders 3* (tenements) comprises a seventy-four-page document that explains in some great detail how New York firefighters are pre-assigned to pre-determined locations on a structure, from the moment the fire service arrive on-scene. In the actual document the detail is very precise and clearly explained. The thirty-seven pages in *Ladders 4* (private dwellings) are equally in-depth in their coverage. However, it would require a well-trained, adequately staffed and broadly experienced fire department to implement these measures in their entirety.

In New York there is no micro-managed style of fire command and the battalion chief, responding a few short minutes behind crews on the initial primary response, will know that the scene is generally staged on arrival for firefighters to readily fulfil roles as needed. These documents give directives for **precision, coordination and communication** in the implementation of any venting actions, and these form the basic foundation upon which safe and effective venting operations are based. Assignments are held accountable and levels of responsibility are devolved, based on the experience of individuals.

In the author's opinion, this places the FDNY response at a clear tactical advantage (in comparison) as they are pre-assigned and in position very quickly to undertake key tasks in the firefighting operation. If the strategy of pre-assigning roles and locations to the primary response were overly rigid – to the extent that any sudden transition in deployment would be slow to react to changing circumstances – then I would agree with DAC Adams' (London) suggestion of 'inflexibility'. However, all of the potential needs of a primary response, along with the range of critical tasks they might be called upon to fulfil, are inbuilt into the documented structure of FDNY strategy that is extremely versatile should the need arise:

> *An operational plan is necessary and has to be formed before the fire. The plan must be understood by all and continual training is required. This bulletin presents such a plan. However, as in any operational plan it must be flexible. For example: There may be only one ladder company at the scene or the second unit may not arrive in time to operate according to the plan. Some minor adjustments may be required.*

FDNY Ladders 3 (p6)

The critical tasks of exterior building façade rescue; forcible entry; reconnaissance; interior light-well rescue; interior search and rescue; primary attack hose-line; back-up support or secondary attack hose-line; vertical ventilation; horizontal ventilation; VES etc., are all pre-assigned and accountable in the FDNY primary response, and do not require micro-management by a single incident commander as they should occur **automatically**, or **instantly** on confirmation/request.

Admittedly, the features of 'standard' types of building construction and external fire escapes in New York are greatly in support of such tactics, whereas in London (for example) the twentieth century modernization of nineteenth and early twentieth century structural interiors was not consistent, generally producing a different layout every time. Whilst in many areas of London, building façades generally maintain a Victorian era or 1930s look, the non-standard interior alterations tend to result in large voids and inter-connecting floor spaces, leading to unpredictable fire development and rapid fire spread where deployment needs can change dramatically in just a few brief moments.

However, the author believes that one **key pre-assignment** that London (and many European inner cities in general) has repeatedly over-looked is that of the roof position. It may be that DAC Adams (and several other contributors) is quite correct in stating a minimum requirement of sixteen firefighters on-scene before tactical ventilation can become a realistic consideration in any primary response to structure fires[3]. However, the placement of a two-person roof team at every call to fire in central London, where a flat roof existed, would ensure the following:

- Early placement at a key position on the structure
- Important reconnaissance of roof and surrounds
- Can usually evaluate both rear and front for signs of fire or occupants
- Where interior light-wells or shafts exist can check the same
- Ideally located, if needed, to undertake rescue/access via rope
- Instantly available for ventilation at stair-shaft access hatches

This one role would have undoubtedly saved lives and assisted interior firefighting operations throughout the inner-city areas of London during the author's time there. In fact the author did begin a pilot project during the 1990s (working with local officers), to set up a two-person roof team for access via adjacent structures or by aerial ladder. During this short-term tactical research project there were nothing but good experiences, where stair-shafts were cleared of smoke and trapped occupants were assisted to safety via interior routes much earlier in the firefighting operation.

This is one pre-assignment that should be occurring in every inner-city area where flat roofs, terraced (row) construction, interior open light-shafts and roof access from adjacent structures may exist. Such early reconnaissance and the placement of pre-assigned roof-teams provide key information to the Incident Commander and interior crews, and assure that critical areas of a structure are immediately and automatically checked for occupants, and signs of fire. Where vertical ventilation over a stair-shaft is viable and needed at an early stage, this is often easily achieved without special tools.

The potential for using PPV as an attack tool is secondary only to the **basic core principles** of natural venting operations, but should not be underestimated. However, where using PPV in pre-attack, the application of 'forced drafts' into a fire-involved structure means that if things are going to go wrong, it's going to happen much faster! Similarly, where firefighters are inexperienced in reading fire conditions, they may not be able to adapt the strategy fast enough and implement safe actions where conditions begin to deteriorate. Therefore, the training need in pre-attack PPV (PPA) is far more detailed than many will currently realize. A solid foundation in practical fire behavior training (**CFBT**) should precede the introduction of PPA as a fire attack strategy. If this training need is not delivered then it is highly likely that the use of PPA by poorly educated firefighters will lead to some situations where buildings are burned down and firefighters are severely injured, or even killed, through the inappropriate use or misapplication of such a strategy.

---

3. Limited staffing SOPs and protocols exist that may allow limited venting tactics, within the context of critical task prioritization at fires (see Chapter Five).

Chapter 4

Important European and US case studies

4.1 INTRODUCTION

New York City – 1975–77
The South Bronx was the epitome of poverty, deprivation and decay during the late 1960–70s. Many have tried to convey the period, the area, the situation, the mood, the atmosphere and the people through words and pictures. However, one thing is certain, you really had to be there to see it for yourself, to believe how a city as large and influential in worldly affairs as New York could ignore such a tragedy within its own backyard.

The term 'South Bronx' was first coined in the 1940s by a group of social workers who identified the Bronx's first pocket of poverty in the Mott Haven neighborhood, the southernmost section of the Bronx. The deprivation in the South Bronx extended up to the Cross Bronx Expressway and just beyond by the 1970s, encompassing Hunts Point, Morrisania, Highbridge and Tremont. It was an area ravaged by fire to the extent that comparisons were often made with the World War Two bombings of London and Dresden.

Operationally and geographically, the FDNY is organized into five borough commands for the five traditional boroughs of New York – Manhattan, Brooklyn, the Bronx, Queens and Staten Island. Within those borough commands exist nine divisions, each headed by a deputy chief. Within each division operate four to seven battalions, led by a battalion chief and typically consisting of 180–200 firefighters and officers. Each battalion consists of four to eight companies, with a company being led by a captain. He or she commands three lieutenants and twenty-five firefighters.

The 7th Division in the Bronx back then was under the command of Deputy Chief William Bohner, a giant of a man who at around 6 ft 4 in towered above his firefighters as he regularly paced the frontage of blazing tenements. (I met Bill along with his wife in London in 1974 and he set up my detachment to the FDNY). His distinct white helmet always stood out and as he was almost as broad as he was tall, there was no doubt he was in command. Inside and around the structures he knew he had other experienced observers in the form of battalion chiefs, who had surmounted extensive experience in fighting fires in these massive brick and timber structures. The command and control was always organized and communication between the various sectors, even back then, was impressive. As soon as an operations request was made, the need was fulfilled, providing the resources were on-scene. If they weren't, then people adapted very quickly.

I was fortunate enough to spend eighteen long months working the area whilst on assignment to the FDNY. The period from 1975–77 was the busiest in the history of the New York City Fire Department and the centre of the fire epidemic at that time was in East Tremont. The 18th Battalion then comprised three engine companies and three ladder companies and in those days, there were additional satellite engines working from the busy firehouses to relieve the workload. At East Tremont's firehouse (Engine 45 and Ladder 58) the quarters were also shared for a time with Squad 1 who would crew the rig between the mid afternoon and the early hours, during the busiest response periods. Every third night the busiest units would interchange with quieter engine companies on the outskirts of the Bronx, to allow some rest and rehabilitation for the battered crews.

Each night shift in the South Bronx was busy beyond belief and brought periods of constant fire action where there were rarely any engine companies available. Anything from abandoned cars, to street and yard rubbish, to false alarms, where kids would 'pull' the street corner alarm boxes on hot summer evenings just for fun – the calls were constant. There was also a lot of death and destruction in this area. The tenement blocks were vast and very closely spaced. They had certain construction features that caused fires to spread and develop very quickly. Often, this rapid fire development was enhanced by gasoline, set by the arsonist. There were also booby traps left for firefighters where arsonists would hang gasoline filled balloons from the ceiling, just waiting for the moment when they burst through the heat. Sometimes the arsonists would cut holes in the floors and then cover them over so that firefighters would fall through. On other occasions a rubbish fire would be set in an alleyway between two buildings and as firefighters started to extinguish the fire, they would be bombarded with bricks and other rubbish from the rooftops. The engines had wire mesh grilles fitted to the windows as protection from stones and bottles that would regularly be thrown from the street as vehicles responded to, or returned from, emergencies. Oh people loved to 'hate' their firefighters in the South Bronx! The firefighters were mostly white and to them, represented a uniformed city authority that appeared to have forgotten their plight. Yet these very men would give their lives regularly, every year, in their attempts to protect the people who lived in the South Bronx. They considered everyone an equal and never based decisions to risk their lives in saving someone on race, colour or creed.

The weapon of choice at the time was predominantly the 1.75 in (45 mm) high-flow attack line that used poly ethylene oxide (PEO) as a water additive. The system, entitled '**Rapid Water**', offered 40% increases in flow-rate and doubled nozzle pressure through decreases in friction loss in the hose-line.

However, it has since been reported[1] that:

> *During the mid-1970s, when New York City underwent financial difficulties culminating in federal bankruptcy, the fire department was notified that layoffs were possible. [1,600 firefighters were laid off in 1975 although 700 were re-hired within three days of the lay-off]. In the midst of ensuing labor disputes, the firemen's union [reportedly] viewed this innovation as a threat to manpower requirements, due to each engine's augmented firefighting effectiveness, and allegedly sabotaged the expensive blending equipment, though this rumor was never substantiated. Moreover, the complex equipment was prone to unpredictable breakdowns, and maintenance problems were severe and ongoing. A strong factor terminating the project was that the term "slippery water" [used by some] conjured up misperceptions of personnel hazards such as unsure footholds in large, slippery areas. Accordingly, interest in the potential of PEO as a viable firefighting agent died, and no meaningful resurrection has since been attempted.*

On some nights, there would be two or three fires burning at the same time in the same street, involving very large tenement blocks. Due to the limited number of engines available, commanders would often simply make do. A single 'all hands' response would handle all the fires, and adaptable crew deployment reflected this. It was almost a nightly event to see very large five or six-story tenement blocks alight on several floors. From a high point you could generally look around the area at night and witness several large glows within a few streets of each other. The Bronx was definitely 'burning'!

Many of these structures were open to the elements and were in fact generally abandoned, with windows and doors missing. Where windows were intact there was a repeated tendency for the firefighters to open up and ventilate wherever they could. A mass assault on the structure seemed common, where all glass would be taken in an effort to remove smoke, heat and dangerous fire gases. This sometimes enabled the fires to take a hold on the building and high quantities of water were often needed from the street to deal with the escalating fire fronts.

One of the worst fires I worked in New York was up at 179th St., where two tenement blocks raged side by side. As the evening wore on the fire erupted from what must have been forty-eight windows on each frontage, and both roofs were completely engulfed in flames. In total there were nearly 100 windows with flames issuing on six levels into the street front. Never in my life had I seen anything on this scale and I think the fire had us beat! We were there for several hours and I remember thinking what it must have been like for London firefighters back in the World War Two era, with limited resources being put under pressure in this way.

1. Chen, E.B., Morales, A.J., Chen, C.-C., Donatelli, A.A., Bannister, W.W. & Cummings, B.T., (1998), Fluorescein and Poly(Ethylene Oxide) Hose Stream Additives for Improved Firefighting Effectiveness, *Fire Technology*, 34(4) p. 291–306

4.2 LEARNING FROM THE PAST

If there is one thing the human race is not good at it is learning from our past. This is a clear human failing and history supports this view. Pessimists and optimists abound among historians[2]. Pessimists argue that the study of history gets us nowhere, because it is impossible to make sense of the past, and man is incapable of learning and improving. Hegel expressed this view when he said, 'What experience and history teach us is that people and governments have never learned anything from history, or acted on principles deduced from it.' Less eloquently, but equally defeatist, Henry Ford characterized history as 'bunk'. Sadly, defeatist views of history can be used to explain and even justify our failure to act on behalf of humanity.

But there are also optimists. Their view of history gives us energy and encourages us to do what is possible to heal old wrongs, to reach out to victims, and to try to prevent the repetition of the same old mistakes. The author fits into this group because he is always of the hope that we, as firefighters, will learn from our mistakes (and we all make them)! *More often than not a firefighter's errors are the result of a failure by hierarchical management to meet their training needs.*

There are a vast number of documented case histories of structure fires that represent a learning ground for us all. The lessons of others are there for us to learn. Sometimes it is necessary to 'read between the lines' to grasp the key learning points, and researchers or investigators may not always address the real issues that are important to fire service tacticians.

Here are some of the critical issues, the root causes of tragedies that are frequently repeated throughout history:

- Complacency
- Poor knowledge or misapplication of SOPs
- Lack of awareness or experience of practical fire behavior
- A lack of knowledge or in-depth experience of venting tactics
- Poor command and control demonstrating inadequate leadership
- Inadequate communications (technological or human failings)
- Poor accountability
- Poor air management
- Inappropriate or poorly coordinated tactics
- Inadequate firefighting flow-rates
- Failure to address gas-phase hazards and combustion

> Author's note: It is not the intention to criticize any particular fire department, or individuals concerned, when reviews of past case histories are undertaken. The overriding objective is to learn what we can from the experience of others. By not doing so, we may dishonor their bravery and service. They would surely want us to learn how we may prevent future firefighters from suffering the same fate. 'With hindsight' is a privilege that just does not exist at the time, on-scene at an emergency. We do our best within the limitations of our own knowledge, experience, understanding and awareness.

2. Bindenagel, J.D., (2001), Speech: *The Return of History*, US Department of State, Forum Alpbach, Austria

There are several websites online which provide up-to-date information on fire-fighter fatalities, near-miss accidents and general safety issues. You can join the mailing lists of these excellent website services and receive regular updates on relevant safety issues that affect firefighters:

- http://origin.cdc.gov/niosh/fire/ **NIOSH firefighter fatality reports**
- http://www.firefighternearmiss.com/ **Near miss accident reports**
- http://www.firefighterclosecalls.com/ **Firefighter safety issues**

> Note: The following case study reports are provided in general abstract form only and the reader is advised to download the complete reports from relevant websites for full review.

4.3 CASE STUDY – STUDENT EXERCISE

The NIOSH reports provide a most useful tool for CFBT and Tactical Deployment instructors. Each case study provides a review that may be taken apart and analyzed from different learning perspectives:

- Fire behavior
- Firefighting tactics and SOPs
- Fire command and control
- Staffing and critical assignments
- Fire-ground risk management (including Risk Control Measures, account-ability, and air management etc.)

One way of doing this is to take a case study fire report (for example from NIOSH) and remove all the investigator's recommendations/conclusions from the front and rear of the report. This leaves a sequence of events as they occurred, and supplementary information, plans, images etc.

Then ask the students to study the report and provide their own list of recommendations. This can be either an individual or group task and may also be a collaborative classroom exercise to promote debate. Give the students a brief as to which specific topic (see example list above) should provide the foundation upon which to base their recommendations and conclusions. Following on from this, you can summarize by comparing the students' conclusions with those of the NIOSH (or other) investigators.

Blaina, UK 1996 (example)

Two male career firefighters died while trying to exit a residential structure as the fire suddenly developed. The first of eight calls to fire initially involving the ground floor of a two-story town house was received at 0603 hours. A single engine, with a crew of six volunteer firefighters, was mobilized to the property. However, the predetermined response was upgraded to two additional engines on receipt of a further call that stated that children were still inside the property.

The first engine to arrive was confronted with a heavily smoke-logged house with no signs of fire visible. A team of two firefighters wearing breathing apparatus

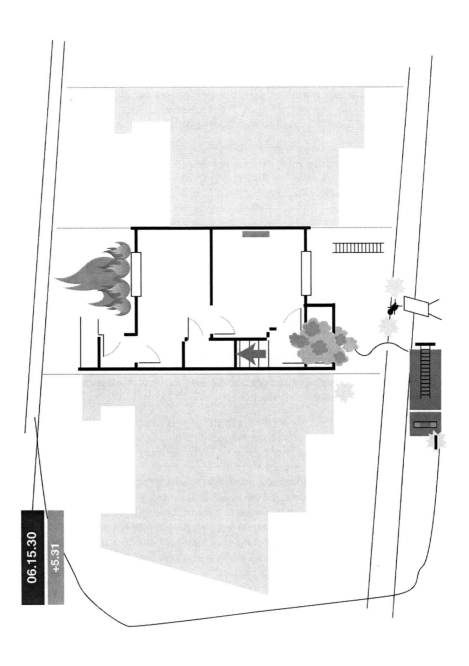

Fig. 4.1 – Blaina, UK 1996

immediately entered the property and proceeded upstairs where they located and rescued a young child. On re-entering the property to continue their search the two firefighters were caught in a backdraft that engulfed the whole house in flames (0615 hours). On trying to make their escape they were unable to open the front door that had jammed shut on the hose-line as the pressure from the backdraft had blown it closed. In their efforts to find an alternative escape route they crawled into the ground floor living room. Both firefighters remained trapped and died from their injuries, despite subsequent efforts from colleagues to advance a 45 mm hose-line into the fully involved structure to rescue them.'

Initial deployment (six firefighters):

- One incident commander
- One pump operator
- One BA entry control officer (ECO)
- Two search and rescue to upper floor to search for known life hazard
- One firefighter to rear of structure in an effort to lay attack hose-line

Initial conditions on arrival at 0610 hours:

- Heavy smoke issuing from front door
- Ground and first floor windows heavily blackened at front
- Dark smoke issuing under pressure from the eaves
- *Two children also died in this fire*

Incident Time	Actions
0610	First engine arrives on scene
0611	Two firefighters wearing BA enter at front with 19 mm hose-reel
0611	Flames reported issuing from ground floor rear
0612	Attempt by firefighter to run second hose-reel to rear of building
0613	First BA crew out of property with one child found
0615	First BA crew return inside to locate second child reported missing
0615	Backdraft occurs engulfing entire house in flames
0617	Second BA crew enter property in attempt to rescue colleagues
0619	Second engine arrives on scene – five further firefighters
0620	One line of hose run from hydrant to augment engine tank supply
0620	Third BA crew enter property to assist rescue of trapped firefighters
0625	Third BA crew exit and re-enter to advance a 45 mm hose-line in
0627	First firefighter victim removed from ground floor to street
0629	Second firefighter victim removed from ground floor to street

Note: The official report on this fire concluded that the door between the kitchen at the rear of the property and the living room at the front was closed by an adult occupant on discovering the kitchen fire. The investigators further surmised that the backdraft was actually a 'smoke explosion' that was caused by the post-flashover fire in the kitchen, breaching the ceiling and igniting an ideal pre-mix of fire gases existing on the upper floor.

UK Fire Investigator John Taylor put forth an alternative theory to this official view and the author is in total agreement that the fire development and 'rapid fire' phenomena were most probably not as the official reports concluded.

The reasons for this are:

- Photographs of a remaining edge of the kitchen door suggested that it may have been in the open position throughout the fire
- There was a heavy smoke layer reported as 'hammering out' of the front door on arrival and very early on in the firefighting operations
- Heavy dark smoke 'hammering' out of the front door suggests a fast moving gravity current (air exchange with hot fire gases and smoke) was in existence
- Very heavily stained windows at the front of the house on both floors
- Firefighters (victims) reported extremely hot conditions at high level as they ascended the stairs on the first entry
- For a heavy smoke layer to come 'hammering' out of the front door on fire service arrival, the post-flashover fire in the kitchen would have had to have breached the ceiling with some heavy flaming combustion into the rear of the upper floor, if the kitchen fire was isolated behind a closed door
- This point to point air-track, from entry door to kitchen fire to upper floor, back down stairs and out entry door, would create extreme heat conditions on the stairway where firefighters may not have been able to advance in
- Also note that a child was rescued from the rear upper floor bedroom and did not appear severely burned but rather overcome by smoke. If flaming combustion had entered this bedroom, from the floor below, some minutes before fire service arrival, then there would most likely be obvious and severe burn injuries.

Present and debate this fire with students, discussing the sequence of events and asking for their conclusions and recommendations.

Template for debates of all case studies:

1. What **fire behavior indicators** were present and how would they affect the tactical approach?
2. Discuss the effects of **staffing** in line with the prioritization of critical tasks and discuss how increased staffing levels will affect deployment.
3. Discuss the **deployment** and **command** processes of this fire.
4. Discuss how the principles of size-up and risk management might be applied in this case and discuss various control measures that could be used to **reduce risks** to firefighters and **secure team safety**, whilst still achieving rescue objectives.
5. Discuss **tactical ventilation** tactics as applied (or not applied), including potential VES approaches.

Additional points worth consideration and further debate in this case:

- Should the rescue attempt have preceded the firefighting action, or vice versa?
- Could both fire attack and rescue have been coordinated together?
- In John Taylor's investigation of this incident it was believed that flaming combustion (rollover) was in progress at the ceiling in the kitchen and living room on the ground floor and was likely extending, but hidden in the dark smoke, out into the hallway and probably curling round and up the stairs, as firefighters made their first entry.
- What simple action would have prevented this rollover from extending into the hallway? (Close the door to the lounge).
- Would VES have been a viable alternative option?
- Would zoning off the fire compartment (closing the living room door) and removing smoke from the remaining areas have assisted?

4.4 FIVE MINUTES ON THE FIRE-GROUND

At this point it is important to recognize how fire operations can take tragic turns within the first five-minute period following arrival on scene. Battalion Chief Ed Hartin (Gresham, Oregon) once presented a theory that twelve minutes was the critical timescale. He was correct, but his timeline started from the time of the first call to the fire department. In the author's view the **'deadly timeline'** begins as the first fire units arrive on-scene. Take a close look at timelines of past fires and note the similarities in events that led to such tragic circumstances. In the Blaina fire (above) the firefighters arrived on-scene at 0610 hours and 'rapid fire' progress occurred at 0615 hours – five minutes.

Another fire that demonstrated striking similarities to the Blaina fire occurred hundreds of miles and three years away. Again, keep the **deadly five-minute timeline** in mind when you review this case.

4.5 KEOKUK, IOWA 1999[3]

On 22 December 1999, a forty-nine-year-old shift commander (Victim One) and two engine operators, thirty-nine and twenty-nine years of age respectively (Victims Two and Three), lost their lives while performing search and rescue operations at are sidential structure fire. At approximately 0823 hours, the three victims and two additional firefighters cleared the scene of a motor-vehicle incident. One of the firefighters (Firefighter One) riding on Engine 3, joined the ambulance crew to transport an injured patient to the hospital. At approximately 0824 hours, Central Dispatch was notified of a structure fire with three children possibly trapped inside.

At approximately 0825 hours, Central Dispatch notified the fire department, and a shift commander and an engine operator (Victims One and Two) were dispatched to the scene in the quint (Aerial Truck 2). At 0827 hours, Engine 3 (lieutenant and Victim Three) responded to the scene.

---

3. NIOSH, USA, http://origin.cdc.gov/niosh/fire/

An adult occupant, sleeping upstairs in the front bedroom, awoke to the cries of a child. The adult opened the front bedroom door to the hall and found hot smoky conditions[4]. The adult returned to the front bedroom, opened a window on the front side of the house and called for help, alerting several neighbors. It is believed that the calls to 911 began shortly after this. The adult returned to the smoky upstairs hallway, found the crying child and exited the residence by the front bedroom window onto the roof of the front porch. Approximately two minutes later, at 0826 hours, firefighters and police began to arrive at the fire-scene. The smoke plume was visible as firefighters approached the scene and there was little if any wind disturbing the plume. Firefighters radioed Central Dispatch, reporting 'white to dark brown smoke' showing from the residence. The adult occupant was outside with the child and explained that there were three children still in the house. The front door to the residence was forced open by a police officer at approximately 0827 hours. The officer discovered heavy smoke conditions. He could not make entry into the house. At 0828 hours, the first arriving crew of firefighters prepared to enter the residence and placed a call requesting additional firefighters.

At approximately 0831 hours, the fire chief arrived at the fire-scene with an additional firefighter. Three firefighters entered the house and brought two infants from upstairs bedrooms to the front door. Two police vehicles were used to transport the infants to the hospital. The fire chief was administering CPR to the second infant and was transported to the hospital. Based on radio transmissions, the first infant was en route to the hospital at approximately 0834 hours and the second infant was en route to the hospital at approximately 0835 hours. According to witness statements, the full fire involvement of the living room, leading to a fully-involved fire condition in the stairwell, occurred as the infants were being transported to the hospital (approximately five or six minutes after fire service arrival).

A hose-line had been advanced into the entry foyer of the house. The 'dry' hose-line was placed on the floor, while the firefighter returned to the engine to charge the line. When the hose-line was 'charged' (pressurized with water) it was discovered that the hose had burned through and flames were coming out of the doorway to the house.

At approximately 0848 hours, as a second fire crew made entry into the house and began to attack the fire with a hose-line, a firefighter was discovered on the floor of the living room. Later the other two firefighters from the first crew were found on the second floor: one on the landing at the top of the stairs with a child victim, and another in the doorway of the front bedroom. All three firefighters and the one child found in the house, as well as the two children taken to hospital, died from injuries caused by the fire.

The critical event in this fire was the onset of flashover conditions in the kitchen. Within 60 seconds after the flashover occurred in the kitchen, the flames had spread through the dining room, living room and up the stairway.

Again, we see a clear opportunity for firefighters to close an internal (hall to lounge) door as they pass it, on their way up to the bedrooms. This might have effectively sealed off the fire spread, protecting their means of egress, and have bought them some more time whilst they rescued the children trapped upstairs. (Note: The end doors from the hallway to the dining room were permanently closed and inaccessible from either side).

4. Madrzykowski, D., Forney, G.P. and Walton, W.D., (2002), *Simulation of the Dynamics of a Fire in a Two-Story Duplex, Iowa, 22 December 1999*, NISTIR 6854

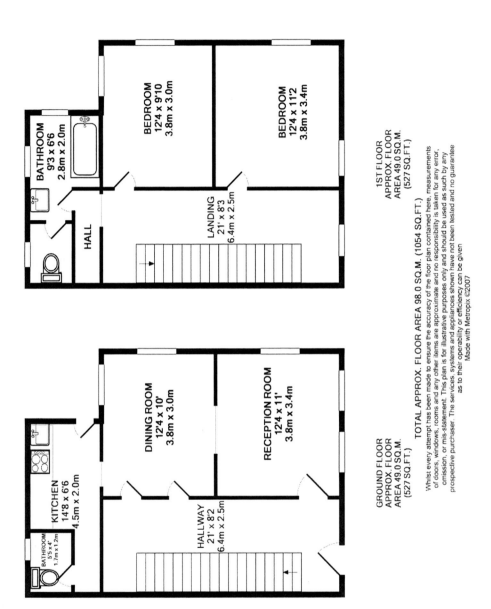

BATHROOM
9'3 x 6'6
2.8m x 2.0m

BEDROOM
12'4 x 9'10
3.8m x 3.0m

BEDROOM
12'4 x 11'2
3.8m x 3.4m

HALL

LANDING
21' x 8'3
6.4m x 2.5m

1ST FLOOR
APPROX. FLOOR
AREA 49.0 SQ.M.
(527 SQ.FT.)

KITCHEN
14'8 x 6'6
4.5m x 2.0m

BATHROOM
5'5 x 4'
1.7m x 1.2m

DINING ROOM
12'4 x 10'
3.8m x 3.0m

RECEPTION ROOM
12'4 x 11'
3.8m x 3.4m

HALLWAY
21' x 8'2
6.4m x 2.5m

GROUND FLOOR
APPROX. FLOOR
AREA 49.0 SQ.M.
(527 SQ.FT.)

TOTAL APPROX. FLOOR AREA 98.0 SQ.M. (1054 SQ.FT.)

Whilst every attempt has been made to ensure the accuracy of the floor plan contained here, measurements of doors, windows, rooms and any other items are approximate and no responsibility is taken for any error, omission, or mis-statement. This plan is for illustrative purposes only and should be used as such by any prospective purchaser. The services, systems and appliances shown have not been tested and no guarantee as to their operability or efficiency can be given
Made with Metropix ©2007

Fig. 4.2 – Keokuk, Iowa 1999

Incident time	Actions
0824	First call reporting fire.
0826	Firefighters arriving on-scene.
0827	Front door open.
0828	Firefighters on-scene requesting back-up.
0831	Fire chief arrives on-scene.
0833	Second infant removed from house by this time.
0834	First infant en route to hospital.
0835	Second infant en route to hospital, hose-line burned.
0848	Discovered firefighter on first floor.

Again, using the template for debates of all case studies:

- What fire behavior indicators were present and how would they affect the tactical approach?
- Discuss the effects of staffing in line with the prioritization of critical tasks and discuss how increased staffing levels will affect deployment.
- Discuss the deployment and command processes of this fire.
- Discuss how the principles of size-up and risk management might be applied in this case and discuss various **control measures** that could be used to reduce risks to firefighters, whilst still achieving successful rescues.
- Discuss tactical ventilation tactics as applied (or not applied), including potential VES approaches.

Interview with Chief Mark Wessel

Chief Mark Wessel is a fire chief who, like many of us, started off at the bottom rung and worked his way up. He has responded to numerous fires, rescues and related emergencies, and has reacted like many of us from the good to the bad, from the happy to the sad. And, like many of us, he has tried to do the best he can with what he has to work with, from the budgets to the equipment to the firefighters. Mark is just another hard working fire chief in the USA.

Things changed drastically for Chief Wessel and the members of the Keokuk FD in 1999 when not only three children were lost in a fire, but three of his firefighters as well[5].

What follows is an extract from a deeply moving but highly educational interview which appeared on FirefighterNation.com, by Chief Art Goodrich[6] (AG) with

---

5. Chief Billy Goldfeder – Firefighterclosecalls.com
6. The author's thanks and appreciation to Chief Art Goodrich for allowing reproduction of his interview with Keokuk's Chief Mark Wessel

Keokuk's Fire Chief Mark Wessel (MW), relating to the tragic events that occurred during this fire. If you want to know what it's like being the on-scene chief at a multi LODD here it is, right from the heart:

AG: 'Let's talk about the NIOSH report and especially, the recommendations. Staffing was an issue. It is obvious that your resources were stretched by the motor vehicle accident (MVA) and then the report of the residential fire. Is it safe to say that your initial response to the fire was a quint, engine and four personnel. Was this SOP?'

MW: 'Yes, that was the initial response. Whenever you have a total shift of six personnel, a five man minimum and answer 850 to 900 calls for service a year, you are going to have times when you respond to an emergency with three, four, or five personnel on the initial response. This is what we learned: It's not how many you respond with, it's what you do with them when you arrive. If you lose perspective of the whole picture, it doesn't matter how many you have.'

AG: 'What do you believe NIOSH considered an appropriate staffing level for a city like Keokuk?'

MW: 'I think this will also better explain the previous question. I don't think NIOSH actually stated how many personnel would be an appropriate staffing level for a community like Keokuk. If you were to take into consideration NFPA and all of the evolutions that need to be accomplished, I would think that number would be somewhere between thirteen and sixteen personnel. Now, that would be for a single-family dwelling. Next, take into consideration the age and condition of the community. How about all of the commercial structures in the community? And, the industrial base that Keokuk serves? I guess one might easily estimate the need for twenty-four to thirty personnel on duty ready to respond. But, the $700 question. How do we pay for it? We don't. We make do with what we can afford. With that comes responsibility to formulate SOPs that can be effected safely. If you can't do that, then stand back and become defensive in your attack of the emergency. It's much easier to stand in front of the media and say we had to let it burn because we did not have the resources to use a reasonable amount of safety to protect the firefighters, than it is to conduct a memorial service. It's much easier to look at a reporter with rubble in the background than to look into the faces of the grieving family of a firefighter. That I can say with certainty, and anyone reading this should take it to the bank.'

AG: 'The report recommended that the IC does initial size-up before initiating firefighting efforts and then continually evaluating risk versus gain as the incident continues. AC McNally was the highest rank initially. Wouldn't he have done a size-up before starting search and rescue? And would you not take command once on-scene under "normal" circumstances?'

MW: 'Under normal circumstances, yes. **TUNNEL VISION** *played a huge role in the way that fire was approached. Mother, with a four-year old in hand, screaming, "MY BABIES ARE INSIDE" was key to the deviation from normal operations. I believe being keyed up from the MVA that morning just prior to the call – in fact they were called off of that incident to this one – played a part in the initial operation. Having no medical transport available played a key role. One might say that this fire was routine. ROUTINE is no longer a word in our vocabulary. Other than pulling into the fire-scene and seeing smoke from a residential structure, there was nothing else routine about it. There was nothing normal about that day.'*

AG: 'Do you think too much emphasis or not enough is put on an ICS? What would it have done for you on this day? You had to get the kids out. In retrospect, break the incident down to what might have been done differently.'

MW: 'I truly feel ICS is the most important aspect of firefighter safety we can have on the emergency scene. Good command should reflect control, coordination, goals and communication. I guess I could beat myself up indefinitely over the operation. Some may even say I should. Trust me, I have. Through this I have gained nothing. What has been most effective is dissecting the incident into small enough pieces to calculate. Also, dissecting the department so that the task is not so overwhelming in the development of good SOPs, SOGs.'

AG: ' "Defensive search" was mentioned. I don't mind telling you that it put a silly look on my face. The only thing that I could think that it meant was to take a long stick and poke it through a window and maybe someone would grab it. How close am I?'

MW: 'Actually Chief, you're not too far off. What defensive search actually refers to is the idea of not over-committing. Do not place yourself in a position that you might become part of the problem. I know we train to rescue people. I know we all have learned the right hand rule and left hand rule on primary search and rescue. Let me just say this: If you have firefighters who have not had this training, they should not be your rescue team. If you are a firefighter who has not had this training, then you should refuse to perform interior search and rescue. I was teaching a basic breathing apparatus class and was asked the question about CEUs for HAZMAT Tech. I asked if the student was a Tech and he replied, "Yes". This particular student had never worn breathing apparatus. Maybe over the years things have changed that much, but I always thought you needed to wear breathing apparatus to train to the HAZMAT Tech level. Don't put yourself or your people in an over-committed environment. When and if other resources arrive, then and only then, might you consider further commitment. Stay next to a door or window to do your search. Do not commit further than your resources or training allow for a reasonable amount of safety.'

AG: ' "Maintains close accountability for all personnel at the fire scene." This would suggest that you didn't know where your FIVE people were, when it is painfully clear that you knew exactly where they were and what they were doing. Was this meant to address communications issues? Who had radios that day?'

MW: 'I did in fact know that they were performing rescue operations on the interior of the structure. When you have this few personnel on the scene, you can track everyone without too many problems. As the incident grows, you must then utilize a formal accountability system to track all the operations that are simultaneously occurring. Having a good accountability of your personnel will help to stabilize a scene, reduce freelancing and provide a safer more proficient operation. Having an established accountability program will reduce the impact of Murphy's Law.'

AG: 'NIOSH addressed communications. Were there difficulties with radio transmissions, radio equipment, and no back-up channels? What caused your radios to be a focus for their review?'

MW: 'At the time of this fire, only the officers had portable radios. Today, all personnel carry radios. There was very little communication occurring at the scene that morning. In fact, it would be reasonable to say little or none, except for initial communications with dispatch. I think NIOSH focused on this mainly because communications seems to be a common denominator in LODDs. It would seem to me that whenever a team is focused on search for a known victim, the radios become very quiet. We have worked on our communications quite a lot. We continue to have a long way to go. With radio communications there is always room for improvement. I think for me the lesson in emergency scene communications was not what was communicated but more of what was communicated.'

AG: 'RIT is a biggie. A lot of discussion over the years. At what point in this incident did you actually have enough manpower to assign RIT? And honestly? Knowing Iowa OSHA like I do, I would have bet on a citation for violating Two In/Two Out. Was RIT part of the equation early into this incident?'

MW: 'No, RIT really wasn't a consideration. Actually the Two In/Two Out rule is negated in Iowa if a known rescue is in progress. Two In/Two Out never played a role in any of the investigation. My only observation towards Two In/Two Out is: Why is it better in OSHA's eyes to perform a rescue with only one person if you know someone is trapped than if you are assuming someone may be trapped? I thought OSHA was about employee safety. If that is the case, even they make an exception to the rules (SOPs).'

AG: 'The last NIOSH recommendation addresses Personal Alert Safety System (PASS) devices. Your firefighters each wore two; one integrated into the SCBA and the other attached to their coats. Yet, no one could recall hearing any audible alarms from any of the stricken firefighters. Could it be speculated that a thermal event inside the structure rendered the devices inoperable?'

MW: 'The third party testing revealed that, due to the extreme thermal event, the electronics failed in all the audible devices. One more lesson: If it is man made, it can, and most probably will, fail at the worst time.'

AG: 'Could you talk about relationships and their importance when dealing with a traumatic event?'

MW: 'Considering I've been fortunate to have not had prior experience with a LODD, I would say we had to learn how to deal with the trauma. Fortunately, the firefighters respected each other through the entire ordeal. There were so many different emotions being experienced, you just had to wonder how the department would make it. I guess the Good Lord stayed with us through to the end. Although I'm sure we remain far from the end. Each person experiences grief in a different way and at different times. Knowing that you are going to have all these different emotions occurring, you have to stay on top of the game. We were able to come through this with little animosity and hurt feelings. It's all about RESPECT.'

AG: 'The last time you and I spoke, you told me about the McNally boys and I saw that gleam in your eye and that smile stretch across your face. Tell our readers about them.'

MW: 'All three of our men had kids at home. Some were rather young and would need to analyze all of this at a later age. Some were older and could, for as well as can be expected, experience the pain and suffering of the loss of their father immediately. I really could not relate to them very well as I had never experienced a loss of this type. All I could do is sit back and pray that the children could rationalize the loss and continue to move forward. Fortunately, to the best of my knowledge, all has gone well. As for the McNally boys; they are doing well. Pat, the oldest son of Dave, was in college working towards a degree in law enforcement. He wised up, changed his mind and moved towards an education in fire science. Pat decided he wanted to be a firefighter. Of course, I was pleased with his decision. Any father would be excited about his son or daughter following in his footsteps. The difference is, Pat had experienced the worst of times. Then Pat came to my office and said he wanted to be a firefighter in Keokuk. Well, you can imagine the mixed emotions I had. We talked quite extensively regarding the reasons he wanted to be a firefighter. Pat had the right answers, the right attitude. Pat has been with the department for over a year now, and is doing very well. I just see so much of his father in him, sometimes he'll do something or the look on his face will remind me of Dave, and I have to walk away. Usually with tears moving down my cheeks. Pat's desire to be a firefighter in Keokuk also in some way makes me feel very good inside. Dave's youngest son has also expressed an interest in the fire service, and he too would like to be a firefighter in Keokuk. I only hope I have the opportunity to make that a reality for him also.'

AG: 'That is a fitting ending to this interview, but your story of that day will continue, won't it? You have such a passion for this that I can tell that you never want anyone else, be it firefighter, family or friend to have to experience it. Your final thoughts, please, Mark.'

MW: 'As it is written in Job, "Should we accept the good that is given and not accept the bad?" Life sometimes throws a curve and we take it on the chin. I knew even as a firefighter I had a responsibility to others. My partner was relying on me for his safety. Then as I was promoted, others were relying on me as well. Eventually the department became my responsibility, and things went bad. I had always thought that I operated safely.

Sometimes your eyes get opened unexpectedly. You don't have to experience what Keokuk experienced. Why is it, we all know if we are punched in the nose, it is going to hurt like hell? Yet some of us still have to pick a fight to believe it. Let Keokuk be your punch in the nose. Let our incident be your incident. Study it. Pick it apart. Plug it into your Operating Procedures. Not just what is written, but how you actually operate on the scene. For most, you will probably find there are some major discrepancies in your written procedures and your everyday, take-it-for-granted, on-scene operations. You have the ability to "Make The Changes". Do you have the desire? If not, let someone else lead. From the bottom to the top, you must be willing to step forward. Not stand back, not stand still. This is not a social club. If you think it is, ask your family if the social pleasure is worth the risk? If you are not willing to train, then get out. Fishing is much more relaxing, but learn to swim first.

Many people have touched my life and supported my department and me through this tragedy. I can only say "Thank you" to all of them. To the firefighters of Keokuk, my hat is off to them. They exemplify the definition of firefighter. They have supported me through this, when often lines are drawn in the sand.

As long as my mind, body and soul can summon the strength, I will continue to carry the message of firefighter safety. Listen to my pain and understand how important it is for

"Everyone to go home". Keep that thought in the forefront of all you do. Do not buckle to the pressures of peers or politicians. If you can do this, you may just find yourself sleeping better at night. Stay safe.'

4.6 FAIRFAX COUNTY, VIRGINIA 2007[7]

At 0059:58 hours on 23 May 2007, Fairfax County's Department of Public Safety Communications (DPSC) received a 911 call from an occupant reporting that there was a fire in her house. Fire and rescue department units were immediately dispatched and were supplied with information indicating that people were trapped in the house. Two occupants exited the house without the assistance of firefighters. One occupant who was on the phone with alarm office died at this fire.

A fire started in the microwave oven in the first-floor kitchen. The occupants (two on the third floor and one on the second floor) awoke to find smoke throughout the house. There were no working smoke detectors as they had been disabled some years prior to this event. The occupant of the second floor went to the first floor and discovered a fire in the kitchen. He opened the front door in an attempt to remove smoke from the house and then tried to extinguish the fire utilizing the spray hose in the kitchen sink, but he was unsuccessful. This occupant also removed one dog from the first-floor bathroom. He heard glass breaking and the fire intensified, causing him to evacuate.

One occupant on the third floor used her cell phone to call the occupant of the second floor prior to calling 911. The third-floor occupants retreated to the bathroom and attempted to escape the smoke by closing the door and blocking the gaps with towels. The situation eventually forced one occupant to seek fresh air by leaning out the third-floor bedroom window where he lost consciousness and fell to the ground. The remaining occupant stayed in the bathroom, called 911 a second time and remained on the phone with the call taker until she lost consciousness, approximately two minutes after the first units arrived on-scene.

Unfortunately, despite a known location and several searches by firefighters, the trapped occupant was found at a very late stage and in a deceased condition.

Fairfax County Fire and Rescue internal findings:

- The response map did not provide an accurate depiction of the address. Specifically, it did not indicate this address was part of a back-to-back style town house.
- The first arriving engine selected and advanced an attack line that was too long. Additionally, the excess hose was not properly flaked out. These actions resulted in a delay in applying water to the fire due to kinks and a lack of pressure at the nozzle.
- The first-due engine did not provide a situation report to include a command statement. Critical information such as the back-to-back town house feature, confirmation from an occupant of a person trapped, initial actions and assignments, command statement, etc. would have provided other responders with critical information and a foundation upon which to manage the incident.

7. Fairfax County Fire and Rescue Department, Virginia, USA

- Initial ventilation operations were uncoordinated. R419 broke out a first-floor window without orders before the attack line and crew were prepared to make entry, which resulted in fire issuing from the front door and window. Ventilation must be coordinated and serve a purpose. At this event, in addition to coordinating ventilation with the attack line, upper windows could have been vented earlier to possibly provide relief to occupants remaining inside (venting for life). Subsequent ventilation by Truck 441's outside crew was coordinated through the unit officer and command and it was quite effective.
- Units failed to follow their initial assignments based upon the order of dispatch. These actions resulted in confusion as to the location and tasks of several units. This includes the fact there was no initial RIT in place. At this event, the fourth-due engine company arrived second, initially positioned too close and had to back up, and then entered the building to conduct a primary search. No apparatus (engine or truck) positioned to cover side C. Although the second truck handled all of the side C duties and deployed the ground ladders, the truck was not positioned on side C.
- Primary search efforts were uncoordinated at the company and command level, which resulted in overlapping primary searches. The result was that several crews repeatedly searched the same areas, often at the same time. This town house, like many structures, is too small to accommodate more than one crew per floor searching at a time.
- Search efforts on the third floor were ineffective. The victim who died at this fire was found in a relatively small bathroom on the third floor. Despite statements from several personnel that they searched the bathroom – both physically and with thermal imagers – the victim was not located for over forty minutes. The victim was located just inside the bathroom. There were no obstructions or obstacles to hinder locating her.
- Crews inaccurately interpreted the fire conditions. R419 entered the front door with E419's crew, but then determined the conditions were too hazardous and the rescue crew withdrew; the engine crew was not withdrawn. Fire investigators determined during their investigation of the incident that the fire was in the free-burning stage with relatively low heat above the first floor. Heat and smoke demarcation indicators were approximately 3 ft off the floor outside the first-floor kitchen area.
- Crew integrity was not maintained. R419's crew was split into two teams to conduct searches of the exposures after the officer determined that the fire building was too dangerous; these teams did not remain intact. Following searches of the exposures, three members of the crew re-entered the fire building. However, one crew member did not enter with the rest of the crew. R419's officer was unaware that R419's full crew was not present. The fourth crew member, after realizing the crew was no longer in front of the building, decided to enter the building to search for them. The crew member located another crew on the first floor and remained with that crew until the remainder of R419's crew was observed exiting the structure following their search of the third floor.
- Not all SCBA voice amplifiers were turned on which impaired voice communications.

The findings from this incident will be incorporated into future firefighter training. (Note: The Fairfax County Fire and Rescue Department are to be congratulated in establishing such an open and thorough approach to the investigation of this incident, in their efforts to ensure that such tactical errors are not repeated).

4.7 PITTSBURGH, PENNSYLVANIA 1995[8]

Three Pittsburgh firefighters died on 14 February 1995, when they ran out of air and were unable to escape from the interior of a burning dwelling. The three victims were all assigned to Engine Company 17 and had advanced the first hose-line into the house to attack an arson fire in the basement. When found, all three were together in one room and had exhausted their air supplies. Three other firefighters had been rescued from the same room, which caused confusion over the status of the initial attack team.

This incident illustrates the need for effective incident management, communications, and personnel accountability systems, even at seemingly routine incidents. It also reinforces the need for regular maintenance and inspection of self-contained breathing apparatus, emphasizes the need for PASS devices to be used at every fire, and identifies the need for training to address firefighter survival in unanticipated emergency situations.

This incident also reinforces a concern that has been identified in several firefighter fatality incidents that have occurred where there is exterior access to different levels from different sides of a structure. These structures are often difficult to size-up from the exterior and there is often confusion about the levels where interior companies are operating and where the fire is located. In these situations it is particularly important to determine how many levels are above and below each point of entry and to ensure that the fire is not burning below unsuspecting companies.

Summary of key issues

Issues	Comments
Incident command	The first arriving company did not establish command. The acting battalion chief was coming from another call and had a delayed arrival. All first alarm companies had self-committed before the acting battalion chief assumed command of the incident.
Accountability	Accountability procedures were not implemented. The locations and functions of companies operating inside the house were not known to the incident commander. It was not realized that members were missing.
Crew integrity	All crews did not function as single tactical units. Some of the individual members from these companies performed unrelated tasks and were not under the supervision of their company officers. Most of the personnel were working in temporary assignments for that shift.

8. Routley, J.G., (1995), *Three Pittsburgh Firefighters Die in House Fire*, USFA Report 078

Issues	Comments
Emergency survival actions	The actions of the three victims when they realized they were in trouble are not known; however, they do not appear to have initiated emergency procedures that could have improved their chances of survival or made other firefighters aware of their need to be rescued.
Rapid Intervention Teams	Some fire departments have adopted procedures to assign a Rapid Intervention Team at working fires. The objective of this team is to be ready to provide immediate assistance to firefighters in trouble.
Communications	There was a lack of effective fire-ground communications at this incident. There was no exchange of information with the interior crews after they entered the dwelling. All of the first alarm companies were operating before the acting battalion chief arrived and assumed command. The incident commander did not receive any progress reports from these companies.

4.8 COOS BAY, OREGON 2002[9]

On 25 November 2002, at approximately 1320 hours, occupants of an auto parts store returned from lunch to discover a light haze in the air and the smell of something burning. They searched for the source of the haze and burning smell and discovered what appeared to be the source of a fire. At 1351 hours they called 911. Units were immediately dispatched to the auto parts store with reports of smoke in the building. Firefighters advanced attack lines into the auto parts store and began their interior attack.

Crews began opening up the ceiling and wall on the mezzanine where they found fire in the rafters. Three of the eight firefighters operating on the mezzanine began running low on air. As they were exiting the building, the ventilation crews on the roof began opening the skylights and cutting holes in the roof. The stability of the roof was rapidly deteriorating, forcing everyone off the roof. The IC called for an evacuation of the building. Five firefighters were still operating in the building when the ceiling collapsed. Two firefighters escaped. Attempts were made to rescue the three firefighters while conditions quickly deteriorated. Numerous firefighters entered the building and removed one of the victims. He was transported to the area hospital and later pronounced dead. Approximately two hours later, conditions improved for crews to enter and locate the other two victims on the mezzanine.

4.9 MICHIGAN 2005[10]

On 20 January 2005, a thirty-nine-year-old male career captain (the victim) died after he ran out of air, became disoriented, and then collapsed at a residential structure fire. The combination department involved in this incident is comprised of

9. NIOSH, USA, http://origin.cdc.gov/niosh/fire/
10. NIOSH, USA, http://origin.cdc.gov/niosh/fire/

sixteen career and twelve volunteer firefighters operating out of two stations. The department serves a population of approximately 22,000 residents in a geographic area of about 26 sq miles.

The victim and a firefighter made entry into the structure with a hand-line to search for and extinguish the fire. While searching in the basement, the victim removed his regulator for one to two minutes to see if he could distinguish the location and cause of the fire by smell. While searching on the main floor of the structure, the firefighter's low air alarm sounded and the victim directed the firefighter to exit and have another firefighter working outside take his place. The victim and the second firefighter went to the second floor without the hand-line to continue searching for the fire. Within a couple of minutes, the victim's own low air alarm started sounding. The victim and the firefighter became disoriented and could not find their way out of the structure. The victim made repeated calls over his radio for assistance but he was not on the fire-ground channel. The second firefighter 'buddy breathed' with the victim until the victim became unresponsive. The second firefighter was low on air and exited. The fire intensified and had to be knocked down before the victim could be recovered.

4.10 CINCINNATI, OHIO 2003[11]

On 21 March 2003, Firefighter Oscar Armstrong III died in the line of duty after becoming trapped in a flashover while battling a residential structure fire. The fire started in the first-floor kitchen of a two-story, single-family residence. The Cincinnati Fire Department had not experienced a LODD since 28 January 1981.

The two-story house was approximately ninety years old and of ordinary un-protected construction with brick exterior walls and wooden interior members. The structure contains two stories and a basement. The entire structure contains six rooms and one bathroom. There are entrance and exit doors on the front A side, left B side, and rear C side of the structure. The entrance on the B side led directly to the main stairwell, which allowed access to the first floor, second floor, and basement. Additionally, the interior walls were covered with a thin wooden panelling throughout the areas on the first floor where the flashover occurred.

The fire originated in the kitchen of the two-story single-family residence. It was determined to have started on the stove-top from a burner that was left on with grease in the cooking pot. There was heavy fire showing from the first floor rear (side C) of the structure. The fire progressed to the flashover stage approximately 3 minutes 40 seconds after the arrival of Engine 9, the first engine company on the scene. One firefighter, Oscar Armstrong III, was killed during the flashover event and two other firefighters were injured, as they were a few feet inside the front door of the structure when the flashover occurred.

Occupant status was unknown to the responders during response and upon arrival at the incident scene. The caller reported to the dispatcher that all occupants were out of the building during his conversation with the 911 operators. This vital information was not relayed to responding companies. This information was also not obtained by first arriving companies. Therefore, the first arriving companies began aggressive interior fire operations.

11. Laidlaw Investigation Committee in cooperation with the Cincinnati Fire Department and Cincinnati Local Firefighters 48, 2004

The initial attack hose-line consisted of 300 ft (100 m) of 1.75 inch pre-connect. This is a very long hand-line and as the engine was sited just a few feet from the front door of the property there was inevitably a large amount of hose that coiled and kinked. This became worse as firefighters laid to the side and rear of the structure before returning to enter at the front. Photographic images taken from above clearly show the hose-line problem as laid to the side of the house.

This caused water problems with low pressure and flow-rate experienced at the attack nozzle being advanced in when the flashover occurred. Just prior to the flashover, there were several horizontal ventilation openings being created via windows at the side of the structure.

4.11 WORCESTER, MASSACHUSETTS 1999[12]

On 3 December 1999, six career firefighters died after they became lost in a six-floor, maze-like, cold storage and warehouse building while searching for two homeless people and fire extension. It is presumed that the homeless people had accidentally started the fire on the second floor sometime between 1630 and 1745 hours and then left the building. An off-duty police officer driving by called Central Dispatch and reported that smoke was coming from the top of the building. When the first alarm was struck at 1815 hours, the fire had been in progress for about thirty to ninety minutes. Beginning with the first alarm, a total of five alarms were struck over a span of 1 hour and 13 minutes, with the fifth called in at 1928 hours. Responding were sixteen apparatus, including eleven engines, three ladders, one rescue, and one aerial scope, and a total of seventy-three firefighters. Two incident commanders (IC One and IC Two), in two separate cars, also responded.

Firefighters from the apparatus responding on the first alarm were ordered to search the building for homeless people and fire extension. During the search efforts, two firefighters (Victims One and Two) became lost, and at 1847 hours, one of them sounded an emergency message. A head count ordered by interior command confirmed which firefighters were missing. Firefighters who had responded on the first and third alarms were then ordered to conduct search and rescue operations for victims one and two and the homeless people. During these efforts, four more firefighters became lost.

Two firefighters (Victims Three and Four) became disoriented and could not locate their way out of the building. At 1910 hours, one of the firefighters radioed command that they needed help finding their way out and that they were running out of air. Four minutes later he radioed again for help. Two other firefighters (Victims Five and Six) did not make initial contact with command nor anyone at the scene, and were not seen entering the building. However, according to the Central Dispatch transcripts, they may have joined Victims Three and Four on the fifth floor. At 1924 hours, IC Two called for a head count and determined that six firefighters were now missing. At 1949 hours, the crew from Engine 8 radioed that they were on the fourth floor and that the structural integrity of the building had been compromised. At

1952 hours, a member from the Fire Investigations Unit reported to the chief that heavy fire had just vented through the roof on the C side. At 2000 hours, interior command ordered all companies out of the building, and a series of short horn

12. NIOSH, USA, http://origin.cdc.gov/niosh/fire/

blasts were sounded to signal the evacuation. Firefighting operations changed from an offensive attack, including search and rescue, to a defensive attack with the use of heavy-stream appliances. After the fire had been knocked down, search and recovery operations commenced until recall of the box alarm eight days later on 11 December 1999, at 2227 hours, when all six firefighters' bodies had been recovered.

4.12 CHARLESTON, SOUTH CAROLINA 2007

The furniture warehouse store fire that occurred in June 2007 in Charleston, South Carolina killed nine firefighters who became disoriented inside the structure as the fire suddenly escalated. As the heart-wrenching 'maydays' were called in from multiples of firefighters lost and trapped within the structure, there were drastic attempts made from the exterior to ventilate some of the smoke out and ease their escape path. The smoke had suddenly dropped down from the ceiling and a progressive flashover ensued across the large floor space within minutes.

The fire occurred at the Sofa Super Store, which was composed of a 42,000 sq ft (3,902 sq m) single-story steel-trussed showroom building with a 17,000 sq ft (1,579 sq m) warehouse building located behind the retail space, located in the West Ashley area of Charleston. The fire started at approximately 1900 hours in a covered loading dock area built between the showroom and warehouse buildings that were attached to both buildings. At the time, the business was still open and employees were present. Charleston firefighters arrived on the scene just three minutes after the alarm, followed soon after by firefighters from the St. Andrews Public Service District.

The initial attack focused on extinguishing the fire in the loading dock area, with a secondary effort to search for and evacuate civilians, and to prevent the fire from spreading to the showroom and warehouse. Crews entering the showroom reportedly initially encountered clear visibility with only very light puffs of smoke visible near the ceiling at the back of the showroom. Shortly thereafter, the fire department opened a door to the exterior, near where the fire was raging. Efforts to close the door failed, allowing the fire to enter the showroom. Firefighters were then ordered to stretch two hose-lines into the showroom to attack the spreading fire, however the pre-connected hose-line from one of the units was reportedly too short, requiring some firefighters to again exit the building to bring in additional sections of hose and leaving only one small hand-line to hold back the growing fire.

At about this time, fire dispatchers advised the crews on-scene that they had received a 911 call from an employee who was trapped in the warehouse, which required some firefighters to direct their attention to the rescue. The trapped employee was eventually rescued by firefighters who breached an exterior wall to reach him.

Despite efforts to confine and extinguish the fire, it continued to spread into the structure and ignited furniture in the showroom, growing more quickly than the few operating hose-lines could control before additional water could be applied to the fire, however efforts to stretch and begin operating additional hose lines continued.

At 1941 hours the showroom area of the store experienced a flashover while at least sixteen firefighters were still working inside. The flashover contributed to the rapid deterioration of the structural integrity of the building, leading to a near-complete collapse of the roof some minutes later. Several calls for help were made by trapped firefighters and efforts to rescue them were commenced. These efforts proved unsuccessful. By the time the fire was brought under control, nine Charleston firefighters had lost their lives. According to Charleston County Coroner Rae Wooten, the firefighters died of a combination of smoke inhalation and burns, but not from injuries sustained from the collapse itself.

This fire is subject to extensive investigations and carries major legal implications. However, there are known facts as reported that are worthy of debate:

- The initial call was to a 'structure fire'.
- The first on-scene chief observed an exterior rubbish fire and radioed this in.
- The primary response of two engines and a ladder arrived within a few seconds of two chiefs being on-scene.
- One of these engines should have obtained a water supply according to department SOP but both reported directly to the structure.
- Both engines were supplying attack hose-lines within the first five minutes from tank water supply.
- The nearest hydrant was 500 ft from the involved building.
- At 1913 and again at 1917 hours, chiefs were calling additional engines in to lay supply lines to feed the two on-scene engines.
- A 2.5 inch attack line had been laid into the structure but could not be flowed for fear of running tanks dry before supply lines were connected.
- By 1924 hours Engine 11 was down to a quarter (tank water).
- When original supply lines were finally laid in to feed attack engines they were single 2.5 inch hose supply lines which were unable to provide adequate flow-rate in relation to the speed and intensity of the developing fire.
- When the chief of department arrived on-scene at 1916 hours the fire was developing rapidly in a large volume structure housing an extremely high fire load. There were water supply problems that prevented the required amount of water reaching the attack hose-lines and a large number of firefighters (at least sixteen) were occupying the structure.
- At this stage there was a report of a trapped occupant who was quickly reached and rescued.

Timeline

- **1908 hours** – First call reporting a possible 'structure fire' is received. The units dispatched include Charleston Fire Department Engine 10, Engine 11, Ladder 5 and Battalion 4, while Engine 16 responded for standby.
- **1911 hours** – Engines 10 and 11 arrive on-scene and B4 reports a trash and debris fire that is up against the wall in the loading dock area, but that they have not yet entered the building to check for extension.
- **1912 hours** – Ladder 5 arrives.
- **1913 hours** – (Approximate). Fire crews enter the showroom building and find no obvious fire, however the incident commander reports some light smoke is visible near the ceiling tiles.
- **1913 hours** – (Approximate). A door leading from the showroom to the loading dock area is opened by the incident commander, and the force of the fire pulls the door out of his hand. Fire enters the showroom.
- **1913 hours** – Additional engines are being assigned to water supply.

- **1916 hours** – The chief of department arrives on-scene as the fire is developing rapidly in a large volume structure housing an extremely high fire load. There are water supply problems that prevent the needed amount of water reaching the attack hose-lines and a large number of firefighters (at least sixteen) are occupying the structure.
- **1917 hours** – Additional engines are being assigned to water supply.
- **1924 hours** – Engine 11 reports that their tank water is down to a quarter full.
- **1926 hours** – An employee of the Sofa Super Store calls 911 and reports that he is trapped in the warehouse building. A crew from St. Andrews is notified of the trapped employee and attempts to locate him from the outside.
- **1929 hours** – (Approximate). The trapped employee is rescued when firefighters breach an exterior wall and pull him out of the building.
- **1932 hours** – The first firefighters in the building may have been breathing compressed air for approximately eighteen minutes and may soon run out of air. Conditions in the showroom continue to worsen while at least sixteen firefighters continue to work inside.
- **From 1932 hours** – A firefighter inside calls 'Mayday!' over his radio. Soon after, another voice on the radio is heard to say 'Car One (Chief Thomas). Please tell my wife that ... I love you.' Another firefighter inside is heard on the radio saying '... in Jesus' name, amen'. Chief Thomas orders his commanders to account for their crews and is told that some firefighters remain inside. One firefighter attempting to escape is trapped behind the large glass window in front of the showroom, and is freed when someone smashes it as other crews prepare to enter the building to rescue firefighters in distress. An emergency alert is activated on the radio of Ladder 5's engineer, who is inside, but calls to that radio go unanswered. Several PASS devices worn by firefighters are heard, meaning that firefighters in distress have manually activated them or have been motionless for at least 24 seconds. Firefighters begin smashing all of the glass in front of the store to allow escaping firefighters out and rescuing firefighters in, but this allows large amounts of oxygen to reach the fire, which quickly begins to grow in intensity.
- **1938 hours** – Chief Thomas orders a full evacuation.
- **1941 hours** – (Approximate). A flashover occurs. Virtually the entire showroom building erupts in fire within seconds. Chaotic radio traffic now ties up the radio channels, but calls about water supply problems continue. A final, unsuccessful attempt at rescue is made but quickly forced back by the intensity of the fire.
- **1945 hours** – A front section of the showroom's trussed roof collapses.

4.13 TAYSIDE, SCOTLAND 2007[13]

A woman died in a fire in her upper floor apartment when a downstairs neighbor started a fire. She was alive when the Fire Brigade arrived and the Fire Brigade knew of her location. That location information was not given to firefighters sent to search upper floors. The search teams did not search properly. Understandable assumptions

13. Tayside Fire Brigade Investigation Report

were made by fire commanders that the search would be thorough. The consequence was that by the time the woman was found she had lost her life to the fire. Some brigade personnel made mistakes.

Many fire service operations result in initial confusion as to who is accounted for and who may be missing. There is often a regular pattern of misinformation, no information and contradictions. The incident commander, regardless of rank, experience and pressure, has to make judgements and take decisions.

It was considered that the first seven minutes following arrival on-scene were the most critical to the potential for saving life (2338–2345 hours) (see gray area on timeline).

On arrival the incident commander assessed the situation. There was a serious fire in one apartment, on the left at ground floor level, and a number of persons were at windows calling for assistance. Within one minute the IC had ordered a hose-line through the front window into the ground floor left flat, which was on fire. Firefighters could not pass through the **close** (common hall) of No. 13 because flames were coming out of the door of the ground floor left flat and the close was full of smoke. The IC ordered a second hose-line through the close of No. 11 and into the rear of No. 13, to assist with the firefighting.

It was during these vital early stages that evidence given by members of the public states that a firefighter spoke to the female victim. However company commanders were adamant that information regarding the victim's predicament never reached them from any firefighter or member of the public. Had it done so they were both equally adamant it would have altered their priorities. One of the firefighters, who did subsequently enter the building in BA, had been in the back court and gave evidence of generally acknowledging various occupants at upper floors who called for assistance. He did not, however, find the victim. That occurred much later after the fire was extinguished and some other occupants had returned to their own or another flat.

Throughout these early operations the IC and sub-officer continually assessed what was happening at both front and rear of the structure and frequently spoke at the front of the building to occupants, to help reassure them that the situation was being brought under control. Naturally some residents were concerned and vocal; others were calm and quiet watching the operation. The IC in particular describes how he remained fairly constantly at the front of the structure, both to ensure he had a good command and observation position and to remain in contact with those occupants he could see.

The IC took the strategic decision to extinguish the fire as the best method of ensuring the safety of those occupants that he could readily identify at the various windows. It was, in fire service terms, very much a normal tenement incident. This normality is also probably one of the reasons for the public concern that surrounds the way the incident was managed on that evening, i.e. how could the fire service get it so wrong on a typical fire.

The speed of events was again nothing unusual. Frequently in tenement fires it can be anticipated that the priority will be the need to extinguish the fire to avoid ladder rescues. The strategy of the IC therefore reflected what was in effect a routine incident.

However, the occupant of the ground floor apartment that was on fire caused such a distraction and excitement amongst the crowd that the police present ultimately held him under arrest. The incident commander found his attention diverted by this individual and by the subsequent need to ensure that sufficient evidence was gathered surrounding this individual's actions relating to the fire.

The fire itself was therefore fought in a conventional way, albeit that the hose-line into the front window had the impact of driving some of the fire, and more particularly the products of combustion, into the stairwell (since the door had been left open by the occupant when he exited the flat).

Crews made good progress in this firefighting endeavour. However, initial deployments came under review during the investigation in to the fire.

Time	Action
2329	This is the estimated time that the fire started.
2334	First call saying that there was a fire in an apartment block in Dundee. Two engines, an aerial and an ADO (battalion chief) were dispatched to the scene.
2325	Second further call. The caller stated that, 'The windows have blown and smoke is coming out.' Third further call received. The caller said, 'There is a fire underneath me.' Fourth further call reporting the fire.
2336	Four further calls received including one from an occupant, saying, 'We are on the second floor and we can't get out because there is smoke belching through the corridor.' The control room radioed the incident commander to say, 'For your information, the occupants of the flat above are unable to exit due to the smoke.' Ninth further call received to fire.
2337	Tenth further call from female victim stating her address and saying, 'It's my house.'
2338	Two engines arrive on-scene.
2339	The aerial arrives and the incident commander sends back the message 'Make pumps 3.' (This is an assistance message used in the fire service to request a third engine). Fire Control seek confirmation: 'Is this "person reported" or just "make pumps 3"?' A11.1 radios back: 'Make pumps 3.'
2339	Eleventh further call received in which the caller refers to someone 'screaming for help.'
2340	Twelfth further call received from an occupant saying, 'I am trapped at the top of a close (hallway/stairs).' The caller goes on to say 'I cannot get out, the smoke is that thick. I cannot breathe or open my door.'

Time	Action
2343	Ambulance Control informed this is a 'persons reported.'
2344	The third fire engine arrives.
2345	Officer in charge sends a radio message saying, 'Ground floor well alight, two jets [attack hose-lines], four × SCBA in use, persons reported.'
2349	Incident commander sends a radio message: 'Three people removed from the first floor by ladder.' One man is led to safety by breathing apparatus crew.
2351	ADO (battalion chief) arrives at incident and takes command.
2354	Further radio message reporting that, 'One further male removed from second floor flat by ladder. Six breathing apparatus sets are in use.'
2355	Divisional officer (deputy chief) arrives at incident.
0000	Radio message from divisional officer saying that ADO will remain in charge of the incident and that divisional officer will undertake health and safety monitoring.
0004	Radio message from ADO that 'All persons are accounted for.'
0014	Radio message from ADO indicating, 'Stop', (i.e. no further resources required – fire extinguished).
0031	Divisional officer (deputy chief) leaves scene.
0040	ADO (battalion chief) leaves scene.
0041	Original IC (captain) resumes command.
0045	Female victim located in her apartment.

Chapter 5

Limited staffing – Three-person crews

5.1 INTRODUCTION

I was in Johannesburg lecturing at a conference when a firefighter came up to me:

Paul, all the textbooks on firefighting tactics seem to take it for granted that there are going to be adequate resources and staffing on-scene at every structure fire. Furthermore, just about all of the Standard Operating Procedures I have seen are written for staffing of five and above. In our little town we will get three firefighters responding to a structure fire on a single engine and they will be there for around thirty minutes before aid arrives from surrounding districts. Believe me when I say, things aren't going to improve in this respect. How should we approach fires? How can we base our documented guidelines (SOPs) that you talk of on a common risk-based approach?

The stark reality of limited staffing and fire-ground resources sometimes means an initial response of firefighters is restricted to a three-person crew. Further still, in some rural areas it is common for this crew to be alone on-scene, without immediate support or back-up for quite some time. It is surprising perhaps that the three person crew is a 'standard' response in many parts of the globe, including parts of rural and even urban USA. One thing is certain, and that is that adequate crewing standards should always be hotly pursued through labour relations where possible. Past staffing studies clearly demonstrate that critical tasking on the primary and secondary response to a structure fire (and a wide range of other incident types) is dramatically affected where staffing is inadequate. Key tasks just don't get carried out and firefighters are sometimes morally forced into situations where their safety is recklessly compromised.

However, where there are three-person crews operating at fires then we must surely offer clear risk-based guidelines from a safety perspective. With this in mind, the concept of 'quick-water' attack using '3D firefighting' techniques – to conserve and maximize the limited water supply on the first-arriving engine – is a strategy that has gained enormous popularity with fire departments that respond with limited resources.

The Critical Tasking Performance Index (CTPI)[1] demonstrates three-person crews are only able to guarantee 23% of the critical tasks necessary on arrival at a structure fire. An initial response of **at least** ten to fourteen firefighters must be assembled on-scene to achieve a CTPI of 100% at even the most basic of 'low-rise' structure fires. It is clear that critical tasks need careful prioritization where crews and resources are limited in numbers. However, the document also demonstrates how a limited-staffing CTPI may be dramatically improved by using the three-phased tactical approach described in the Fire2000.com staffing bulletin.

Reduced property damage, improved viability of retrieving live casualties, and safer firefighting operations for limited-staffed responses are the result of careful deployment of three-person crews following risk-assessed firefighting concepts.

In 1983–84 a study was undertaken in Dallas, USA that measured the impact of various staffing levels on the effectiveness of using three, four and five firefighters on fire apparatus responding to structure fires. The research included ninety-one full-scale fire simulations and three full-scale fire tests, where performance was measured. Prior to this there had been several other studies that measured the effect of varying crew sizes on the efficiency of existing fire strategy and tactics. However, few, if any, studies have actually attempted to optimize structural firefighting strategy and tactics in line with pre-existing *reduced* or *limited* staffing response.

5.2 CRITICAL TASK PERFORMANCE INDEX (CTPI)

The Fire2000.com research (2005) is based upon a Critical Task Performance Index (CTPI, see Fig. 5.1) and approaches the problem of limited fire-ground resource management by recognising reduced crewing as a pre-existing state. The CTPI serves as a competence grading of a first-arriving structural fire response. It is based on qualified estimations, supported by the vast experience of an international team of operational fire officers and compartment fire training specialists. Whilst the CTPI is clear to point out that fire-ground performance and firefighter safety is severely compromised in situations of limited resources, it is proposed that a risk-assessed Standard Operating Guideline can be structured in such a way that the efficiency, performance and safety of limited-resourced crews can be greatly increased.

The Dallas research of 1984 was quick to highlight that in some (a few) situations a three-person crew operated more effectively than a four-person crew. It was suggested this was due to a combination of key factors such as leadership, planning, attitude, skill, congestion, coordination, experience and motivation. However, the

1. Grimwood, P., SOG4242 Limited Staffing CTPI, http://www.fire2000.com/

general consensus of the research was quick to point out that critical tasking on the fire-ground was directly related to *time versus crew size* and key tasks were delayed where crews were under-resourced. It was proposed in the Dallas study that crew sizes below four firefighters were literally unable to achieve effective performance in laying out hose-lines, placing ladders, and augmenting water supplies to the attack pumper. The research also acknowledged that performance of crews varied depending on the differing types of risk, occupancy and levels of fire protection existing therein.

The Fire2000.com research project fully endorses the findings of the Dallas research. It is directed at both the management and deployment of resources and offers a three-phased approach to optimizing and increasing performance of limited-sized fire crews. The Fire2000.com SOG is based upon a simple twelve point **Incident Action Guide** (IAG, see Fig. 5.2) that clearly defines situations when a limited-resourced crew might commit to an interior offensive attack or when it is safer and more effective to function in a defensive mode – attempting fire confinement and/or exposure protection. A day of Visual Pattern Recognition (VPR) training can be used to assist firefighters in forming a thought process that enables decision-making based upon the IAG.

The CTPI takes into account **nine essential features**[2], or critical objectives, that require effective implementation on first arrival at a structure fire. The grading index suggests that fires on the upper floors of tall buildings, or those at large commercial or industrial risk, will place greater demands upon the responding fire force, and these are not addressed directly in the supporting IAG. As an example, the CTPI recommends that a minimum complement of ten to fourteen firefighters are needed on first response to achieve a 100% grading. A team of three firefighters is only able to guarantee 23% of the CTPI – that is one quarter of the critical tasks they may be faced with – at a small working fire in a low-rise residential structure. As with the Dallas research, the CTPI does not apply to large fires in large volume structures, where fire-ground resources are generally stretched beyond the limits of an initial response. A basic 'working fire' in a tall open-plan office building (above the sixth floor) would require at least thirty-six firefighters just to implement the fundamentals of an incident management plan, by ensuring safe and effective crew deployment and fire-ground resource support.

Critical task analysis – Nine critical core tasks (objectives)

1. Peripheral (visible) rescue (windows, ledges etc.)
2. Fire confinement (protecting exposures – defensive mode)
3. Primary fire attack (offensive mode)
4. Fire isolation (closing interior doors etc.)
5. Primary interior search and rescue
6. Continuous augmented water supply obtained
7. Provision of incident command (IC)
8. Provision of motor pump operator (MPO)
9. Provision of two outside firefighters ('Two Out' RIT)

These nine critical objectives are based upon assessed needs at hundreds of fires. They are also often perceived as contributory factors to the error chain in multiple

2. These nine core tasks may vary according to personal assessments and local review (see CTPI form).

life loss incidents, including LODDs. It may also be seen that national, federal and local health and safety regulations are acknowledged within the scope of the CTPI. In contrast to the Dallas research, the CTPI addresses 'tasks' as 'objectives' or 'roles' and does not apply the principles of physical competence aligned to first-response actions – as in placing a ladder or laying a hose-line. The Dallas research itself acknowledged that several factors such as attitude, skill, experience, coordination and motivation would directly influence such tasks.

The nine objectives are then graded individually in the CTPI, depending on their importance or relevance, to the effect that crews of one to six firefighters are able to achieve the objectives in order of priority. It should be noted here that the OSHA Two In/Two Out rule (see section 5.6) is legislated in the USA, and this may affect the percentage applied in the CTPI, though local interpretations[3] of OSHA apply. Although a three-person crew is only able to **guarantee** 23% of achievable objectives in the CTPI, they are graded at 44% overall effective in contrast to other sized crews, where working to the IAG, and if adequately trained and equipped under the three-phased approach.

Critical Task Performance Index (CTPI) *(Initial response to small working fires in low-rise, low volume structures of average fire load)*.

One firefighter – 13% effective
Two firefighters – 31% effective
Three firefighters – 44% effective
Four firefighters – 61% effective
Five firefighters – 65% effective
Six firefighters – 74% effective

Ten to fourteen firefighters – 100% effective (Where more complex tactical venting actions are required, then at least fourteen to sixteen firefighters are needed to achieve 100% grading on the CTPI).

5.3 THREE-PHASED TRAINING APPROACH

The training begins with the twelve point Incident Action Guideline (IAG), which is supported by a one day Visual Pattern Recognition (VPR) program. This will provide crews with the knowledge to make risk-assessed decisions based on sound tactical principles, which encourage structured offensive or defensive modes of attack. Following on from this, a three-phased approach is used to improve performance of limited-staffed crews by introducing a range of strategies and tactics that are ideally suited to their situation.

- **Phase One** – The use of CFBT nozzle 'pulsing' or 'bursting' techniques will conserve the apparatus water tank supply and increase the tank's working duration without an augmented supply. These techniques will optimize the available water supply, effectively cooling the overhead and gaining some rapid knock-down of fire in a 'fast attack' mode.

---

3. Some US States interpret the OSHA ruling to allow the IC to form one of the 'Two In' members, leaving the sole outside firefighter operating the pump to take temporary command.

- **Phase Two** – The use of Positive Pressure Ventilation (PPV) and anti-ventilation tactics, to create 'safety' zones, is explored, and strict protocols are followed to assure that firefighters implement the strategies safely and effectively. The result may be a safer and more comfortable working environment in which firefighters can advance to search for trapped occupants and locate the fire. This strategy is supported by the use of thermal imaging cameras. An alternative option to PPV attack is **VES**. This is most definitely a viable option for three-person crews where 'quick hit', 'get in and get out' search tactics will enable a search pattern by taking one room at a time.
- **Phase Three** – Finally, for exterior attacks, fire confinement and exposure protection, the use of water additives is explored to increase the duration of the apparatus water tank supply. Both class A foam or Compressed Air Foam Systems (CAFS) are known to extend a limited water supply's suppressive performance by up to six times, increasing the capability of a limited-staffed crew with limited resources.

It can be argued that no primary response to a residential structure fire is truly 100% effective unless first responders are able to:

- Begin or complete exterior (visible) rescues; and
- Attack the fire; and
- Undertake an immediate primary search of the interior where occupants are 'known' or 'believed' to be trapped.

5.4 INCREASING PERFORMANCE OF LIMITED-STAFFED CREWS

The objective of a three-phased approach operating according to strict risk-based protocols is to increase the performance of a three-person crew whilst maintaining their safety on the fire-ground.

- Conserving available tank water by using nozzle 'pulses' and short 'bursts' will increase tank duration by three to four times and ensure water applications are optimized by reducing run-off and increasing efficiency
- Use of Positive Pressure Ventilation (PPV) and anti-ventilation techniques, according to safe working protocols (see Chapter Two), allows approach routes to clear of smoke, heat and dangerous gases and enables the fire room to be closed down (close door) and the fire itself to be isolated, whilst the structure is searched; or for the fire to be extinguished by direct attack. Note: The ideal ventilation outlet may already be in existence and a firefighter may not therefore be needed for this task.
- Use of Class A foam or CAFS (or similar) from an exterior position, gains rapid knock-down of extensive flaming combustion and protects exposures.

Note on the IAG: Ensure compliance with OSHA (Two In/Two Out), NFPA, and other local directives at all times, and follow your own departmental procedures. The IAG is a model procedure that may be adopted or adapted where such compliance is not stipulated or applicable.

	Task	Task	Task	Task	Task	Task	Task	Task	Task
1 FFR									
2									
3									
4									
5									
6									
7									
8									
9									
10									

Critical Tasking Performance Index (CTPI)

Primary Response To:_____

√ = Yes, can complete the task with the resources available.
? = May possibly complete the task depending on task priority and resource deployment.
X = Cannot complete the task with the resources available on-scene.

Prioritize importance of nine tasks by percentage %

√ = Full percentage achieved
? = Allotted percentage divided by three
X = Zero percentage achieved

Fig. 5.1 – The Critical Tasking Performance Index

RISK-ASSESSED INCIDENT ACTION GUIDE Three-Person Crewing Structural Response	MODE OF ATTACK
1 Take control of the situation from the outset through the implementation of effective **actions** and clearly **establish achievable objectives.**	Offensive/defensive
2 **Do not commit** to an interior offensive attack or interior primary search without ensuring any obvious 'rapid fire' development is dealt with. This may entail an exterior covering line, or the closing of an interior door.	Go defensive **unless** able to 'counter' rapid fire development.
3 **Do not commit to interior alone!** Always work in crews and stay together until returning to the outside of the structure, unless using **VES** tactics where one stays at the head of the ladder whist the other enters a room.	Go defensive **unless** in crews of two at minimum.
4 **Do not commit interior** to a fire that has spread beyond the compartment of origin.	Defensive.
5 **Do not commit interior** to a compartment of origin that exceeds 25 sq m (270 sq ft), where it is fully involved in fire unless access is immediate, it is clearly safe to do so, and there is sufficient flow-rate at the nozzle	Defensive.
6 **Do not commit to interior** where more than a single length of hose (15 m or 50 ft) is needed to reach the fire from the street entry doorway	Defensive.
7 **Do not commit interior** to a smoke condition where visibility is below 1 m (an arm's length).	Defensive.
8 **Do not commit interior** into a fire demonstrating a fast moving 'air-track' or backdraft-like conditions.	Defensive.
9 **Do not commit to interior** where the structural elements: walls, floors, ceiling have been breached.	Defensive.
10 **Do not commit to interior** where the ceiling is higher than 3 m, due to the potential for accumulations of dangerous fire gases in the ceiling reservoir.	Defensive.

RISK-ASSESSED INCIDENT ACTION GUIDE Three-Person Crewing Structural Response	MODE OF ATTACK
11 **Do not commit to interior** where hot or uncomfortably developing conditions are experienced.	Offensive/defensive.
12 **Do not commit to interior** for longer than ten minutes to tackle a fire. If the fire is not extinguished within ten minutes of entry, then evacuate out to safety.	Offensive/defensive.

Fig. 5.2 –The Incident Action Guide (IAG)

5.5 EXTERIOR ATTACK STRATEGY

It should be clear that the prime purpose of the Incident Action Guide (IAG) is to protect firefighters. On arrival at a structure fire a three-person crew should take every opportunity to approach a fire from the interior, but in all cases, should conform with the IAG as well as local interpretations of the OSHA Two In/Two Out rule[4]. Where there are 'known' occupants trapped then the OSHA ruling does not apply and an interior search may be undertaken with any number of firefighters on-scene.

Another tactical option that may be considered is **VES** (see Chapter Two) where 'known' or 'suspected' occupants may be involved. A two-person crew may operate as a team checking rooms (especially bedrooms at night) from the exterior, with one remaining at the window whilst the other enters to search the room.

However, where the location of a fire compartment is obvious from the exterior, an attack from the outside of the structure must always be a consideration, even if just to gain some knock-down of flaming combustion and to stop fire spreading unchecked. If the fire is not developing to involve other parts of the structure, is fairly well confined, and the interior access routes conform to the IAG, then an interior approach may be made, providing OSHA rulings are complied with.

5.6 COMPLETING THE CTPI (OPERATIONAL REVIEW OR CLASSROOM EXERCISE)

The CTPI may be completed as an operational review of an individual fire department's objectives and capabilities. It may also be used as a classroom tool to get firefighters thinking. The nine boxes across the top are for entering what the student believes are the most critical core tasks, or objectives, that should be fulfilled at a fire of any particular occupancy type (refer to box at top of form). These 'tasks' do not necessarily relate to individual actions such as breaking windows, throwing

---

4. Some US states interpret the OSHA ruling to allow the IC to form one of the 'Two In' members, leaving the sole outside firefighter operating the pump to take temporary command.

ladders or forcing doors, but are more general **'objectives'**. The nine examples given earlier may offer some idea as to what is required of a primary response to a structure fire. It may be that you consider tactical ventilation and forcible entry should be on the list, so put them in and also increase the number of boxes (or give each student two forms) if needed.

Having then identified a list of critical core tasks (objectives) for the primary response, the students should attempt to **prioritize** and **grade** by percentage the importance of each task. For example, the need for interior fire attack **may** be seen as a higher priority over interior search and rescue, or obtaining a continuous water supply. Each task should be listed and graded on the basis of ten objectives being equal to 100%. Therefore, the average grading for each task would be 10% but some tasks will be graded higher or lower in percentage, according to their importance.

Having completed this part of the exercise, students should then be asked to consider how many of these objectives are likely to be achieved by the primary response during those vital first few minutes following arrival on-scene. It can be seen that primary response levels up to ten firefighters are included, but you can increase or reduce this, as needed. The students should enter a **tick** where the task can be viably implemented by each specific number of firefighters on-scene; a **question mark** where it is doubtful; or a **cross** where the task cannot possibly be achieved.

Having reached this stage, students can then roughly estimate a performance grading for each number of firefighters on-scene by following the guide at the base of the form:

- $\sqrt{}$ = Full percentage achieved
- ? = Allotted percentage divided by 3
- X = Zero percentage achieved

Therefore, any box achieving a tick ensures the full grading percentage given to that task is added to the overall percentage achieved for that number of firefighters. Any box with a question mark will only receive a third of the graded percentage (e.g. 3%, if 9% was the graded percentage); and any box with a cross will not receive any figure to add to the final total.

This way a grading can be concluded for three-person crews in achieving the listed tasks or objectives and then compared to other primary responses ranging from one to your own choice.

5.7 LIMITED STAFFED STRUCTURAL FIRE RESPONSE
OSHA TWO IN/TWO OUT AND NFPA 1500 STANDARDS
FIRE2000.COM GRIMWOOD

Objective: A review of OSHA, NFPA and other local standards that may present legal implications affecting the strategy and tactics of limited-staffed crews.

Date: 6 January 2006

Training: PowerPoint Presentation – One hour

OSHA (USA) Two in/Two out Regulations

The OSHA standard *29 CFR 1910.134* specifically addresses the use of respirators in Immediately Dangerous to Life or Health (IDLH) atmospheres, including interior structural firefighting. OSHA defines structures that are involved in fire beyond the incipient stage as IDLH atmospheres. In these atmospheres, OSHA requires that personnel use SCBA, that a minimum of two firefighters work as a team inside the structure, and that a minimum of two firefighters be on standby outside the structure to provide assistance or perform rescue.

The standard is badly worded and poorly written but some would argue that this incomplete text offers a range of legal 'loopholes' and options that might be exploited to bypass any rigidity in the Two In/Two Out rule when applied on the fire-ground. What is far more likely is that the standard has been badly written, period! The existence of **loopholes** leaves those that use them open to the potential for legal test cases through the courts.

The *NFPA 1500 Standard for Fire Department Occupational Safety and Health Program* also recommends a minimum of four firefighters be on-scene prior to an interior fire attack being initiated. However, both standards provide exceptions and recognize critical-tasking needs where lives may be saved or serious injuries averted.

OSHA 29 CFR 1910.134:

- At least two firefighters are to enter the IDLH atmosphere and remain in visual or voice contact with one another at all times;
- At least two employees are located outside the IDLH atmosphere; and
- All employees engaged in **interior structural firefighting** use SCBAs.
- One of the two individuals located outside the IDLH atmosphere may be assigned to an additional role, such as incident commander in charge of the emergency, or safety officer, so long as this individual is able to perform assistance or rescue activities without jeopardizing the safety or health of any firefighter working at the incident.
- Nothing in this section is meant to preclude firefighters from performing emergency rescue activities for a **'known life hazard'** before an entire team has assembled.
- The term **'known life hazard'** is defined (by the International Association of Firefighters [IAFF]) as seeing or hearing an occupant or being told directly by a witness that there is **definitely** an occupant (or occupants) inside. The normal prompts of toys on the lawn, cars in the drive, or tot finder stickers in a window, are not sufficient to bypass the OSHA standard and commit to an interior search without at least four firefighters on-scene.
- Where the IC forms part of the Two Out, and where the MPO is **NOT** documented by state OSHA as one of the acceptable Two Out, **FIVE** firefighters are needed on-scene to comply with the OSHA standard before an interior attack can begin!

Legal compliance

It is stated that compliance with federal or local OSHA (and other) regulations (where in force) is beyond debate. However, it is most likely that in practice there are fire departments that are utilizing loopholes in the standards to implement 'urgent' measures on the fire-ground, in the belief they are still legally conforming to applicable standards.

The purpose of the standards is to improve the safety of first responders but it can equally be argued that in some situations, the standards serve to hinder a safer approach. Two outside firefighters might be better employed in some circumstances in attacking either the fire itself, or in protecting those that have committed in advance, in search of trapped or remaining occupants. This statement is neither in support of the standards nor against them, but merely recognizes that there are definitions and statements that appear therein that are ambiguous in any practical application.

Loopholes
(Open to the legal test process)
In some local state OSHA definitions it has been documented that the pump operator (MPO) cannot form one of the Two Out if the IC remains outside. In other states the pump operator has been approved to act as one of the Two Out providing the pump is set and running. *NFPA 1500* itself states the IC can delegate the IC to the MPO if he/she commits as one of the Two In.

In some states (e.g. Oregon) it has been documented that a team committed for 'investigation' is not involved in 'structural firefighting' and can therefore work outside the scope of the regulations – a Two In/One Out for example. Other potential loopholes exist where a team is committed to an interior fire that is defined as being in the 'incipient' stage. The term 'incipient' can be defined as 'a fire not having reached the free burning stage,' and this definition itself is open to challenge in the courts.

The State of New Mexico (OSHA), with Federal acknowledgement, have further interpreted the rules as follows:

The standard does not require the Two In/Two Out provision if the fire is still in the incipient stage and it does not prohibit firefighters from fighting the fire from outside before sufficient personnel have arrived. It also does not prohibit firefighters from entering a burning structure to perform rescue operations when there is a **reasonable belief** *that victims* **may** *be inside. It is only when firefighters are engaged in the interior attack of an interior structural firefighting that the Two In/Two Out requirement applies. It is the incident commander's responsibility to judge whether a fire is an interior structural fire and how it will be attacked.*

Another legal loophole (and a good strategy) might be the use of **VES** tactics whereby a team of two firefighters (incident commander and firefighter) work from the exterior windows serving bedrooms and other parts of the structure. With the IC remaining at the head of the ladder he/she is still in visual or voice contact as the other firefighter enters for a quick sweep search of individual rooms. As each room is searched the firefighter returns to the ladder and they then re-site from the exterior to search another room. The legal argument is interesting as can the IC (at the head of the ladder) be considered both 'In' and 'Out' for the purposes of the OSHA ruling? Nowhere does OSHA stipulate four firefighters are needed on-scene for compliance, although this is implied. The legal wording requires Two In and Two Out and this point (in this scenario) may be arguable in a court test-case.

NFPA 1500
It is stated within the text of this Standard that if immediate action(s) might serve to prevent life loss or serious injury, then the need to act prior to four firefighters arriving on-scene is acceptable if based on an effective and justified size-up and risk assessment.

British Columbia, Canada

What seems to offer a compromise ruling to the OSHA Two In/Two Out approach may be seen through regulations applied in British Columbia, Canada.

Regulation 31.23 Entry into buildings

1. When self-contained breathing apparatus must be used to enter a building, or similar enclosed location, the entry must be made by a team of at least two firefighters.
2. Effective voice communication must be maintained between firefighters inside and outside the enclosed location.
3. During the initial attack stages of an incident, at least one firefighter must remain outside.
4. A suitably equipped rescue team of at least two firefighters must be established on the scene before sending in a second entry team and not more than ten minutes after the initial attack.
5. The rescue team required by Subsection (4) must not engage in any duties that limit their ability to make a prompt response to rescue an endangered firefighter while interior structural firefighting is being conducted.

In Summary:

- Firefighters utilizing SCBA in IDLH, potentially IDLH or unknown atmospheres shall operate in a buddy system with two or more personnel.
- Firefighters using the buddy system are required to be in direct voice or visual contact or tethered with a signal line. Radios or other means of electronic contact shall not be substituted for direct visual contact for employees within the individual team in the danger area.
- Identically equipped and trained firefighters are required to be present outside the IDLH, potentially IDLH or unknown atmospheres prior to a team entering, and during the team's work in the hazard area in order to account for, and be available to assist or rescue, members of the team working in the IDLH, potentially IDLH or unknown atmospheres.
- A minimum of four individuals is required, consisting of two individuals working as a team in the IDLH, potentially IDLH or unknown atmospheres and two individuals present outside this atmosphere for assistance or rescue at emergency operations where entry into the danger area is required.
- OSHA allows for one of the two individuals outside the hazard area to be engaged in other activities, such as incident commander in charge of the emergency incident or the safety officer. However, OSHA does state that the assignment of operators of heavy equipment as standby personnel, could clearly jeopardize the safety and health of the workers in the danger area.
- If a rescue operation is necessary, OSHA requires that the buddy system be maintained by the rescue team while entering the IDLH, potentially IDLH or unknown atmospheres and that this team shall be properly equipped and trained for this operation.

Chapter 6

Primary command and control – Tactical deployment

6.1 INTRODUCTION

London 1986

We had a series of experiences over a hectic few months that taught me a lot about tactics, but perhaps more about human psychology.

It was the 1980s and London was in a transition stage. We were just coming out of the 'iron-lung' era where firefighters, who often chose to 'eat smoke' as a way of proving they 'had balls', were facing the conflicting enforcement of occupational health and safety legislation. For years I had battled against the forceful opinions of those who were either too lazy or too incredibly complacent in their tactical approaches. Those who chose not to wear their SCBA provided a good example. There was also a collection of company commanders (junior officers) who were of the typical 'reactionary' belief that SCBA wasn't needed until it was needed! In fact our procedures were so rigid that in some situations these officers refused outright to allow SCBA to be removed from an engine until ordered. Then, one after the other, there were some serious lessons to be learned!

1986 – The Water Gardens are a series of high-rise apartment buildings situated in London's Edgware Road, Paddington. One day we were called to a smell of smoke at the 12th floor level of one of the buildings. As we approached the fire floor in the elevator we had no SCBA or firefighting equipment with us. The IC hit the button for the fire floor despite our procedure that stated we go two floors below and take SCBA and hose-lines with us. Hell, it was only a smell of burning right? Wrong!

When we arrived at the 12th level the lift lobby was full of smoke. The fire had taken hold in an apartment along the hallway and we were told there was a family trapped inside. There were frantic radio calls to get SCBA equipped firefighters up with a hose-line but we needed to do something. I entered the apartment with Hughie Stewart and we began to crawl past the fire down a long corridor. The smoke was as bad as I had ever experienced and at first I sucked the remaining air out of the carpet by placing my nose flat on the floor. Then when that failed, I took my fire helmet off and breathed what dead air was inside the 'head space'. All the time we were moving forward and you know what, I could hear the women and children screaming ahead of me. Suddenly, I started to feel the smoke get to me and that dizzy feeling kicked in. I just took a breath, held it, and crawled as fast as I could out of there! Hughie was already lying prone on the landing and I joined him.

The IC informed me that there were two women and two children who were threatening to jump from the balcony of the flat, and that all attempts to reach them by aerial ladder had failed. The crew coming with SCBA had taken the wrong elevator and were trying to find their way down to us but had become confused as to our exact location. There were no ventilation or roof assignments and, in this case, no opportunity to utilize rope access/rescue from the balcony above. In the end, we got them! But it was a close call and we could have lost an entire family, along with a few firefighters, simply through complacency.

1987 – King's Cross underground railway was a fire that took the lives of thirty-one people, including a colleague, Station Officer Colin Townsley of Soho fire station. The standard approach to just about any fire in those days was to (1) investigate; (2) locate the fire; and (3) call for SCBA and a hose-line. I asked the question: Why do we wear all our PPE, helmets, boots and a whole weight of clothing to investigate? If we are going to find fire then we need our SCBA as well! Or why not just investigate in our shirtsleeves and then call for PPE, SCBA and a hose-line?

That's what happened at King's Cross. The first crews down onto the ticket hall concourse of the underground station, located just a few feet below the surface, found a developing fire. They had no equipment and no SCBA. Colin Townsley remained at the heart of the developing fire, calming and controlling hundreds of people exiting from the train platforms below, whilst his crews returned outside to collect SCBA and a hose-line.

Within a few short seconds, before the firefighters could return, the fire suddenly erupted from what was a relatively small fire to a raging inferno, trapping all nearby. Colin was found some distance from the fire on an exit route, just a few feet from the base of the stairs leading to the street. Close to him lay another victim, a woman, whom he had apparently attempted to bring out with him.

I say to this day, if Colin had his SCBA on his back he would still be here. It would have taken him twenty to thirty seconds to travel from the concourse to the location he was found. It takes less than five seconds to get air into the mask and the mask over your face. Colin's cause of death was smoke inhalation.

In **1989** I started a national campaign to encourage the donning of SCBA for fire reconnaissance purposes. In my station area alone we had five or six calls to fire alarms actuating every shift. In some instances we would descend some hundreds of feet sub-surface into the tube network to investigate a smell of burning or a call to an automatic fire alarm. In 99% of cases these surmounted to nothing, and to carry in a weighty SCBA seemed hard work to many. There was a clear feeling of overkill, and complacency crept in, in opposition to my proposed strategy. I countered this with the view that as long as we were responding with lights and sirens, we were responding to an emergency and that level of emergency response remained until we were able to confirm for ourselves that a 'non-emergency' existed. If you're going to jump red lights and break speed limits to respond, why downgrade the level of urgency because you don't see anything from the street on arrival? The relevant procedures stated SCBA should be worn in any situation where entering smoke, but only with a directive from the incident commander. I was proposing nationally that SCBA should be taken in to any situation where a firefighter may encounter smoke, and from which he/she might be unable to escape to safe air; or, in deep reconnaissance situations where SCBA might be urgently needed for the purposes of deploying a rescue team. The Chief Officer of the London Fire Brigade[1] himself stated that, as he understood written procedures, it was not necessary to receive a directive from the incident commander to simply have SCBA on your back, but it was necessary to receive such a directive if you were 'going under air'. However, it was clear that a 'gray' area existed in written procedure.

What did I learn during these few months? I confirmed in my mind, from a tactical perspective, how critical it was to approach every emergency that was showing 'nothing' on arrival as if it was likely to be the worst-case scenario waiting to be uncovered within the depths of a structure. I would assign only limited credibility to information passed to me by persons not part of the London Fire Brigade and would wish to check for myself that everything was in order.

Also, from a psychological point of view, I learned that complacency was rife, not only throughout the fire service but also in everyday life. What is needed is a certain type of person who will repeatedly maintain a level of self-discipline and conform to safe practice, even when a short cut might make things quicker and easier, despite an element of risk being attached. How many times do we take that short cut in life? In the emergency field, you just cannot afford to, because one day it will catch you out! Believe me when I say **'Complacency is the firefighter's worst enemy!'** Approach every single response with a strong element of professionalism and base your approach as if the worst-case scenario is about to happen. Act ahead – don't 'react' – and make sure you follow your procedures (SOPs), unless there is a genuine and viable reason not to.

The last thing I learned (but I really already knew) is that there are good fire commanders – those that are conscientious and care about their crews – and then there are the complacent ones, who haven't had anything bad happen to them yet. You'll know the difference when the time comes.

6.2 MILITARY RULES OF ENGAGEMENT AND STRATEGIC PRINCIPLES

In a speech entitled *U.S. Forces: The Challenges Ahead*, Powell said, in part, *'We owe it to the men and women who go in harm's way to make sure that . . . their lives are not squandered for unclear purposes.'* He was challenging leaders to make strategic

---

1. Clarkson, G., (1988), *The Fennel Public Enquiry into the King's Cross Fire*

decisions based on a core ethic: **Don't waste human life**. Implicit in his speech and in the Powell Doctrine is that committing troops to combat should be neither an easy nor an automatic decision. In fact, such a decision should be made only if there is a significant advantage to be gained.

Fire Service veteran Eric Lamar writes[2]:

The 21st century battleground is dynamic, chaotic and complex, and so is the fire-ground. As with the military, we have gone to great lengths to employ organizational systems and technology to instil a degree of order and predictability to the working fire environment. Both line firefighters and infantry soldiers now have an array of modern protective gear, surveillance equipment and offensive tools to achieve rapid victory. The uniform application of command and control systems is designed to ensure coordinated and effective action and to strictly limit casualties. In reality, our systems, protocols and technology often fail us with disastrous results. Why?

He continues:

Almost without exception, our firefighting forces are most vulnerable during interior structural firefighting. This operational environment most closely resembles the combat setting to which Colin Powell refers in his famous Doctrine. In his view, committing forces requires four imperative strategic considerations.

- Committing troops must be an absolute necessity;
- There must be a compelling risk posed by not acting;
- Overwhelming resources must be applied;
- A clear exit strategy must be in place.

Mr Lamar continues:

Do fire officers and firefighters routinely commit to interior operations where the objectives are fuzzy and the strategy is unclear? Are firefighters routinely killed in interior environments where the responses to these four strategic considerations should suggest completely different tactics?

A review of LODD reports will suggest that this is undoubtedly the case, and that fire officers should carefully review their own fire-ground strategy and tactics, within a cultured view of justifying their tactical decisions on risk-based approaches supported by the practical application of all necessary risk control options.

6.3 FIREFIGHTING RULES OF ENGAGEMENT

As stated in Chapter One, when considering risk management fire departments should consider the following *NFPA 1500 Rules of Engagement*:

- What is the survival profile of any victims in the involved compartment?;
- We WILL NOT risk our lives at all for a building or lives that are already lost;
- We may only risk our lives a LITTLE, in a calculated manner, to save SAVABLE property;
- We may risk our lives a lot, in a calculated manner, to save SAVABLE LIVES.

2. http://www.firehouse.com/ – Firefighter Safety (August 2007)

Further to this, the International Association of Fire Chiefs (IAFC) present their view on Rules of Engagement.

IAFC ten Rules of Engagement and risk assessment (see Figs 6.1 and 6.2).
Acceptability of risk:

1. No building or property is worth the life of a firefighter;
2. All interior firefighting involves an inherent risk;
3. Some risk is acceptable, in a measured and controlled manner;
4. No level of risk is acceptable where there is no potential to save lives or saveable property;
5. Firefighters shall not be committed to interior offensive firefighting operations in abandoned or derelict buildings that are known or reasonably believed to be unoccupied.

Risk assessment:

6. All feasible measures shall be taken to limit or void risks through risk assessment by a qualified officer;
7. It is the responsibility of the incident commander to evaluate the level of risk in every situation;
8. Risk assessment is a continuous process for the entire duration of each incident and the incident command system should ensure this occurs from the moment firefighters first arrive on-scene;
9. If conditions change, and risk increases, change strategy and tactics;
10. No building or property is worth the life of a firefighter.

	HIGH				LOW
RESCUE	1	2	3	4	5
	EARLY				**ADVANCED**
FIRE STAGE	1	2	3	4	5
	HIGH				**LOW**
SAVE PROPERTY	1	2	3	4	5
	LOW				**HIGH**
FIREFIGHTER RISK	1	2	3	4	5

	4–9	10–14	15–20
Strategy	**Offensive/Interior**	**Marginal Rescue**	**Defensive/Exterior**

Fig. 6.1 – IAFC Risk Assessment/Rules of Engagement – Grading of risk.

Firefighter Safety/Life Safety HRisk	High Probability of Success	Marginal Probability of Success	Low Probability of Success
Low risk	Initiate offensive operations – continue to monitor risk factors.	Initiate offensive operations – continue to monitor risk factors.	Initiate offensive operations – continue to monitor risk factors.
Medium risk	Initiate offensive operations – continue to monitor risk factors – employ all available risk control options.	Initiate offensive operations – continue to monitor risk factors – be prepared to go defensive if risk increases.	DO NOT initiate offensive operations – reduce risk to firefighters and actively pursue risk control options.
High risk	Initiate offensive operations only with confirmation of realistic potential to save endangered lives.	DO NOT initiate offensive operations that will put firefighters at risk of injury or fatality.	Initiate offensive operations only.

Fig. 6.2 – IAFC Risk Assessment/Rules of Engagement – Guidelines 2007.

For example, a fire officer can use the notebook charts in Figs 6.1 and 6.2 to assess an incident on a scale from one to five, for concerns like rescue possibility and savable property (high to low), fire stage (early to advanced) and firefighter risk (low to high). If the incident has a total risk rating of four to nine, an offensive interior strategy is a good option. A rating of ten to fourteen requires a marginal rescue, and a fifteen to twenty rating warrants a defensive exterior strategy.

6.4 CRITICAL TASKING AND THE PERFORMANCE INDEX (CTPI)

The safe and effective deployment of firefighters into a fire-involved structure relies on several factors. Needless to say, we have covered the theory of risk-based approaches and the use of Risk Control Measures already. Other factors will include staffing levels, command provision, water supplies, accountability factors and other resources etc.

We are generally able to divide essential fire-ground tasks on the initial response into 'primary' and 'secondary' actions and place them into an order of priority. In doing so we are then able to match the various tasks with a minimum number of firefighters needed to achieve or action these tasks. This provides us with a Critical Task Performance Index (CTPI) that may be varied for different scenarios (see Chapter Five). The CTPI provides a simple method of evaluating resource potential on the primary and secondary response, which can be extended to all types of incident, or even be used for extended operations. The index is used to optimize staffing on the primary/secondary response, ensuring that staffing availability is assigned most effectively to fulfil fire-ground critical tasks.

6.5 PRIMARY RESPONSE (FIRST ALARM) SYSTEMS

There are several types of primary response systems that will optimize the deployment of firefighters as they arrive on-scene. Two of the most common are:

- Reactive 'conditions-based' primary response;
- Pro-active pre-assigned task-based primary response.

In simple terms, both offer advantages and disadvantages. The **'reactive' conditions-based approach** sees firefighters arriving on-scene under the control of a designated incident commander from the outset. This may be a lieutenant or captain riding one of the rigs, or a chief who responds ahead of, or alongside, the primary response units. Using this system places a lot of responsibility on the first arriving IC who may have to communicate his/her plan across the fire-ground as he/she issues directives and assigns tasks and roles. This ensures the response is extremely flexible but may be slow to react to good effect. This type of response system may also see elements of freelancing creeping into the early approaches being made.

The **pre-assigned task-based response** is, on the other hand, very efficient in automatically assigning roles and tasks to pre-determined positions on the rigs. The New York City Fire Department (FDNY) is perhaps the master of such a response system that suits their company logistics, staffing arrangements, structural typing, extensive documented SOPs and incident management system. The battalion chief, who may be responding with or just behind the primary response units, will know that on his/her arrival the strategic scene will be set in the same way at each structural type. Teams of firefighters will automatically be in position at key points in/on the structure and the levels of responsibility are based on experience in the assigned task. Where a critical task is to be undertaken without a verbal directive, this decision to act will be taken by an experienced firefighter. Other roles, requiring confirmation from the IC or an interior company officer, can be filled by lesser-experienced firefighters. Most importantly, the concepts of a pre-assigned response must not be so rigid that the plan cannot be changed or redirected instantly, to account for any developments in fire conditions or operational needs. The FDNY ensures that assignments are flexible and easily adaptable and effective incident management and good communications support this.

Throughout this training manual the 'pre-assigned' response system is described in some detail, particularly in Chapters Two and Three. Although this form of response is well suited to inner-city fire departments, where the majority of staffing is on-scene within sixty seconds of each other, the smaller rural or suburban department may still adapt the fundamentals of pre-assignment for use in their own jurisdiction.

To utilize a pre-assigned task-based response system in your department the following is required:

- A documented SOP for structural fire response, pre-assigning tasks and roles as needed;
- Written directives that devolve responsibility to company officers or individual firefighters assigned to fulfil roles;
- Effective training in procedures so that all personnel acknowledge the importance of locating, communicating and coordinating any pre-assigned tasks with other core roles being fulfilled.

There are several 'tasks' that your department may already be fulfilling automatically on the primary response. In some locations a department will pre-assign the first responding engine to the fire, whilst the second engine will reverse lay to the hydrant. In other situations the first engine will go straight to the water supply. These are pre-assignments, and their objectives are established through written directives. The first arriving ladder company may automatically fulfil the roles of forcible entry, search and rescue, and ventilation, again conforming to pre-assigned directives.

'Company assignments' may not always be the most effective means of deploying staffing. It has been shown that pre-assigned response systems are more effectively utilized by breaking companies down into smaller work-group assignments, in order to optimize critical fire-ground needs. Rather than assigning an entire five-person company or unit to a role, it is often more productive to assign two or even three critical tasks to a company. Look at fire case studies and review how many times entire companies were 'assigned fully to the C side to ventilate' or to advance a back-up hose-line, without even considering the staffing arrangements.

6.6 PRIMARY RESPONSE

The moments following arrival of the primary response on-scene at a fire and the decisions made/actions taken will be absolutely critical in the overall firefighting strategy and outcomes. Past experience demonstrates that the most critical periods are:

- The first sixty seconds following arrival on-scene;
- The first five minutes following arrival on-scene; and
- The first five minutes following the arrival of a senior fire-ground commander on-scene.

The first sixty seconds on-scene

As the first responders arrive on-scene, it is absolutely essential that they follow their established (and documented) procedure in all cases. They should also listen to any specific instructions from their crew commanders or the incident commander. It is equally important to communicate and coordinate the actions of crews/firefighters. It is vital that both company officers and incident commanders **assert their command and take control** during these critical opening stages of the response, and ensure that the SOPs are being followed. Any deviation from normally accepted procedure should be held accountable at a later stage, where sound reasons must be given for any such deviation. Any tactical errors or bad decisions made during this initial stage may turn out to be life threatening to both occupants and firefighters alike.

Where company officers or incident commanders fail to take control right from the outset of operations, unless firefighters are operating under a *pre-assigned task-based response system*, or unless they are very well trained and responsive, the tactical approach will most likely rapidly deteriorate to one of 'freelancing'. This may quickly lead to uncoordinated actions and result in a tragedy.

The first five minutes on-scene

It is well known that a large number of events associated with rapid fire progress occur within the first five minutes following arrival on-scene. Looking over past fire reports it is certain that the sequence of events that takes place during this initial 300 second period are often the result of firefighters' actions, or non-actions. In general, it has been shown that at least 25% of working structural fires worsen following fire service arrival, before they are finally suppressed. That is to say that the area of fire involvement grows larger between the time firefighters arrive on-scene and control being achieved. This is easily understood, since firefighters are likely to create an opening or openings (entry points) in an effort to locate the fire and gain information on structural layout, even before they have equipment ready to get water on the fire. Such openings will most likely provide a path for vital air to enter the building, allowing the fire to increase in size and intensity. Therefore, one of our objectives should be to **prioritize incident stabilization** during this vital first five minutes. This calls for a more in-depth appreciation of how ventilation profiles in a fire-involved compartment or structure may be widened or narrowed through our own actions. A better understanding of fire behavior and air dynamics is critical where incident stabilization is to be effective.

Firefighters should take actions that will serve to stabilize conditions and slow fire development, rather than actions that may destabilize the interior fire environment. Don't vent without a directive or an objective and communicate and coordinate your actions with the interior crews. Take viable actions that will ensure the security of crews working the interior of a structure such as laying in a back-up support hose-line, providing them with some much needed lighting, or controlling the door opening at their point of entry, ensuring that any development or changes in conditions are communicated immediately both to the interior and the IC.

Reconnaissance
Firefighters arrive on-scene to reports of a fire deep inside a large structure. It might be a high-rise building, factory, large warehouse or sub-basement of an underground railway system. From the street there is nothing showing and no signs of fire. Sound familiar?

They send an investigation team to locate the fire and report on status. On arrival they find a reasonably small fire that could be immediately extinguished – if only they had some suppressive agent with them! By the time they get hose-lines laid in and a fire attack initiated, the fire has developed into a conflagration.

This scenario happens daily! In many cases it has led to the deaths of firefighters and many trapped occupants. The concept of 'fast attack' uses special strategies, tactics and highly innovative equipment to achieve immediate knock-down on arrival of a small fire that is threatening to spread out of control.

Fire confinement may not be possible and it may take several minutes to establish water supply and lay in attack lines deep into the structure. An on-the-spot suppressive action with just a few liters of water might be all that it takes to prevent a conflagration.

The tactical IFEX 3000 backpack is the most mobile support unit to be used in combination with the fog gun. The unit holds a 13-liter water/agent cylinder, a 2-liter air cylinder and a pressure regulator with two outlets giving air pressure to the gun and to its water support. Water and any additive can be filled directly into the water cylinder. The concentration of additive should be reduced to 0.5–1% rather than the 3–6% normally recommended. The harness is provided with an additional bracket for mounting an extra air cylinder to support a breathing apparatus.

Air resistance acting on the water stream breaks the water droplets down and reduces the normal mean droplet size from about 700 microns to an average of 100 microns. So the cooling surface of 1 liter of water is increased from the normal

5.8 sq m to 60 sq m. This small amount of water is applied in rapid series high-velocity bursts, to knock back small fires before they become big problems!

Water/agent tank capacity	13 liter
Size air cylinder	2 liter
Overall width × depth × height	360 mm × 260 mm × 625 mm
Weight empty/overall weight	10.3 kg/23.3 kg
Cylinder material	Stainless steel 1.4301-SS304
Harness material	Flame resistant synthetic
Filler cap unit/handle material	Brass/chrome plated steel
Release valve	Manual valve; 6.3 bar
Water and air hoses	Snap on connections
Water inlet filter	Optional, mesh size 0.6 mm
Operating/test pressure	6 bar/7.8 bar
Recommended additive concentration	0.5 to 1%

Note: Further information on the IFEX 3000 is available from the Euro Firefighter website http://www.euro-firefighter.com (use reader's key link code provided at the front of this book).

- Protocols for reconnaissance:
- Purpose of recon is to locate and confine/isolate the fire;
- Do not enter the compartment unless for 'known' life hazard;
- Recon is not an attempt to tackle the fire unless it is not possible to confine the fire behind a closed door, or the amount of fire is small and within the limits of the portable backpack or suppression system;
- Where the fire is already confined within a closed compartment, do not commit firefighters in SCBA without the protection of a fire service hose-line.

The first five minutes of senior command
As soon as the first senior fire commander arrives on-scene, it is essential that he/she undertake a rapid assessment for the **safety of firefighters** who are committed to dangerous positions in the 'hot' zone[3].

As well as ensuring that the firefighting and rescue operation is effectively underway, the incident commander's first role is to undertake an immediate risk assessment and balance risk versus gain in relation to the task assignments, firefighter locations, fire development, structural layout and integrity, and resource availability on-scene. Again, past experience has shown that the path the firefighting operation is taking can normally be redirected within these brief few minutes but that after that, a sequence of events will ensue that are beyond the control of even the most knowledgeable and efficient fire chief. Playing 'catch-up' with the fire or overall operation is not something that any chief would want to undertake, but that is quite often what happens where initial decisions or actions were inappropriate for the situation to hand. On occasions, this style of command (playing catch-up) may lead to fatalities.

The first-arriving on-scene chief must get into the operation as quickly as possible and this first five-minute slot is perhaps the only opportunity he/she will have to turn things around to a tactical advantage or even save firefighter lives. Is the fire

---

3. From a structural firefighting perspective, the 'hot' zone refers to any area occupied by firefighters where they are forced to wear SCBA and full PPE.

developing beyond the capability of hose-lines in place? Is water even on the fire? Is the water supply continuous and augmented? Quickly scan the structure for warning signs. Do your firefighters need to come out now? Are you convinced that the size-up the captain just gave you on hand-over was an accurate assessment of the situation? Have they really got water on that fire? Are you in direct communication with the attack team(s)? Do you know how many firefighters are committed and roughly where they are?

You have five minutes Chief – It's your call!

Water supply
It is absolutely critical that a continuous flow of water is provided to the fire-ground at the very earliest opportunity. Rural fire departments that may have immediate supply problems must be well practiced in locating and transporting water to the fire-scene. If an interior fire attack (offensive tactical mode) is implemented, then the minimum attack hose-line flow-rate of 100 gallons/min will only provide a few minutes of unit tank supply where flowed. Where an exterior defensive mode is in operation and exposure protection is the strategy, then higher flow-rates are almost certainly needed.

One of the biggest tactical errors possible is to deploy firefighters to an offensive interior attack and have them run out of water. Make sure your primary response (first alarm) SOPs provide clear directives of who is responsible for supplying the fire-ground with a continuous flow of water.

Never underestimate fire-ground needs in this respect. Exterior dumpster or other outside fires may appear controllable with a single tank supply but if there is potential for exposures, get that water supply heading in.

Forward lay – The supply hose is run from hydrant to fire building. This may entail the first engine going straight to the hydrant and laying its own supply onto the fire-ground (may be time consuming). In other situations a pump is sited at the hydrant as well. Sometimes the first engine goes straight to the fire and the second due takes the hydrant and forward lays to the fire-ground engine.

Reverse lay – The first due may drive by, take a look at the fire and complete a quick visual size-up, before dropping a supply line and running this in a reverse lay to the hydrant. The second due will respond straight to the fire building and connect to the supply line as left by the first engine. Where hydrants are closely spaced, the first engine may also drop two attack lines before reverse laying from fire to hydrant.

Fast attack – The first due (or arriving) engine responds direct to the fire building and runs direct off the tank, leaving the second due engine to provide a continuous supply by using either a forward or reverse lay. This strategy will enable rapid attack and search operations where staffing permits. The danger is that the tank water supply may be exhausted before a continuous supply arrives on-scene.

Water relay – Several engines are used to boost pressure in a very long run from water supply to fire building.

Water shuttle – Special water carriers (or engines) are used to collect from a supply source and transport water and dump this into a portable on-site container or dam left at the fire-ground.

Not considering an immediate need for laddering the exterior for a **visible** or **'known' life hazard**, or an obvious **exposure** problem, the optimum tactical objectives at any interior (offensive) fire operation will be to implement three basic tactical operations as follows:

1. Primary attack hose-line;
2. Interior search and rescue (or RIT after second line in operation);
3. Secondary support (back-up) hose-line.

A minimum staffing of twelve firefighters, along with a minimum on-scene flow-rate of 200 gallons/min (750 liters/min) and three engines are needed to safely meet these tactical objectives providing the hydrant is close by. The first engine goes to the fire, supported by the second due to the hydrant (forward or reverse lay), and the third arriving mainly for staffing (few engines carry six firefighters). Even in this situation a staffing of twelve may not be able to meet an efficient first alarm CTPI rating depending on the structure, occupancy, fire load and stage of fire development.

6.7 INCIDENT COMMAND SYSTEM (ICS)

What makes a good fire commander? There are a wide range of skills, attributes and traits that are coupled with experience that will serve to make a good officer. These skills and attributes include:

- Knowledge
- Desire to learn
- Leadership
- Experience
- Management skills
- Ability to formulate strategy and make decisions under pressure
- Self discipline
- People skills
- Ability to assert command.

However, a fire-ground commander will only be as good as the organization and the systems employed to achieve strategic, tactical and operational objectives. ICS is one such system that resulted from the obvious need for a new approach to the problem of managing rapidly moving wildfires in the early 1970s. At that time, emergency managers faced a number of problems:

- Too many people reporting to one supervisor;
- Different emergency response organizational structures;
- Lack of reliable incident information;
- Inadequate and incompatible communications;
- Lack of a structure for coordinated planning between agencies;
- Unclear lines of authority;
- Terminology differences between agencies;
- Unclear or unspecified incident objectives.

Designing a standardized emergency management system to remedy the problems listed above took several years and extensive field testing. The Incident Command System was developed by an inter-agency task force working in a cooperative local, state, and federal inter-agency effort called FIRESCOPE (Firefighting Resources of California Organized for Potential Emergencies).

Basic and essential features of an effective fire-ground ICS:

- A documented pre-plan;
- Effective departmental SOPs;
- Adequate command staffing on response;
- IC responsibility to be fulfilled on the primary response;
- Tactical mode established and communicated to all;
- Effective fire-ground communications;
- Effective radio procedure followed (clear, concise and acknowledged);
- Assertive and experienced commanders;
- Training in ICS procedure;
- Size-up and risk assessment processes built into the plan;
- Command emphasis on the 'first sixty seconds' on-scene;
- Command emphasis on the 'first five minutes' on-scene;
- Command emphasis on the senior commander's first five minutes on-scene;
- Rapid but detailed handover from company commanders;
- Effective briefings and debriefings of crews and commanders;
- All essential roles fulfilled including that of 'safety officer';
- Clear decisions by commanders, with **firefighter safety** the primary consideration at all times (risk versus gain).

6.8 TACTICAL DEPLOYMENT INSTRUCTOR'S (TDI) COURSE[4]

The concepts of Compartment Fire Behavior Training intend to provide firefighters with the ability to read fire conditions, enter fire compartments safely, advance into smoke-logged and super-heated environments, and take actions to counter 'rapid fire' development. However, CFBT training does not equip firefighters with the ability to estimate fire-flow needs, deploy effectively into a wide range of situations, deal with heavy fire, or fully appreciate the pros and cons of primary reconnoiter and fast-attack tactics. Fire brigades commit large amounts of money towards training their firefighters in CFBT skill sets but are the actual skills effectively transferred onto the fire-ground?

There has been a distinct learning curve informing us that in some (a few) situations, the CFBT concepts are left on the training ground and the transfer of skills and knowledge to 'real world' fires never took place, or was misinterpreted. As typical examples, we have seen firefighters in London (Telstar House) and Madrid (Windsor building) inappropriately attempt to apply 'pulsing' water-fog applications against major fire development in high-rise open-plan floor space. In other situations we have seen CFBT trained personnel caught in 'smoke explosions' for failing to follow safe-zoning protocols. Why does the transfer of CFBT knowledge out onto the fire-ground sometimes appear to have failed?

- CFBT learning objectives include: fire behavior, reading fire conditions, gaining safe entry to compartments, stabilizing conditions, countering rapid fire phenomena whilst dealing with small fires (1–2 MW) in the gas-phase, or burning under controlled conditions.

4. http://www.cfbt-eu.com/

- CFBT is **not** about: firefighting tactics, deployment of crews, command functions and decisions, dealing with rapidly developing fires involving large fuel loads, tackling forced draft fires, dealing with multi-compartments, or evaluating needed flow-rates for varying fire size etc.

If a CFBT-trained fire officer, or a CFBT instructor who has specialised in fighting 1.5 MW training fires, is responsible for making risk-based tactical deployment decisions at 'real fires', then they should also gain a full understanding and appreciation of how such fires are likely to be different from training scenarios, where compartment **temperatures** may be similar to real fires, but actual energy (**heat**) release is likely to be far more severe.

In some cases, the TDI course[5] is probably the most important training link that will ensure the concepts of CFBT will make an effective transition from training ground to fire-ground. A critical part of this **transitional training** involves specific crew command skills, deployments into a variety of multi-compartment fires, hallway approach tactics where heavy fire loads and exterior winds are involved, and high-flow direct attacks into heavy flame-fronts.

CFBT Tactical Deployment Instructor course – Learning objectives

1. Establish safe and effective tactical objectives based on the limitations of available resources and staffing levels likely to be available on the fire-ground.
2. Operate according to established guidelines or directives provided within local (or CFBT model) Standard Operating Procedures (SOPs).
3. Learn to effectively 'assert command', ensuring incident command becomes a primary and influential function from the moment crews arrive on-scene at an incident.
4. Establish the difference between 'size-up' and 'risk assessment' at fires and initiate both processes immediately on (or even before) arrival on-scene.
5. Utilize safe and effective primary approaches and reconnoiter where needed.
6. Ensure safe and effective deployment and placement of primary and secondary attack hose-lines, cover hose-lines and support (back-up) hose-lines.
7. Gain an appreciation of situations where 'quick hits' and rapid 'knock-down' of fast moving flame-fronts may prove effective, and utilize fast-attack 'rapid water' tactics where needed.
8. Learn to gauge resource requirements at an early stage and implement an effective strategic plan that offers the most productive outcome through optimized resource deployment.
9. Utilize safe and effective risk-based tactical modes of attack and be sure to communicate the 'mode of attack' to the fire-ground immediately.
10. Prioritize exposure protection where necessary and utilize the most effective means of achieving this within the limitations of on-scene resources.
11. Establish every available means of securing team safety within the confines of a fire-involved structure.

5. http://www.cfbt-eu.com/

12. Establish the minimum needed flow-rates to deal with a compartment or structure fire, based on simplified but reliable fire-ground formula in line with visual observation. Then advance and apply high-flow direct attacks into heavy fire fronts, establishing the practical limitations of various firefighting flow-rates.

13. Utilize safe and effective risk-based deployments of search and rescue teams working behind, ahead of and above attack hose-line positions.

14. Utilize effective interior search patterns, including the use of safety ropes and thermal image cameras (TICs).

15. Implement safe and effective 'snatch rescue' procedures for interior search operations.

16. Gain a more in-depth working appreciation of VES (Vent-Enter-Search) tactics within the context of crew and firefighter deployments.

Pre-course requirements[6] are that students are qualified CFBT instructors. Training is undertaken in multi-compartment, multi-level CFBT tactical units as well as acquired structures, where available.

6.9 'TAKE THE FIRE FIRST'

A *Fire Engineering* 2002 round table review[7] discussed the prioritization of tactical firefighting objectives. The contributors were mixed in their views in prioritizing tactical options and the discussion touched upon strategic areas first proposed in 1991 by the author, in his book *Fog Attack*[8].

The *tactical objectives* for first-arriving firefighters have historically placed life-safety as the number one priority in the strategic plan at structure fires. In definition, *life-safety* has also been taken to mean the safety of firefighters, but this concept has rarely placed firefighters' lives ahead of those trapped inside burning buildings. It is common for firefighters to place themselves at great risk in an effort to remove victims to safety as the priority, and this act of selflessness has frequently cost them their lives.

In 1991 the author first proposed that the priority in tactical objectives should shift in situations of limited crewing. Where an initial response of six to eight firefighters arrive together, then there is every likelihood that *fire attack* along with *search and rescue* objectives may be implemented jointly. However, with a single engine arriving on-scene, a choice often has to be made – fire attack or rescue? Which is the priority?

If there are building occupants visible at windows or balconies from the exterior, and they are within reach of a ladder, then this almost certainly is the priority. A rapidly escalating fire that threatens multiple occupants may be the only exception to this rule. However, under limited crewing situations the priorities are these:

- Rescue visible occupants by ladder;
- Isolate the fire (close doors) and stabilize interior fire conditions;
- Place a hose-line that will protect escape routes;

6. http://www.cfbt-eu.com/
7. *Fire Engineering Magazine* USA, (2002), December edition
8. Grimwood, P., (1991), *Fog Attack*, DMG/FMJ International Publications, Redhill Surrey, UK

- Extinguish the fire if easily accessible;
- Ensure Two-in/Two-Out rule is followed.

These are primary actions and should take priority over all others.

The author's 1991 assessment of fire-ground primary and secondary actions placed *interior search* strictly into the list of **secondary actions,** and this was considered highly controversial at that time. Ever since then we have operated under constant threats of cuts in on-scene staffing that have perhaps made this approach even more topical. If our staffing and resources are to be stretched then perhaps our strategy and tactics should be influenced by such reductions in the weight of attack. If the second due engine will take an additional two or three minutes to support the first due, or where there are reductions in staffing, it is certain that critical tasking on the fire-ground will become more complex and this will affect the order of priorities if our risk assessment is undertaken correctly. There will be greater delays in effecting primary operations where we must implement additional Risk Control Measures. In some situations, it will become necessary to change the order of core objectives in order to maintain safety standards. Where before we were able to immediately coordinate fire attack with search and rescue, or search and rescue with ventilation etc., we may now have to put in place some Risk Control Measures to support a single objective – an example might be to take a hose-line to support our search, in situations where much needed fire-ground support is not immediately available.

In 1994 a retired Los Angeles Fire Chief John Mittendorf claimed[9] that the priority between fire attack and search and rescue was changing, and that controlling the atmosphere and conditions within a fire-involved structure was increasingly being viewed as more important than carrying out search and rescue. He stated his belief that fire attack rather than search and rescue was the first-crew job and that this view was spreading across the USA. He further stated that a more efficient use of limited manpower could be achieved by redirecting efforts towards controlling and relieving interior conditions.

This proposal became a tragic lesson when, in 1996, two UK firefighters were killed by a backdraft that occurred a few minutes after they, and four other fire-fighters, arrived on-scene as the initial response to a house fire. They faced the moral dilemma of several children being reported trapped upstairs, and opted to take the interior search prior to taking the fire, failing also to initiate any form of confinement or isolation strategy. The fire escalated suddenly, producing a massive fireball and subsequent flashover inside the house.

Deputy Chief (Toledo, Ohio) John 'Skip' Coleman proposed that:

Unless you can effectively do several things at the same time [on the initial response] – PUT THE FIRE OUT [first].

Chief Tom Brennan (FDNY, retired) made some important points, and went on to say:

Tactical objectives used to isolate the fire and account for human life are as valid for one as the other.

---

9. Fire Research and Development Group (UK), (1994), *Report 6/1994*

He continued:

> *Some of our strategy and tactical texts of the past have put the stamp of approval on fire control being put on hold if the life exposure is too severe and must take priority.*

Chief Brennan could see a need to reverse this rule. Ron Hiraki is an Assistant Chief in Seattle, Washington. He said:

> *We should always remember that the best way to accomplish the rescue objective is to take the danger away from the victims or put out the fire. Even if the fire is not immediately controlled or extinguished, a quick attack can slow the spread of the fire and buy other firefighters additional time to take the victims away from the danger.*

Lieutenant Bob Oliphant of Kalamazoo, Michigan, suggested that rescue should be the first consideration, but not necessarily the priority. He said,

> *I am truly saddened when I read accounts of firefighters who died trying to effect rescue when there was only a remote chance of finding anyone.*

Frank Shapher, Chief of St. Charles Fire Department, Missouri made his point,

> *Rescue is always our highest priority at a structure fire, but it should not be the first thing we do unless, of course, we are determined in getting ourselves injured or killed! Therefore I always maintain that the best way to rescue people from a burning building is to put out the fire.*

Chief Shapher challenged those who disagreed with him to read the NIOSH reports to see how firefighters get injured or killed whilst making rescue attempts. Chief Rick Lasky of Lewisville Fire Department, Texas similarly suggested looking at the same contributing factors causing losses on the fire-ground, and proposed a switch of rescue for fire control in larger commercial structures.

The author's original proposal in 1991 recommended essential 'primary' and 'secondary' actions ('Golden Rules') to be followed by firefighters on arrival at a structure fire. As a Standard Operating Procedure, these rules placed **fire attack** ahead of interior search as a primary action.

The plan was described as 'comprehensive but by no means complete'. It remained flexible in as much as 'tactical options' may be either up or down-graded in the hierarchy to suit specific circumstances – but a sound basis of risk analysis must be put forward to support any such decisions. The simple action of thinking laterally and closing a door or restricting airflow towards the fire may be enough to prevent fire spread and save lives!

6.10 SEARCH AND RESCUE

We have discussed search and rescue assignments extensively, in line with risk control and tactical priorities, throughout this book. However, one area of concern is demonstrated by three fires – the fire in the CCAB Chicago high-rise (see Chapter Eleven), and the fires in Fairfax County and Tayside, Scotland (see Chapter Four).

It is an area we can all learn from because there is a common thread of tactical error amongst these three fires. In all cases:

- There were occupants still alive in fire service arrival;
- They were in phone contact with the central alarm office;

- They were passing information as to their location;
- This information was not relayed effectively to the IC;
- The multiple victims were not found until sometime after fire control; and
- Search skills, techniques or procedures were inadequate.

In all of these fires the occupants were saying, 'We are here – come and get us,' but in some way or other, in each case, we failed them. Make sure you review these case studies, download the reports and learn the lessons. Here are the main learning points:

- Primary and secondary searches must be documented by SOP;
- They must occur at the earliest opportunity;
- They should be clear as to who searches where and when;
- Stairshafts may be key areas for primary search – top to bottom;
- A reliable method should be used by the IC (or aide) to record and account for multiple units (apartments), rooms and areas;
- Secondary searches should be undertaken promptly by different personnel to the primary search, where possible;
- The search plan should be trained, coordinated and well briefed to prevent crews searching the same areas;
- Where there are phoned reports from trapped occupants, these must be of highest priority (they are 'known', 'confirmed', and still alive);
- Communications are absolutely critical here – there may be two or three exchanges before the message is passed to the fire commander who can initiate action. If one of these exchanges fails to pass accurate information, lives may be lost;
- This is an area of fire-ground operations that may overwhelm the IC where information transfer and critical tasking is extensive;
- The IC must remain focused on the rescue objectives and needs.

In London, the author was assigned to a district in the 1970s where serious hotel fires were commonplace. It wasn't unusual to have over fifty occupants requiring aid, and there were always over 100 different areas to be searched. These firefighters became experts in undertaking search operations in large buildings with great speed and various methods were used for marking searched rooms along hotel corridors. The searches were thorough, and a target time of ten minutes from arrival was always the objective. Where exterior rescues were obvious and ladder access was possible (it usually was) then these were prioritized.

As with any fire service operations, an effective pre-plan, adequate staffing and regular training (six monthly) are the keys to success.

Chapter 7

Operations – Tactics – Strategy – 'Back to Basics'

7.1 INTRODUCTION

Experience – Tradition – History
The fire service in every country, in every locality, is rooted in great tradition. Our ancestors, the brothers and sisters who have gone before us, have all attained a level of experience through their time served fighting fires. This experience may be personal, reserved only for those who were there. This type of experience base is short-lived and dies with those directly affected.

However, in many cases this broad experience base becomes 'tradition' in the finest sense, through a sharing of experience. This 'sharing' may come in the form of direct transfer through verbal instructions and advice. It may be that firefighting tips and tactics are handed down from generation to generation, from battalion chief to battalion chief, from senior firefighter to junior firefighter. This is how solid tradition is formed and evolves, as each generation adapts the tactics slightly to conform to changes in the way the modern urban environment is fast developing around them. They may also be forced to address changing provisions in resource allocation and align or counter this with various advances in technology. Where events have occurred, but any opportunity to learn something valuable that may be added to our experience base is lost, then instead of becoming tradition, it simply becomes history.

Another way that experience can become established tradition is through documented SOPs. The author has reviewed and studied the development of various SOPs over the past five decades, and witnessed how changes therein have been related to specific events. We should follow our SOPs without question, providing they are based on a solid foundation of prior knowledge, experience, and tradition, and that they fit the circumstances of any particular situation during the dynamic changes we may face at a fire or emergency. On review of these SOPs the author

169

notes that experience from past events is often recorded in the form of short paragraphs or even a simple sentence: just a few words, giving a directive of something that should be done, should not be done, should be avoided, or should be carefully risk assessed and controlled.

It is through these few words that we often fail to spot **critical information** based on past fire experience. A simple sentence may appear as sound advice that we may consider at some time, but most of us are generally unaware that this information is the result of a tragic event that occurred many years, or decades, before.

It is therefore essential that we provide a means for our experience base to develop, broaden, expand and update. This should be in the form of written documented directives. The author holds strong views that the directives should **not** serve solely as guidelines but rather as definite orders or procedures to be followed. These directives (SOPs) should only be varied, or deviated from, where a clear tactical decision is made on the basis that a **safer** or **more effective** outcome will likely result. This decision to deviate must only be taken by those who can be held responsible and accountable for any such digression, and sound tactical reasoning should be presented in subsequent debriefings to support any decisions which depart from directives.

Good legal argument has traditionally presented the view that SOPs are directives and SOGs are guidelines. Where clear directives include the words 'will', 'must', etc., the orders are there to be followed. The words 'should' and 'may' are there to give leeway to those who deviate from procedure. But why should there be doubt here? The SOP should be clear in its directives, although deviation is acceptable where sound argument can be tendered as to why such deviation occurred. Where words (should/may) allow too much flexibility in a SOP, then deviations will frequently occur without reasoning or justification. If for example, a SOP states specifically that a 24 mm nozzle **'should'** be taken up to a high-rise floor for the primary attack line, then anyone who takes up a combination fog nozzle has not effectively broken with directives in the procedure. If a SOP states you **'must'** take the elevator to a point at least five floors below the reported fire floor, but you decide to take the stairs because the reported fire is on the sixth floor, then that is a perfectly sound tactical reason to deviate from procedure. If a firefighter breaks with procedure that states 'Do not create a ventilation opening without a clear directive and objective, and without approval from the interior crew(s)', then questions need to be asked as to why that opening was made. These instructions are giving a clear order: Do not do it unless three boxes can be ticked. If the firefighter makes the opening anyway, without ticking all the boxes, then there may be a tactical error or the firefighter might have made a sound tactical decision based on the fact that communications were down and in his/her extensive experience the situation was obviously in need of venting. Perhaps the vent opening was authorized under a pre-assigned directive or a VES operation. Either way, we need to be clear in the way we word our SOPs.

7.2 FIREFIGHTING TIPS AND TACTICS

7.2.1 THE FIREFIGHTER:

1. 'Take the fire first', unless visible exterior rescues are of a greater priority. Placing the primary attack hose-line and getting water onto the fire, before interior search and rescue takes place, may well save more lives in the process.

2. Where staffing permits, coordinate fire attack with interior search and rescue – but entry to the structure should be made behind a hose-line until/ unless fire conditions are stabilized.

3. Only enter a structure ahead of the primary attack line being placed under the following circumstances:
   - A 'known' life hazard exists
   - To locate and isolate the fire where conditions appear stable
   - A suspected life hazard may exist, where fire conditions appear stable.

4. Isolate the fire wherever possible by 'zoning down' the structure (anti-ventilation) in an effort to control unchecked fire development and to stabilize fire conditions. Begin all operations from an anti-ventilation stance.

5. Only create ventilation outlets with:
   - An assigned directive; and
   - A clear objective; and
   - Confirmation of the interior crew(s);
   - Consideration given to wind speed and direction.

6. Ensure such vent openings are **precise** (correctly located in line with the fire and interior crew's positions); **coordinated** (in line with fire attack); and **communicated** and approved by the interior crew(s).

7. As you enter a structure for interior operations, look up and get a view of as much of the exterior as possible. Are there any signs of fire? What stage is the fire development? Are there pre-existing vent openings? Are there security bars on windows? Take in what information you can with a quick visual scan.

8. Stay low, stay oriented and at all times be very aware of what is occurring at the ceiling. In high-ceiling compartments, or heavy smoke conditions, this may not be possible. However, it is essential to gauge temperatures in the overhead to prevent fire getting behind you. Use brief bursts of water into the ceiling to gauge conditions and listen for rapid conversion to vapour. Be very aware of dangerous smoke or fire-spread behind false ceilings.

9. Always follow your SOPs and/or direct orders from commanders at a fire and avoid freelancing at all costs.

10. At building fires there will be very few occasions when you need to run. Always move and act quickly, but take time to pause, take a breath, think about your surroundings, promote a calm fire-ground demeanour and always be in control of yourself as well as the fire!

11. Always buddy up and work with at least one firefighter, particularly when entering a fire-involved structure. Remain in close proximity to each other at all times until exited from the structure. Conform to SCBA air management or accountability procedure.

12. If proceeding into the structure for search or rescue purposes, in advance of the primary attack hose-line being in place, or without water being applied to the fire, close all doors and make attempts to isolate the fire wherever you are going beyond or above the fire.

13. When forcible entry is required for an inward-swinging door behind which there is intense heat and fire, the inward swing must be controlled. A firefighter or officer should hold the doorknob closed with a gloved hand or short piece of rope while other firefighters force the lock open (Vince Dunn FDNY).

14. Wherever procedure is not followed, be prepared to give a professional account as to the reasons why this deviation occurred.

15. Crouch down and keep one leg outstretched in front of you when advancing an attack hose-line in a smoke-filled fire room. Proceed slowly, supporting your body weight with your rear leg. Your outstretched leg will feel any hole or opening in the floor deck in your path of advance (Vince Dunn FDNY).

16. Don't open a door leading into a stair-shaft (behind which a fire is suspected) prior to clearing the stair-shaft of both occupants and firefighters who may be situated above.

17. After flashover occurs inside a superheated, smoke-filled room, there is a point of no return after which a firefighter cannot escape back to safety. The point of no return – or maximum distance a firefighter can crawl inside a superheated room and be sure he/she will get back out alive and not badly burned after flashover – is 5 ft. If you are 5 ft inside a room that has flashed over (walking 2.5 ft per second) it takes you two seconds to get out. During this time you are engulfed in 1,000–1,200 °F heat. If you are 10 ft inside and flashover occurs, you are exposed to 1,000–1,200 °F for four seconds. At 15 ft you are exposed for six seconds and badly burned even with protective clothing. Think about it! (Vince Dunn FDNY).

Author's note: Chief Dunn offers some important advice above. However, be aware that firefighters have escaped flashover conditions after thirty second periods of entrapment, or even longer! There are fire-scene videos and firefighter testimonies that prove this. On occasions the interior fire presents full flaming combustion from exterior windows, but incomplete combustion inside the compartment. In other situations the interior flaming occurs at high level and floor temperatures are around 300–400 °F (not full flashover) – severe but survivable. Don't ever operate recklessly and believe you can survive a flashover, but be ready and oriented to escape quickly where interior conditions suddenly deteriorate.

18. Be proactive as opposed to reactive. Always look to prevent bad things from happening rather than having to react to them when they do. That's what makes a good firefighter a great firefighter.

19. When stretching a hose-line to an upper floor of a building, do not pass a floor on fire unless a charged hose-line is in position on that floor (Vince Dunn FDNY).

20. Lay out an exterior cover hose-line at the earliest opportunity wherever firefighters are working on ladders whilst sited adjacent to, or above, windows or openings, or where flame-spread threatens the building façade or the floors above through auto-exposure.

21. Lay in and crew secondary back-up hose-lines as soon as possible, in support of primary attack hose-lines. These should be of at least equal flow or greater than the primary line and their role is to protect the escape route of the primary line, which should be working just a few meters ahead.

22. Notify your officer when going above a fire to search for victims or vertical extension of flame or smoke (Vince Dunn FDNY).

23. Ladder buildings on all sides wherever firefighters are known to be working on floors inside a building, in order to provide rapid access to alternative exit routes.
24. Always ensure an adequate flow-rate (absolute minimum 100 gallons/min [380 liters/min]) is available at the primary attack nozzle(s) prior to initiating any ventilation opening.
25. Be very aware of wind speed and direction before creating any openings in a structure, including at the entry point (might need to create an outlet point before opening the entry).
26. Utilize the optimum attack stream pattern in line with achievable objectives – water-fog for gas cooling or gas-phase attack, or straight stream for direct fire attack.
27. Ensure that the chosen stream pattern has an adequate flow-rate for the objective at hand.
28. Ensure that the flow-rate available at the nozzle is adequate considering the size of the compartment and the potential fire-load involved.
29. Multiple crews must not be deployed into a structure for search and rescue purposes without **at least** one interior hose-line protecting their means of egress.
30. Always try to maintain contact with a wall and remain oriented when searching or advancing a hose-line in heavy smoke conditions. If laying hose across large open floor-space in such conditions, do not lose contact with the hose for any purpose.
31. Hose-lines may become difficult to follow out to the exit in heavy smoke where they become heavily looped or laid across furniture etc. Guide-lines (ropes), or some form of high-intensity lighting beacons, or safety/security crews with thermal imagers should be used to guide firefighters to the exits under such circumstances.
32. In large areas, all fire exit routes/doors should be instantly operable, should they be needed, and exterior crews may be assigned to force them open where considered necessary, being careful not to admit additional air into a developing fire situation.
33. Ensure SCBA cylinder contents gauge checks are carried out at least once every five minutes.
34. Be conscious of SCBA air management 'turn around times' (TATs) – always base the TAT on the lowest cylinder reading (see Chapter Eight)
35. If you enter a smoke and heat-filled room, hallway, or apartment above a fire and suspect flashover conditions behind you, locate a second exit, a window leading to a fire escape, or portable ladder, before initiating the search (Vince Dunn FDNY).

7.2.2 THE COMPANY OFFICER (CREW COMMANDER)

1. Select and communicate the tactical mode immediately on arrival.
2. The role of incident command should be fulfilled from the outset of operations, from the moment the first alarm response arrives on-scene.
3. Only deploy firefighters above, or beyond, the primary attack hose-line where there are 'known' or 'reasonably suspected' occupants trapped.

4. Assert command (be in control from the outset, even before you arrive), and deploy firefighters strictly according to local procedure and documented protocols.

5. Wherever procedure is not followed, be prepared to give logical account as to the reasons why this deviation occurred.

6. Multiple crews must not be deployed into a structure for search and rescue purposes without **at least** one interior hose-line protecting their means of egress.

7. Utilise the Rapid Deployment Procedure for interior 'snatch rescue' attempts.

8. Only deploy firefighters under an effective system of fire-ground accountability and resource management.

9. Ensure an emergency team (or RIT) is available at the earliest opportunity, especially where offensive interior operations are initiated.

10. If flames are discovered still burning at a gas meter or broken pipe after a fire has been knocked down, do not extinguish the flame. Let the fire burn, protect the exposures with a hose stream, and alert command that the gas has to be shut off at the cellar or street control valve (Vince Dunn FDNY).

11. During a fire in a one-story strip store, vent the roof skylight over the fire before advancing the hose-line to prevent injury from backdraft explosion, or flashover (Vince Dunn FDNY).

12. When it is not possible to vent the rear or roof of a burning store quickly and signs of backdraft or explosion are evident from the front of the store, vent the front plate-glass windows and doors, stand to one side, let the superheated combustible gases ignite temporarily, and then advance the hose-line for fire attack (Vince Dunn FDNY).

13. Ensure SCBA cylinder contents gauge checks are carried out at least once every five minutes.

14. Be aware of SCBA air management 'turn around times' (TATs) – base the TAT on the lowest cylinder reading.

7.2.3 THE FIRE CHIEF (INCIDENT COMMANDER)

1. **Three-minute handover and primary size-up** – Time is critical so when you arrive on-scene, don't delay in getting into the operation. Exert a **calm** but **assertive** demeanour and take in as much **information** as you can in the first three minutes. The safety of your firefighters may depend on what you do, and the decisions you make, in the next 120 seconds following this three-minute size-up. If the primary response has self-deployed without clear objectives, there may be too many firefighters operating inside the structure. If there are multiple crews working, without water being applied to the fire, then this needs addressing immediately. Even consider bringing all crews out to the street and redeploying as necessary, with safe working practices and clear objectives.

2. After handover from the IC, communicate with those working the interior. Reports from different sources that describe inconsistent fire conditions might lead an incident commander to question the safety of crews working the interior. One sector might report 'fire under control', while another reports heavy smoke or fire conditions. The discrepancy suggests they are

looking at different areas or that one does not know about fire conditions that are evident to the other. Crews operating in the area where the fire appears to be under control might be in serious danger if they do not know where the fire is still burning.

3. Evidence of a significant interior fire that cannot be located should sound a warning to the incident commander. Crews working in a smoke-filled building might be unable to find the fire. At the same time the continuing or increasing presence of heavy smoke suggests that a significant fire is burning somewhere inside the structure. The risk of a sudden outbreak of fire or a structural collapse increases with time spent on the scene. Be sure to deploy adequate flow-rate and implement Risk Control Measures to reduce the potential of any hazards.

4. Ensure you have adequate communications from the dispatch centre to your aide, to you, to your fire-ground sectors, and to those undertaking operations in the hazardous or 'hot' zones. It is established that unnecessary radio traffic can serve to hinder fire-ground operations. Only allow essential information or requests to be passed, make full use of command and tactical channels, and use effective radio procedure – the receiver must always acknowledge important information. If not, keep sending!

5. Before committing crews, ensure they are given a clear brief on their assignment and objectives. Further, make sure they are kept up-to-date with important information. If they are assigned to search for a missing person who suddenly turns up outside, make sure the crew is evacuated or clearly and effectively re-assigned.

6. When crews exit the structure, make sure they are immediately debriefed and that vital information is exchanged. Briefing and debriefings of firefighters should be precise, clear, accurate, and effectively documented on-scene as they occur. Use bullet points to record vital points.

7. Consider the role of safety officer at the earliest opportunity.

8. Sometimes 'known' life hazards are forgotten! That's a tragic statement but it's true. Most recent fires in Fairfax, Tayside and Chicago have all presented situations where there were 'known' life hazards, but, even more importantly, their locations were known as well. Yet these people all died and we failed them. Yes we failed them.

Chapter 8

SCBA air management – BA control

8.1 THE HISTORY OF BA CONTROL PROCEDURES IN THE BRITISH FIRE SERVICE[1]

In **1943** the UK *Manual of Firemanship* Part 1 recommended that where breathing apparatus is available:

- Men in BA should always work in pairs;
- On occasions it may be desirable to trail a bobbin line to enable men to retrace their steps;
- In some circumstances line signals may be advantageous provided a separate line is used for signals.

In **1945** Part 6a of the *Manual of Firemanship* further recommends in respect of BA procedures:

- If the smoke is 'thick' BA should be worn.
- Burning electrical insulation and fires involving industrial processes may make BA essential due to the noxious atmosphere produced.
- Precautions for moving in smoke and darkness using the hands and feet to feel the way (training techniques).

1. http://www.fire.org.uk/FireNet/ba.php

These pilot procedures served to greatly improve the safety of firefighters working the interior of fire-involved structures whilst wearing BA. However, they were to be of little assistance when the first of two disastrous fires occurred at London's Covent Garden Market in 1949.

The first fire occurred at 1110 hours on 20 December **1949** in the basement of Covent Garden Market. It continued until 1340 hours on 22 December 1949 and was a very difficult and hazardous fire.

The lessons learned from this fire were:

- Hose used to enable firefighters to follow the way out was difficult to trace in the deepening water, which eventually reached 4 ft in depth.
- Men worked alone. In trying to rescue a colleague, one fireman became so exhausted he barely made it back to street level to summon assistance. He did in fact collapse and vital minutes were lost in the rescue attempt.
- There were no recording and supervising procedures for men entering and leaving the incident in BA.
- No method of summoning assistance in an emergency as with present day DSU (PASS) alarms.
- Communications were bad to non-existent. This consisted of signals or, as was often practiced, the mouthpiece of the BA was removed thereby allowing the ingress of toxic products into the respiratory tract.
- No minimum charging pressure for BA cylinders. Many were only two thirds full.
- No low cylinder pressure warning alarm.
- Many firefighters wore BA but did not start up until it was absolutely essential, by which time they had taken in quantities of smoke and gases, which had their effects. It would appear that an ability to 'eat smoke' and the time taken to service sets were contributing factors in this procedure.

It is interesting to note that none of the above points was deemed worthy of further investigation, and it was considered that the brigade's organization was satisfactory, as stated in Chief Fire Officer, Mr F.W. Delve's report dated 24 January 1950 to the London City Council.

In **1950** the London Brigades introduced a 'nominal roll board' that was held in the watch room. All riders were listed by name but at this time these boards were not, it would appear, carried on fire units responding to incidents. Other than the nominal roll board, the procedure for BA did not change between the 1949 fire and the next in 1954.

The second fire at Covent Garden Market occurred in a five-story warehouse at 1500 hours on 11 May **1954** and continued until approximately 2230 hours on the same day. Two London firefighters were to lose their lives at this tragic incident.

The lessons learned from this fire were:

- No recording and supervising of men entering and leaving the incident in BA. In fact one fireman was only unaccounted for when roll calls were taken later at the fire stations which had responded to the incident.
- No means of summoning assistance (RIT) in an emergency. Crews took nearly an hour to locate a trapped colleague after a collapse.
- No evacuation signals to warn men to withdraw if signs of collapse became evident.

- It is obvious that the above lessons were some of the same as experienced at the 1949 basement fire.

Following the second fire at Covent Garden, the Home Office issued **Technical Bulletin No 2/1955**. This document stresses the importance of two fundamental points of good breathing apparatus procedure:

- BA should be donned and started up in fresh air before the wearer enters the incident.
- If the wearer's nose clip or face mask becomes dislodged for any appreciable amount of time he should return to fresh air to avoid the problems associated with the exposure to noxious atmospheres.

Once again it would appear that no other moves were made to provide a more detailed procedure for the operational use of BA.

In the early hours of 23 January **1958**, a fire broke out in the basement of London's Smithfield Market. This fire was to be one of the most difficult London Fire Brigade had faced and two more firefighters were killed. The incident continued for three days.

Once again there were lessons to be learned: The same problems occurred at Smithfield as had occurred at the two previous fires at Covent Garden. The single exception was a **local procedure introduced by the London Fire Brigade(s)** in **1956** following the second Covent Garden Fire. This was the provision of a control point set up in Charterhouse Lane to record the entry of men into the BA incident. The control point consisted of a blackboard and recorded:

- Name;
- Station;
- Time of entry to structure;
- Time due out from structure (based on calculated oxygen consumption rates).

This procedure (the first ever BA Control Procedure) proved invaluable and was to indicate later in the incident that two men were missing and overdue.

Following the loss of life at Smithfield and Covent Garden, January 1958 saw calls for a more comprehensive schedule of BA procedures to be formulated. These calls came from Mr Delve, Chief Fire Officer of London, Mr Leete, Deputy Chief Fire Officer of London, and Mr Horner of the Fire Brigades Union.

Due to the outcry over the recent deaths of firemen, the Home Office set up a Committee of Inquiry into the operational use of BA. This was a sub-committee of the Central Fire Brigades Advisory Council. It appeared from its first meeting that some efforts had previously been made by the Home Office to establish a procedure for the use of BA but nothing had been circulated to brigades on the progress made.

By June 1958 twelve brigades were circulated with a trial procedure and by August a number of observations and recommendations had been received by the Committee of Inquiry who began to prepare an interim report.

In October 1958 **FIRE SERVICE CIRCULAR 37/1958** was issued. It detailed the findings of the Committee of Inquiry and recommended the following:

- Tallies for BA sets;
- A Stage One and Stage Two control procedure for recording and supervising BA wearers;

- The duties of a control operator;
- The procedure to be followed by crews;
- A main control procedure.

Paragraph 4 of the accompanying letter to the above circular requested brigades report their observations and recommendations in light of experience by the end of November 1959. The letter goes on to say that no specifications for the design and use of guide or personal lines would be issued until more experience had been gained. Recommendations were, however, made in respect of a specification for:

- Low cylinder pressure warning device.
- Distress signal device (PASS alarm).

Following submissions by brigades on the interim report and procedures recommended by the Committee of Inquiry into the operational use of BA, a DEAR CHIEF OFFICER letter was promulgated in August 1961. This dealt with the revised procedures:

- A control board as opposed to a nominal roll board that was to be carried on all appliances equipped with BA.
- In Stage Two, a column for the location of crews, and, as a low-pressure warning device was now available, a 'time of whistle'.
- A 'remarks' column.
- Different coloured tallies for different types of apparatus.
- A set of working duration tables (based on average oxygen or air consumption) to be permanently attached to each BA control board.
- A **safety margin** of ten minutes allowed in the calculations.
- An armband to identify entry control officers for both Stages One and Two.

To take account of the newly established recording and supervising procedures for BA introduced into the Fire Service, Part I of *Manual of Firemanship* was reprinted in 1963.

A Home Office survey reported that, as a result of the increasing need to use BA at incidents, the Fire Service had available approximately 3,490 sets on first line appliances. The problems associated with a lack of progress in producing and adopting a specification and procedure for the use of guide-lines (search ropes) was soon to become a major issue.

On 6 February **1966** at 1245 hours a fire broke out at a secret underground radar station at RAF Neatishead in Norfolk. It burned for nine days.

The lessons learned from this fire were:

- **No personal lines (search ropes).** Team members got separated and lost. This led to the deaths of two firemen.
- **No main or branch guide-lines (search ropes).** In the case of the former, the distance between the main entrance to the incident and the fire was some 1,500 ft (500 meters). The hose, being used as a method of locating the way out, was so long and snaked under pressure that it was difficult and sometimes impossible to follow. As a result of the snaking, travel distances were increased dramatically.
- **The SCBA in use were of relatively short duration** (twenty minutes). In taking the extended route in and what was thought to be a more direct route

out, men lost contact with the hose, costing the life of a divisional officer (deputy chief) who ran out of air.

- **The communications equipment used was not successful** as it became entangled with other equipment. Communications were lost in the early stages of the fire. Communication between crews with oxygen breathing apparatus was non-existent but SCBA air sets with face masks allowed good intercommunications between crew members.
- **RAF (Royal Air Force) had no recording or supervising procedures** thus there was a lack of knowledge for responding local fire department crews.
- **1(1)(d) visits were very few** and the sparse information and lack of plans available did not assist firemen as to the best route to take to the seat of the fire.
- **Relief crews sent in five minutes before low pressure alarm.** No appreciation made of the time needed for relief crews to enter and reach the fire and working crews to return to control.
- **Distress Signal Units (PASS) not available.** When men got lost or separated these would have assisted in locating them.
- **Heat problems.** As men were never trained in heat, there were severe operational problems, even for experienced crews.

In 1966, the Home Office issued *TECHNICAL BULLETIN 10/1966*. This included the physical specification for the DSU (PASS) methods of attachments to sets and the prescribed testing procedure.

Following the lessons learned at Neatishead, the Central Fire Brigades Advisory Council issued *FIRE SERVICE CIRCULAR 46/1969* in December 1969 following extensive trials by brigades. The circular dealt with both the specifications and operational procedures for the use of guide-lines, personal lines and branch lines (a system of personal and crew search ropes). To provide more information, a number of diagrams of associated equipment were attached as an appendix. The facing letter of the circular recommended adoption of the procedures.

8.2 UK BA CONTROL – THE SYSTEM BASICS

The British system of BA control (based on London Fire Brigade's original procedure introduced in 1956) perhaps serves as a foundation for risk-managed firefighting operations that has been widely used by fire services around the world to increase firefighter safety at fires. The following offers a simple guide to the system that is explained in some greater detail through the eighty-four pages of *UK TECHNICAL BULLETIN 1/97* (due for revision).

- A BA control system to be used on every occasion firefighters enter a hazard zone whilst wearing SCBA.
- Dedicated BA entry control officer employed.
- BA control system implemented in three stages:
  1. Rapid Deployment
  2. Stage One
  3. Stage Two
- Under Rapid Deployment Procedures special rules apply (exceptional circumstances in limited-staffing situation).

- SCBA wearers have a plastic BA tally that records information such as department (brigade), name, cylinder pressure on entry, time of entry, and estimated time of exit based on tables calculated at 40 liters/min air consumption.
- Entry control officers (ECOs) are equipped with a board where a certain number of tallies (depending on stage of control) can be inserted. In some boards a digital clock is fitted, in others an automatic feature calculates the estimated time of exit.
- Crews work as a minimum of two and maximum of four firefighters and will always remain together, until exit, although with the ECO's knowledge a four-person team may split into two teams of two under specific circumstances.
- All crews enter structure through BA entry control point where briefing is given – this is not a slow process and in practice this normally occurs within less than a minute of arrival.
- The system details circumstances where a RIT (emergency team) is assigned to each entry control point as soon as practical – this is deemed a tactical priority.

8.3 RAPID DEPLOYMENT PROCEDURES

The procedures may be used only when the total number of BA wearers in the risk area does not exceed two, and:

- It is immediately clear that persons are at great risk and in need of rescue, and are either within view or known to be within a short distance of the entry point; or
- Dangerous escalation of the incident can be prevented by immediate and limited action.
- Where possible another crew member should be nominated as a Rapid Deployment entry control officer (ECO), with responsibility for recording the 'Time In' on an entry control board (ECB) (this may be in addition to other essential duties being undertaken). Alternatively, before entering the risk area BA wearers will ensure their tallies are attached to a Rapid Deployment ECB so that the 'Time In' is recorded automatically (as tallies are inserted into the ECB a deployment clock automatically starts).
- As soon as is practicable the Rapid Deployment Procedures shall be replaced by BA control procedures. When transferring to Stage One or Two procedures, care should be taken to ensure that the BA tallies are effectively handled to ensure accurate and prompt recording and monitoring of BA wearers in the risk area.

8.4 STAGE ONE PROCEDURE

To apply control procedures to meet the demands of small or limited incidents and to monitor the safety of breathing apparatus (BA) wearers.

Stage One procedure applies where:

- The size of the incident is small and the use of BA is unlikely to be protracted;
- No more than two ECPs are used; and
- The total number of BA wearers within the risk area does not exceed ten.

NOTE: Branch guide-lines are not to be used under Stage One procedures.

Note: The term 'guide-lines' is used here to describe search ropes. The guide-line (primary search rope) is 60 meters in length (190 ft) and is laid from the entry point. The term 'branch guide-line' refers to secondary search ropes laid in off the primary line, at points within the structure. These lines are marked A or B (main guide-lines) or 1–4 (branch guide-lines). A maximum of two guide-lines and four branch lines may be laid from an entry control point, depending on the stage of BA control implemented. Each crewmember also attaches to these search ropes with his/her own shortened personal line attached to the SCBA.

The ECO shall:

- Take up the position nominated by the officer in command for the ECP;
- Provide an entry control board (ECB), complete with suitable waterproof marker;
- Indicate clearly on the ECB that Stage One is in operation and ensure the ECB is clearly sited;
- Synchronize time on the ECB clock, in accordance with brigade procedures;
- Receive the tallies of BA wearers and check that the name of the wearer and the cylinder content at the time of entry into the risk area are correct;
- Enter the 'Time In' on each tally;
- Place each tally in a slot on the ECB so that the tallies of each team of wearers are together and are indicated as a team by bracketing the tallies using the waterproof marker (the earliest 'time of whistle' [low-pressure alarm actuation] being placed outside the bracket);
- Calculate the **'time of whistle'** for each wearer by using the ECB clock, and enter this in the appropriate section of the ECB, opposite the tally. The 'time of whistle' should be calculated by:
- Carefully referring to the correct section of the duration tables noting both the cylinder pressure reading at entry and the type of cylinder/apparatus in use;
- Acting on the guidance of the IC if necessary, restrict the length of exposure in difficult or strenuous conditions. The BA wearer and team leader must be advised to withdraw from the risk area at a predetermined pressure gauge reading. The ECO should calculate the time of exit and make a note in the remarks column accordingly;
- Where appropriate, taking into account any elapsed time since entry of BA wearers who entered the risk area under the Rapid Deployment Procedure.
- As soon as resources permit, a minimum of two BA wearers should be kept available at the ECP for emergency purposes (RIT).

Emergency procedures
The ECO shall:

- Commit an emergency team(s) (RIT), if available, and immediately inform the IC of the incident if:
- Any team fails to return to the ECP by the indicated 'time of whistle' (indicated outside the brackets);
- A DSU (PASS) is operated;

- It is clear that a dangerous situation is developing which will affect the BA team; or
- It appears that any BA wearer is in distress.

Note: If the IC is not available, the ECO shall initiate a radio assistance message, 'BA emergency'.

8.5 STAGE TWO PROCEDURE

To apply control procedures which meet the demands of 'larger' and more complex incidents and to monitor the safety of breathing apparatus (BA) wearers.

Stage Two entry control procedures normally supersede Stage One procedures and are used where one or more of the following apply:

- The scale of operations is likely to be protracted or demand greater control and supervision than is provided by Stage One procedure;
- More than two ECPs are necessary;
- More than ten BA wearers are committed into the risk area at one time; or
- Branch guide-lines are used.

During transition from Stage One to Stage Two procedures, care should be taken to ensure that the number of BA wearers whose entry control tallies are supervised by an ECO (on one or more ECBs) does not exceed ten (excluding the emergency team).

The Stage Two ECO shall:

- Ensure the ECB indicates that Stage Two procedures are being applied;
- Check the 'time of whistle' calculations of the Stage One ECO being relieved;
- Ensure BA teams are relieved at the scene of operations in sufficient time to allow their return to the ECP by the 'time of whistle';
- Have available (at least five minutes before they are due to enter) sufficient relief teams to allow pre-entry checks and briefing to be completed without delaying their entry;
- Liaise (by radio, runner, etc.) with other ECPs and inform them of the names of BA wearers who leave the risk area other than via the control point at which they entered;
- Liaise with a Main Control, if one is established, and ensure that personnel who have collected their tallies report immediately to Main Control.
- Ensure that if it becomes necessary to use additional ECBs, tallies remain on the initial ECB under the control of the ECO (tallies must NOT be transferred to a second ECB until the wearers collect their tallies and the initial ECB can be disestablished); and
- Synchronize the clock of the Stage Two ECB and the Main Control clock to the clock on the first ECB used.

Stage Two emergency procedures
The ECO shall:

- Have a fully equipped emergency BA team (RIT) rigged and standing by each ECP throughout the period that it is in operation.

- Commit the emergency teams if line communications are lost.
- Stage Two ECO duties are restricted to those directly related to the BA function. It may therefore be necessary to have an officer close by to give direction as to firefighting requirements, equipment supply or casualty handling.

8.6 MAIN CONTROL PROCEDURE

To ensure BA wearer safety by establishing additional control, with the aim of coordinating BA requirements where there is more than one Stage Two ECP or the number of BA wearers is large.

Where there is more than one Stage Two ECP, or the number of BA wearers is large, an additional control to coordinate BA requirements should be established. This control, known as 'BA Main Control', should be set up at the most convenient site for easy access and communication with all Stage Two ECPs and the fire-ground control. A control unit, emergency tender, or other suitable vehicle may be used as the main control point.

A BA Main Control officer (MCO) should be appointed by the officer in charge of the incident. The MCO should have the appropriate command and management skills and have shown proficiency in the responsibility required.

Duties of the MCO
Monitoring duties
   The MCO shall:

- Establish and record the availability of BA, associated equipment and personnel at the incident;
- Identify the location of each Stage Two ECP, record the name of each ECO and establish communications with Stage Two controls and fire-ground control;
- Take account of any time variations between clocks;
- Establish and record the requirements for relief teams of BA wearers from each of the Stage Two ECPs;
- Have available sufficient BA wearers to provide the relief teams required by each Stage Two ECP and dispatch them to arrive at the ECP at least five minutes before required.

Summary of British BA control system

Rapid Deployment	Stage One	Stage Two	Main Control
Exceptional circumstances where urgent action is needed but staffing is limited on the primary response	Up to ten SCBA wearers	More than ten SCBA wearers committed into the incident	More than two entry control points

Rapid Deployment	Stage One	Stage Two	Main Control
Maximum of two SCBA wearers	Maximum of two Stage One entry control points	More than two Stage One entry control points	Large number of SCBA wearers
Trapped occupants are within view or are known to be near the entry point (In practice this is interpreted the same as 'known' life hazard)	Operations are not protracted with large numbers of reliefs	Protracted incident requiring relief crews	
Used to prevent a dangerous escalation of the incident by taking limited action	Branch guide-lines (secondary search ropes) are not allowed	Branch guide-lines (secondary search ropes) in use	
SCBA wearers can self-deploy using a Rapid Deployment control board		BA emergency teams (RIT) assigned to each entry control point	

Fig. 8.1 – Summary of British system of BA Control.

8.7 RAPID INTERVENTION (BA EMERGENCY TEAMS)

In **1970** the London Fire Brigade recognized the need for assigning dedicated firefighter rescue teams at BA entry control points and this concept was introduced through the BA control system. Equipped with Emergency Air Transfer Air Line (EATAL) and later Emergency Air Supply Equipment (EASE), these Rapid Intervention Teams (termed BA emergency teams) would stand by in cases of downed or missing firefighters, or where SCBA teams had not exited by the time their low-pressure alarms actuated, and enter with portable air rescue sets to locate, assist and rescue trapped firefighters.

It can be seen that UK firefighters have been implementing the **'buddy system'** since 1943 – the **RIT** concept introduced and developed by London Fire Brigade is now nearly forty years old – and the principles of **SCBA 'air management'**, as introduced by London Fire Brigade in 1956, served as another innovative development in the field of global firefighter safety.

8.8 SCBA AIR MANAGEMENT

Whilst the basic concept of SCBA 'air management' has been around for some time in the USA, it is only quite recently that firefighters have been encouraged to take measures to avoid getting caught in a 'low-air' situation. These measures include conducting regular cylinder gauge checks; addressing accountability for interior operations; monitoring and controlling the entry and exit of SCBA crews, and calculating reasonable estimates for crew working durations. This risk-based approach, based on the London Fire Brigade's original 1956 BA control system, has been shown to greatly improve firefighter safety at fires and is now widely integrated into fire department operations throughout Europe and in many countries around the world.

Historically, firefighters have worked until their low air alarm, or End of Service Time Indicator (EOSTI)[2], on their SCBA has activated. This alarm had served as the indicator for firefighters to leave the IDLH environment. Initiating egress after the activation of the EOSTI requires the individual to utilize the reserve air supply to exit the IDLH area. This has had tragic consequences. Evidence shows that firefighters do not call for help until they have consumed their reserve air supply. This practice puts the Rapid Intervention Team (RIT) at a severe disadvantage and lessens the likelihood of a successful rescue.

In addition, the sounding of multiple alarms is commonplace, and therefore not seen as an indicator of a firefighter in trouble. Many firefighter testimonials have documented that individuals in trouble, with alarm bells ringing, went unnoticed by crews working in the same area.

To many firefighters, 'air management' still means waiting for the low-air vibration alert or alarm to sound, signaling it's time to leave the building. This occurs when three-quarters of the air supply has been consumed. Many consider that such a procedure is acceptable during a routine room-and-contents fire in a small building. However, take a look at how many firefighters have 'run out of air' in residential fires and lost their lives! In larger structures, or where there are large numbers of firefighters operating, the issue of interior accountability and 'air management' is critical. Where SCBA operations are extended to periods longer than thirty minutes and relief crews are required, a greater element of SCBA control is called for.

During the tragic Charleston Sofa Superstore furniture warehouse fire in South Carolina in 2007 where **nine firefighters lost their lives** as the fire suddenly escalated, a firefighter recounted how several firefighters came running past him in the blinding smoke screaming their cylinder air supplies had almost run out and that they were unable to find the exit. He tried to calm them but they were in a state of panic as all of their low-air alarms were actuating. The firefighter knew the way out and escaped, but tragically all the others died as their cylinders emptied, one by one.

A review of NIOSH reports will also demonstrate how many other firefighters have run out of air before they leave the structure.

SCBA cylinders are rated in **operating pressures** and **capacity**. They may also be marked in minutes. As an example, a 2,216-psi, thirty-minute carbon-wrap cylinder holds 45 cubic feet of air (1,270 liters) when the pressure gauge reads 2,216 psi. However firefighters are generally unaware that this thirty-minute cylinder will rarely last them thirty minutes in a fire. The reason is that this cylinder duration is based upon an air consumption rate of an adult male undertaking a moderate workload, such as walking along at a speed of 4 mph. This rate is based on a respiration rate of 24 breaths per minute with a volume of 40 liters/min (1.41 cubic ft). When firefighters are working hard in structure fires their air consumption rates will

2. Portland Fire and Rescue, (2006), *Operational Guidelines: Air Management*

increase dramatically. In two US studies[3] (Seattle and Phoenix Fire Departments) it was demonstrated that hard work would cause air consumption rates between 130–140 psi/minute (compared to 75 psi/min for moderate work).

As a rule of thumb, firefighters undertaking hose-lays up a stairway and completing a search pattern in a training situation, will reduce the air supply of thirty-minute cylinders to around twenty minutes (to empty) and forty-five-minute cylinders will be reduced to about thirty minutes (empty). Using up 75% of cylinder contents, leaving a 25% time to allow exit, means that thirty-minute cylinders may only allow five or six minutes to exit and forty-five-minute cylinders will allow seven or eight minutes (air reserve to empty).

The British system **enforces** a more stringent protocol whereby firefighters must have already exited the structure before the alarm actuates. This approach allows for a safer margin of error in leaving a reserve of air for the unexpected. European firefighters have developed simple methods to calculate specific 'turn around times' (TATs), implementing cylinder contents gauge checks every five minutes to assist this objective. In 1993[4] the author developed a rule-of-thumb fire-ground formula for estimating the TAT of firefighters working in SCBA, based on the 40 liters/min average air consumption rate. Of course, **where the work-rate increases during the egress from the structure**, say for example because firefighters are carrying a casualty out, then the **accuracy** of the formula will be **seriously affected**. However, this formula has been used for over fifty years by the British Fire Service to calculate fairly reliable estimates for estimated time of exit. It should be noted though that communications technology has greatly improved during this period and may be used to improve air management and monitoring by an exterior ECO.

Cylinder Pressure (Bars)	Cylinder Pressure (psi)	Turn Around Time (TAT) TAT = CP/2 + 25 = Bars	Turn Around Time (TAT) TAT = CP/2 + 500* = PSI
220	3,200	135	2,100
210		130	
200	3,000	125	2,000
190	2,750	120	1,875
180		115	
170	2,500	110	1,715
160		105	
150	2,220	100	2,110
140		95	
130	2,000	90	1,500

3. Morris, G., (2005), FireChief.com
4. Grimwood, P., (1993), *Fire Magazine* (UK), DMG Publications, Redhill

Cylinder Pressure (Bars)	Cylinder Pressure (psi)	Turn Around Time (TAT) TAT = CP/2 + 25 = Bars	Turn Around Time (TAT) TAT = CP/2 + 500* = PSI
120	1,750	85	1,375
110	1,600	80	1,300
100	1,450	75	1,225

Fig. 8.2 – The author's simple TAT formula that may be used by interior firefighters to estimate their turn-around times; i.e. the starting cylinder pressure is halved and then 25 is added (or 1000 psi if using 45 min cylinders or 500 psi if using 30 min cylinders) to the resulting figure. This will recommend the gauge pressure at which a firefighter should turn around and begin to exit. (The 25 bars represents a ten-minute safety margin, at which stage the low-pressure alarm would begin to actuate and by which time the firefighter should be outside with his crew, or buddy. (The 1000 psi represents ten minutes and 500* psi equals five minutes reserve air). These figures are reliable estimates where moderate work is undertaken. Where work is considered heavy (sucking air) then air consumption may almost double and TAT is greatly reduced .(But check your own cylinder pressure and contents ratings before relying on this data.)

One US firefighter's formula employs the **one-half time plus five minutes method**. To accomplish this, subtract 5 minutes from 33, giving you 28 minutes. Half of this is 14 minutes. For this operation, your team would penetrate for 14 minutes and then turn for home. This leaves a five-minute air reserve time. However, this is a **time** estimate and does not account for real time air consumption. A firefighter can still easily run out of air before exit.

A simplified rule of thumb guide to air management, based on reaching the **exterior** with an estimated five or ten minutes of air reserve, can be used to guide firefighters as follows:

- 30 min/cylinder (moderate work) – TAT of 1500 psi (5 min reserve)
- 30 min/cylinder (heavy work) – TAT of 1800 psi (5 min reserve)
- 45 min/cylinder (moderate work) – TAT of 2000 psi (5 min reserve)
- 45 min/cylinder (heavy work) – TAT of 2200 psi (5 min reserve)
- 45 min/cylinder (moderate work) – TAT of 2250 psi (10 min reserve)
- 45 min/cylinder (heavy work) – TAT of 2800 psi (10 min reserve)

Note 'Firefighter 1' guidance figures, and be sure to check the contents (liters) and charging pressures of actual cylinders in use in your department before relying on these charts, guides and 'rule of thumb' estimates:

- Low pressure cylinder – One minute per 100 psi
- High pressure cylinder – One minute per 200 psi

Common pressures are 153 bar (2,216 psi), 207 bar (3,000 psi), and 310 bar (4,500 psi) for 1,800 liters and 2,500 liters of compressed air. Some European fire departments utilize the twin-cylinder concept to increase working duration and provide more comfort to the wearer.

The **Seattle Fire Department** is one to have recognized the importance of SCBA air management and has introduced and modeled its procedure based on the British system. They use the acronym **ROAM** (Rule Of Air Management) that states:

Know how much air you have, and manage that air, so that you leave the hazard area before your low-pressure bell rings.

This is the rule on which UK firefighters base their SCBA control system and since 1970 they have committed Rapid Intervention Teams (dedicated emergency teams) into structures where firefighters fail to exit before the (ten-minute) low air alarm actuates (in accordance with pre-calculated air consumption tables based on the 40 liters/min rate).

If a low air alarm activates in the IDLH environment, it calls for an immediate radio transmission to Command specifying **WHO** you are, **WHERE** you are and **WHAT** your status is.

NFPA 1404 mandates training in air management.

NFPA 1404 (5.1.7, plus appendices) states the following:

'Training policies shall include, but shall not be limited to the following:

1. Identification of the various types of respiratory protection equipment.
2. Responsibilities of members to obtain and maintain proper face piece fit.
3. Responsibilities of members for proper cleaning and maintenance.
4. Identification of the factors that affect the **duration** of the air supply.
5. Determination of the **Point of No Return** for each member.
6. Responsibilities of members for using respiratory protection equipment in a hazardous atmosphere.
7. Limitations of respiratory protection devices.'

Most departments have training policies relating to points 1, 2, 3, 6, and 7.

The **2007 version of** *NFPA 1404* adds three new points:

- Exit before you use your reserve air;
- The alarm indicates use of reserve;
- Alarm activation is an 'immediate' action item.

In an article about 'air management' concepts by Seattle Fire Department trainers, the authors[5] believe that the 'Point of No Return' is not the point where you die but rather that point at which you or your team stops becoming part of the solution and starts to become part of the problem. By crossing the Point of No Return, you are now a part of the problem and, most likely, in need of intervention by resources that might otherwise have been directed toward the initial problem. Crossing the Point of No Return and doing nothing about it can lead you toward death.

The authors of this article concluded:

Firefighters in countries with progressive air management policies have lower per-capita firefighter death rates in structure fires than the United States. The Rule of Air Management and SCBA control is essential to firefighter safety and survival on the fire-ground.

---

5. Bernocco, S., Gagliano, M., Jose, P. and Phillips, C., (2005), 'The Point of No Return' in *Fire Engineering Magazine*, Penwell Publications, USA

8.9 TRAPPED FIREFIGHTERS – AIR CONSERVATION

The rescue of a lost or trapped firefighter is extremely time sensitive, and success may depend on the victim's and rescuer's ability to conserve air[6]. This was illustrated in a report issued by the US Fire Administration after a fatal fire in San Francisco in 1995. The fire claimed the life of a lieutenant and injured eleven firefighters, one critically, and received widespread attention because three firefighters became trapped when an overhead garage door closed behind them without warning. After a frantic rescue effort, all three were removed from the garage. The lieutenant and the firefighter who was critically injured were found to have fully depleted their air supplies. The third firefighter, recognizing their grave situation, attempted to conserve air and remain calm. At the time of his rescue, he had approximately 2,800 psi remaining in a 4,500-psi cylinder. All three firefighters were on air for less than twelve minutes.

The difference in air consumption rates was a significant factor during this fire, which was pointed out in the *Lessons Learned and Reinforced*. The report said:

> *There are many factors that affect SCBA duration. Both physical and emotional stresses cause an increase in the consumption of oxygen, and therefore, air. A person's physical size and conditioning are also major factors in air supply duration. Each firefighter should know how he or she reacts to stress when wearing an SCBA. These reactions affect the duration of the air supply.*

The author initiated a London Fire Brigade research project in 1990 that looked at conserving air consumption rates in entrapment situations. The research demonstrated that SCBA emergency air (ten minutes duration signalled by the start of the low air alarm) could be extended in duration by cycle breathing (slow resting controlled breathing). It was shown that the ten-minute duration could be extended to 63 minutes (6.3 liters/min) and in one test a firefighter equipped with SCBA and full PPE ran for the entire working duration (to alarm activation) of the SCBA on a treadmill and then rested to cycle breath from the start of the low pressure alarm. The ten-minute emergency air supply was then increased in duration to 43 minutes in this case.

This might suggest that a cylinder could be made to last for an hour and a half for each 15 minutes of normal air (yes that's 4.5 hours total), simply by cycle breathing: sitting down, relaxing and breathing as slowly and as little as possible. Such a technique takes practice and in a real situation may not be quite so effective. Even so, the firefighter should learn this life-saving action.

---

6. Carrigan, S., *Training for Air Conservation*, Nashua Fire-Rescue

Chapter 9

CFBT (Fire Behavior) Instructor

9.1 HISTORY OF CFBT AND 3D FIREFIGHTING TACTICS

Throughout the 1990s, a particular set of firefighter life safety initiatives was fast developing throughout Europe, Australia and Asia, where two Swedish fire engineers (Mats Rosander and Krister Gisellson) had introduced the concepts (1983) of 'bursting' or 'pulsing' water-fog patterns into the overhead, through the use of quick on/off movements at the nozzle, in an attempt to offer greater control over gaseous-phase combustion. These techniques were also used to cool the super-heated fire gases that accumulated in the upper regions of a compartment/room fire, as a means of preventing flashover. It should be noted that these methods of applying water-fog were not anything like the indirect applications that were based on the Lloyd Laymen or Iowa research from the 1950s, where the mechanisms of extinction, or fire suppression, were completely different.

Shortly after these theories were published, a fire officer in Stockholm (Anders Lauren) introduced the idea of using standard design steel ISO shipping containers to teach firefighters these innovative methods of fire control and suppression. By lining one end of the container's walls with woodchip board, and by setting a small fire to heat the boards until they produced enough fire gases to simulate repeatable 'rollovers', firefighters were able to practice these interesting new nozzle techniques against live fire.

The ISO containers were further used to enable firefighters to observe how fires developed in an enclosed space, and this provided an ideal opportunity to learn about fire behavior. Over time, the container units were further developed, and incorporated various purpose-built design and safety features such as manually-controlled roof and wall venting hatches. These venting hatches served three main purposes:

- To release excessive amounts of steam where students had been over-zealous with nozzle/water applications;
- To observe the effects of (a) increasing, or (b) reducing ventilation, during a fire's growth stages; and
- To alter the speed of the air-track fire (gravity current) and to demonstrate the benefits and disadvantages in doing this.

The ISO shipping containers were then purpose-built into a wide array of geo-metrical layouts, some with several compartments existing on different levels, and training in this form of structure (Fire Development Simulator or FDS) became very popular, as firefighters were able to work at close quarters to the fire with large amounts of flaming combustion. Essentially, where the original Swedish training and design model was followed, this method of training was extremely safe and took place within measured and calculable parameters. This meant that a standard fuel load, in line with the established design and ventilation features of each unit, provided an almost identical training burn each time.

This form of training became known as (Compartment) Fire Behavior Training (CFBT). However, the training objectives of these FDS units were open to variable definitions and some trainers, particularly in parts of Australia and the USA, adapted the concepts to fit in with local practices and tactics. Different fuels were used and techniques far detached from the original Swedish model were taught. In some situations, FDS units were occupied by crews without the protection of a hose-line! In many situations, the true benefits of CFBT were lost and there were even reports of serious injuries caused through inappropriate methods and techniques being taught therein. It wasn't just that you could obtain steel shipping containers and set fires inside to train firefighters. In order to achieve safe and effective training evolutions, everything had to be in accordance with clearly defined risk-assessed protocols.

Most importantly, this method of training firefighters presented easily repeatable conditions, where each firefighter/student would experience exactly the same levels of fire development under safe and controlled conditions. When held in contrast to live fire training in acquired structures, where much preparation time was needed and each firefighter experienced somewhat different conditions, the training in ISO shipping containers ensured a more uniform approach to training a squad, unit or entire force of firefighters.

Learning curves

As Rosander and Gisellson were developing their gas-phase firefighting theories and nozzle 'bursting' techniques at the start of the 1980s in Sweden, the author had not long returned from a two-year working detachment (1976–77) to the New York City Fire Department. During this period he had been assigned into the (South) Bronx area (Division 7) during the busiest period for fire action in the history of the FDNY. It was common to receive anything up to twenty or thirty calls per shift and, of these, four to five responses per night would be to serious working fires in large structures. In fact, this level of fire action saw occasions where there were just not enough units to handle all the fires, and it was common for a single response to be fighting three or four large building fires in a single street at the same time.

During this period the author learned a great deal about flow-rate! The level and extent of fire action clearly demonstrated how adequate amounts of water would deal quickly and efficiently with fast developing fires. The tactics were impressive, particularly where staffing and resources were readily available. The response system was based very much on a pro-active pre-assigned task based approach, which meant that key tasks were pre-planned according to riding position and company assignments. Prioritizing critical tasks or deployment issues on arrival was not something a commander usually needed to think about as the response of (generally) three 'engines' and two 'ladders' (companies) automatically filled roles on arrival, in accordance with a well documented pre-plan of assignments, delegated to first or second arriving engines and vice versa for ladders.

However, it became fairly easy to stand back, take a look at a fire and instantly estimate the needed fire-flow requirements simply by taking in the fire conditions as they presented. How many windows/floors were issuing fire or smoke? What type of occupancy or building was involved? What color was the smoke and how intense was the fire? How fast was the smoke moving? Was there attic or interior void involvement? What was clear to the author was that the tactics were nearly always to 'open the fire up' (ventilate) to allow smoke and heat to leave the building, so that the pre-assigned response could rapidly advance into the structure in order to overpower or counter any fire development with a superior flow-rate.

In contrast, the 'new-wave' European Rosander and Gisellson techniques were suggesting an optimized application of very small amounts of water droplets (water-fog) to control the gaseous-phase combustion, prior to advancing in and cooling walls and ceiling areas, and finally suppressing the base of the fuel-phase fire (burning surfaces). The concept of ventilating structures was at conflict with their methods, as this would inevitably lead to a greater release of energy from the fire (heat release rate), which may exceed the small amounts of water being used to overcome the gas-phase fire.

However, at this time, the author – having returned from New York – was in the process of introducing a new-wave strategy he termed 'tactical ventilation'. This was a compromise between the aggressive use of venting tactics (USA) and the more conservative use of 'anti-ventilation' tactics (UK) that would see a structure remain tightly closed during the vast majority of fire attack operations. Quite simply, the US approach was intended to relieve smoke and heat from a structure whilst the UK approach was intended to prevent the flow of air in to feed a fire. This latter approach to isolating fires had a lot to do with the fact that the fire attack strategy in the UK and Europe in general was based very much on the principles of 'fast attack' using 1,400–1,800 liter water tanks and high-pressure (40 bar) (500 psi) low-flow hose-reel booster lines on engines for 85% of primary fire attacks.

Nevertheless, the author could see great merit in both approaches. Through his experience of inner-city firefighting, it was demonstrated that in some instances it was a better option to keep the structure closed down (anti-ventilated), whilst in others the creation of vent openings would have greatly assisted firefighting and rescue operations and possibly have saved lives. It was certain, however, that both tactical approaches had been responsible for life loss at past fires, either through the inappropriate use of venting or the failure to create openings when needed. With this in mind, the tactical ventilation solution was being proposed as a compromise. The compromised stance came in the form of a strict range of parameters and protocols with which to work.

In 1984, whilst assigned to units working in the heart of London's busy West End district, the author worked with local commanders to develop a strategy for combining the Rosander and Gisellson tactics with his own tactical ventilation strategy at real fires. Over a ten-year period (1984–94) these combination tactics were used operationally at a wide range of fires with great success. The main objectives were to:

- Begin operations from an anti-ventilation stance, wherever possible;
- Create openings where an obvious tactical advantage may exist;
- Attempt to 'fog' areas prior to entering (door entry procedure);
- Attempt to 'fog' an area prior to ventilating to the exterior;
- Attempt to cool gases in the overhead using brief bursts of water-fog;
- Attempt to gain rapid 'knock-back' of gaseous combustion using brief bursts of water-fog.

Whilst there were some obvious successes achieved locally in London in combining both strategies, there was never an overall national acceptance of such methods, and a platform upon which to change the culture of firefighting practices in the UK seemed non-existent. This was despite the author's constant and extensive publication of articles in national trade journals, as well as a book[1] promoting the benefits of CFBT, tactical ventilation, and the Rosander/Giselsson Swedish fire suppression techniques.

Then, over a tragic three-day period in 1996, things suddenly changed. On the first day of February 1996, rapid fire progress killed two UK firefighters during their

1. Grimwood, P., (1992), *Fog Attack*, FMJ/DMG International Publications, Redhill, Surrey, UK

attempts to rescue several children from a house fire. Then, just three days later, further rapid fire progress caught a female firefighter and her colleague during a fire in a large superstore. Whilst her colleague was pulled from the store, she had reportedly died instantly.

There were suddenly national calls for action as it was apparent that both fires raised concerns over firefighting tactics coupled with a lack of knowledge of fire behavior. It is tragic, but typical, that it took these deaths before any deliverance would finally be acknowledged, and Compartment Fire Behavior Training (CFBT) became nationalized across the UK as a strategy in 1997, along with the introduction of tactical ventilation and the Rosander/Gisellson techniques.

These 'life safety' initiatives would go on to dramatically lower firefighter LODD statistics in the countries where such training was delivered in a continuing modular phased-in approach, and *Fog Attack* (by the author and published in 1992) became the recognized training manual of the period that provided the springboard for starting CFBT. The US Navy in their 1994 research tests, and the Fire Service College UK – as well as several fire brigades around the world – referred to it frequently when writing their original CFBT/tactical ventilation training syllabuses.

'Real world' fires

However, as effective as the CFBT training programs are, there were still some issues in both their delivery and their ability to transfer the entire message across to the 'real world' of fighting building fires.

CFBT is all about:

• Creating a working knowledge of fire behavior and fire dynamics;
• Creating a greater awareness of gas-phase fire hazards (smoke burns);
• Developing the necessary skills needed to 'read' fire conditions;
• Developing the knowledge required to locate air-tracks and understand their likely affects in a structure fire;
• Developing the skills necessary to prevent rapid fire progress;
• Developing the skills needed to safely gain entry to a fire-involved structure or compartment;
• Developing the skills necessary to advance and retreat hose-lines safely in a fire-involved structure or compartment;
• Developing the skills needed to suppress or control gas-phase combustion;
• Developing the skills needed to suppress or control fuel-phase combustion.

Having trained and equipped an entire generation of UK firefighters in the skills needed (above) over a decade of CFBT delivery, it became apparent, in the majority of situations, that the transfer of knowledge to the fire-ground had not taken place in the following respects:

• There was no attention given to firefighting flow-rate requirements;
• The concepts of zoning off 'safe' areas was little understood;
• The concept of compartment fire loading was little understood;
• The view that the 1.5 or 2.8 MW training fires were representative of 'real' room fires was commonly held;
• The skills needed in applying direct attacks against fast-developing structure fires were neglected;

- The basic principles needed to control air-tracks to tactical advantage (tactical ventilation or anti-ventilation) were not taught;
- The practical limitations of gas-phase fire attack were not clearly defined.

These failings clearly led to situations where newly trained firefighters would often implement inappropriate tactics or nozzle techniques for the fire conditions being experienced. In some cases they would use brief nozzle 'bursts' against fast developing fires in large volume structures; neglect the accumulation of dangerous fire gases in rooms, compartments or roof spaces; or fail to demonstrate an understanding of how adequate flow-rate is required to overcome the energy release involved in any particular enclosed fire at any particular time, and that this too was dependent on the amount of ventilation available or provided by firefighters.

In other cases, many CFBT instructors held strong beliefs that low flow-rates would be equally effective against real fires as they worked so well against the gas-phase fires experienced inside the ISO containers. Some even liaised with nozzle manufacturers to develop primary attack nozzles flowing as low as 40–90 liters/min, without seeing a need for higher flow requirements in the 'real' world. When the Rosander and Gisellson techniques were initially introduced, they advised minimum flow-rates of between 100–350 liters/min and there were sound reasons for this!

3D firefighting tactics

The concept of 3D firefighting was born out of the need to address structure fires from the point of view that fires, or occupant status, should not be allowed to deteriorate further following fire service arrival. As a training concept, '3D firefighting' was used to influence any CFBT failings from a 'real world' perspective.

It was established through research data in London that, in general, building fire conditions actually worsened following the arrival on-scene of firefighters in around 25% of occasions. That is to say, the extent (area) of fire involvement actually increased after firefighters arrived, prior to fire control occurring. Whilst it is easy to defend this well-defined statistic from the viewpoint that working fires are sometimes most likely to develop further before firefighters are able to take necessary action, perhaps we should seriously take a look at our tactical approaches first! In many instances, you will note firefighters taking actions, or not taking needed actions, that cause fires to worsen.

Examples:

- Creating an opening (outlet vent) without good reason or logic;
- Selecting an entry point (doorway) without consideration of wind direction or speed;
- Creating an opening at the entry point prior to a charged hose-line, and crew ready for entry, being in place;
- Failure to close doors in an attempt to control fire development and isolate the fire;
- Inappropriate deployment, prioritizing interior fire attack over exterior exposure protection;
- Failure to ventilate essential areas, such as at the head of stair-shafts, where smoke and heat is mushrooming across and back down.

Of course, many of these issues are reliant on adequate staffing and resources (water) but, even so, simple actions by firefighters are so often neglected, and the

25% statistic referred to above was a result of a study involving 307 serious fires in an inner-city area that was considered reasonably well staffed compared to some situations.

The main reasons for deteriorating fire conditions during the first few minutes after arrival may well be a failure to understand the principles of air-track management, practical fire dynamics and basic fire behavior at fires.

Whilst CFBT and 3D firefighting clearly share many of the same objectives, a few of the failings of the early CFBT programs were:

- There was little, or no, integration with tactical ventilation training;
- There was (is) no emphasis placed on minimum safe flow-rates;
- There were (are) no limitations placed on the size of fire where pulsing or bursting fog patterns might become ineffective;
- There was little (or no) attention paid to maintaining firefighting skills in the more traditional methods using straight stream fire attack.

The training concepts of 3D firefighting were used to equip firefighters with a more rounded view of how compartment and structure fires were likely to present themselves outside the training scenario, and attempted to form a stronger bridge, to assist the transfer of knowledge and skills between the training environment and the fire-ground itself. The use of combination tactics, when venting areas that had been pre-water-fogged, was also central to the 3D firefighting culture. The combination tactics had been termed 3D firefighting as the main emphasis was on dealing with the three-dimensional hazards of smoke and fire gases that firefighters so often neglected during their early tactical approaches.

Certainly, 3D firefighting is about getting water into the gas layers, but it is also about getting adequate amounts of water onto the fire, using the optimum methods of water application with the equipment and resources available. This meant that the more traditional methods for suppressing fires, in the form of straight-stream direct attacks, were not forgotten.

The introduction of CFBT training in the UK had seen both positive and negative effects. Whilst the ISO container fires presented challenging training scenarios for firefighters, at just 1.5 MW and 2.8 MW maximum heat release rates (HRR), the fires fell short of 'real-world' post-flashover compartment and structure fires which generally presented a far greater level of fire intensity of 3–15 MW. A whole generation of new firefighters were trained in dealing solely with gas-phase combustion but were not taught how to handle rapidly developing fuel-phase fires. This caused many real fires to be under-flowed, with serious consequences.

The 3D firefighting approach also dealt specifically with firefighting flow-rate, from the perspective of providing firefighters with a minimum amount of water that would enable a safe advance into real fire conditions developing close to, or verging on, the point of flashover. This minimum flow-rate was termed the 'tactical' flow-rate.

It is essential that CFBT instructors understand the difference between a 1.5 MW training burn at its peak of development and a 5 MW room fire that is still developing. They must further appreciate the importance of flow-rate as well as application technique, when applied to intense and rapidly-progressing enclosure fires. This greater depth of knowledge and awareness is critical to the safety of firefighters in the real world, and 3D firefighting manuals have always addressed these very issues.

9.2 COMPARTMENT FIRE GROWTH AND DEVELOPMENT

We shall firstly define the difference between a 'compartment' fire and a 'structure' or 'building' fire. The term 'compartment' refers to any room, space, or confined area which has clear boundaries consisting of walls or sides, ceiling or roof, floor or base, and which may have openings that could provide the option of sealing/closing off (doors and windows etc.). Such areas may be rooms, attics, hallways, corridors, stair-shafts, voids, cellars etc. A 'structure' or 'building' consists of the outer structure that houses all the various compartments within. A 'compartment fire' therefore is one that is restricted to the material contents and surface linings of an individual compartment. Where fire has spread to involve several rooms or compartments then this is a 'multi-compartment' fire. Where any compartment boundary has been breached by fire (walls, floor or ceiling) to involve the additional supporting elements of structure (joists, beams, trusses etc.), then this is now termed a structure (building) fire and the potential for structural collapse is further considered a hazard.

9.3 FIRE DYNAMICS AND FIRE BEHAVIOR

This is always an area that firefighters find difficult to deal with as, quite simply, it is 'boring'! This is one reason why the fire behavior training module should be delivered in the morning, when we are at our most wakeful period of alertness! However, by using small prop demonstrators and relating to real world experiences, this module can actually become interesting to firefighters. It is essential that you grab firefighters' attentions early on, from a tactical view, because it is most often due to a lack of understanding of fire behavior that firefighting operations generally deteriorate.

9.4 FIREFIGHTING FLOW-RATE

It is useful try to use the analogy of a very small fire to explain the concept of flow-rate. Take, for example, a waste-bin fire or a doll's-house training prop. Set it on fire in a safe area and ask one or several of the students to attempt to extinguish or control the flaming fire using the water available from a hand-held plant water sprayer or child's water pistol. It is an exercise that leaves a simple but lasting impression! Then provide something with a bit more water in the stream, such as a larger hand-held portable device like a water fire extinguisher.

Ask them why the fire wouldn't go out when using the water pistol. The answer is obvious to them. However, when they have a hose-line in their hands, are firefighters able to give any reliable guide as to the flow-rate in use? More experienced firefighters may have some idea from the amount of nozzle reaction, or backpressure, the nozzle is exerting. However, it is usually the case that firefighters cannot tell how much water is leaving the nozzle, in terms of flow-rate.

An example of this was frequently demonstrated to the author during a national survey of firefighting flow-rate undertaken in the UK in the period from 2000–2003 where fifty-eight fire brigades were assessed for the flow-rate available on their primary 45 mm attack hose-lines. The research demonstrated that 89% of brigades were actually flowing far less water through their attack hose-lines than they realized – in some cases as little as 16% of their target (manufacturer's nozzle specification) flow-rates! In these cases, the flow-rate being applied to real fires through 45 mm lines was actually reducing (just 80 liters/min) when replacing high-pressure hose-reel booster lines that would flow around 80–110 liters/min on the fire.

When demonstrated to the firefighters, using inline flow meters, what was occurring, they appeared to struggle with the possibility that this could actually be the case. They were sincerely of the belief that because the fire stream 'looked' good and reached a good distance, they were providing a higher flow attack line compared to the smaller booster lines. They had not heard of the 'automatic' nozzle (which they were using), that constantly trimmed the nozzle outlet to achieve an effective throw, at the expense of some trade-off to flow-rate! Depending on the pressure being sent to the nozzle, the nozzle aperture automatically adjusted to provide stream 'throw' over flow. This can be a good thing under circumstances where nozzles are flowed correctly with adequate amounts of pressure in the first place.

There was a general problem right across the UK (and commonly elsewhere, including the USA) that nozzles were often under-pumped. The UK research showed that only 11% of the fifty-eight fire brigades were flowing effectively from their 45 mm primary attack hose-lines, achieving 500 liters/min at the nozzle! In fact, the nation's average flow-rate was shown to be just 290 liters/min, when fire-fighters believed they were achieving almost twice this amount. In reality, with this flow rate, they are only capable of putting out half the amount of fire! Even more concerning was the fact that since the UK Fire Service had begun a transition at the turn of the 1990s from its use of traditional smooth-bore and combination fog/straight 'impingement' nozzles, towards the more modern concepts of nozzles with flow-selectors, automatic internal mechanisms and spinning teeth rings, the actual flow-rate available at the nozzle had halved.

These facts – coupled with the belief that lower flow-rates were just as effective on all fires because 'new-wave' burst and pause techniques had been widely taught in the UK for compartment firefighting since 1997 – saw several instances of fires being under-flowed. During this process, firefighters may have lost their lives.

The coroners' narratives (summary) of four UK firefighter deaths, over two fires in 2004 and 2005, suggested that inadequate amounts of water might have been available at the nozzles to deal with the fires in question and, in effect, were most likely contributory factors in the cause of their deaths. In one case a crew was attempting to gain entry into an apartment fire, in which at least one firefighter was believed to have still been alive following an event of rapid fire progress, but stated that despite applying brief bursts of water from the nozzle into the overhead, the hose-line in use provided an ineffective stream that 'seemed to be having little effect on the fire.' The energy release for this fire was estimated somewhere between 5–15 MW and was wind-assisted into a direction that opposed the advancing hose-team, who were quickly beaten back off the fire-floor. There was a clear need for greater amounts of both pressure and flow at the nozzle and the limits of gas-phase firefighting were clearly surpassed in this situation, meaning alternative approaches were needed, as 'pulsing' droplets into 5 MW fires with low flow-rates and inadequate nozzle pressure becomes problematic.

The science of firefighting flow-rate

There are many flow-rate formulae used around the world by fire protection design engineers that have been produced following painstaking engineering calculations based on theory. In general, the vast majority of these formulae has been designed

with a strong 'fail-safe' approach, ensuring dramatic overestimates of water requirements, and is far too complex to be used by fire officers at fires.

In practical terms, the most viable formulae for fire-ground use have been
provided as follows:

- Iowa University (for indirect attack) – gallons/min
- National Fire Academy (NFA) (for direct attack) – gallons/min
- Tactical Flow-rate (TFR) (for general fire attack) – liters/min

Despite the fact that each of these formulae was provided for practical use by
firefighters, based on empirical research undertaken in different parts of the world,
there are distinct similarities in the various formulae produced. This suggests that
each formula is viable in relation to its specific and intended use.

Iowa University flow-rate formula (Royer/Nelson)	V (sq ft)/100 = gallons/min (US) V (cubic m)/0.75 = liters/min (metric)	Where V = the volume of the fire-involved compartment. This formula was derived solely from 'indirect' methods of suppression.
National Fire Academy (NFA) flow-rate formula – to 1,000 gallons/min	A (sq ft)/3 = gallons/min (US)	Where A = the area of the fire-involved compartment. This formula was derived mainly from direct attack methods of suppression and includes two hose-lines (one as back-up).
Tactical flow-rate (TFR) (author's) formula – Areas between 50–600 sq m	A (sq m) × 4 = liters/min (metric) A (sq m) × 6 = liters/min (metric)* *A × 6 is used where the fire load has breached and spread to structural members; or where the fire is affected by an inflow of wind.	Where A = the area of the fire-involved compartment. This formula was derived from both direct and indirect (gas-phase) methods of fire suppression, although direct attack would be predominant.

Critical flow-rate (CFR)

The CFR refers to the 'minimum amount of water-flow (liters/min or gallons/min)
needed to fully suppress a fire at a given level of involvement' (i.e. during growth or
decay stages of development). The actual CFR for compartment fires of a given size
(sq m or sq ft), existing in different stages of fire development, may be widely
variable. Where a compartment/structural fire exists in its growth-phase, the heat
output (energy release) will be constantly increasing and the amount of water
needed to extinguish the fire effectively will be much greater than when the fire has

progressed beyond 'steady-state' combustion into a decay-phase of burning, when most of the energy release has already occurred.

In theoretical terms of simply meeting a critical rate of flow, this does not offer the best use of resources, as it requires a more or less infinite amount of time. An increase in the flow-rate above the critical value causes a decrease in the total volume of water required to control the fire. However, there exists an optimum flow giving the smallest total water volume. Above this flow, the total volume of water increases again. In practical terms however, a margin of safety, or error, must be designed into the application of any firefighting tactic and this includes methods of fire suppression and flow-rate. An increase in water flow will generally darken a fire quicker. However, there is an upper limit on flow-rate in terms of what is practical for any given size of fire, in line with the resources available on-scene during the early stages of primary attack. The author's tactical flow-rate is the target flow (liters/min) for a primary attack hose-line(s). It is based upon extensive research and empirical data relating to firefighting flow-rates used at real fires in the USA and UK. The tactical flow-rate discussed in this text is for fire suppression during the growth phases of development, or in post-flashover steady state enclosure fires before the decay-phase has been reached. It is always an operational objective to achieve control during the growth stages of a compartment fire's development, rather than during the latter decay stages, to reduce the chances for serious structural involvement and any potential collapse, particularly where an interior approach is made.

The theory of needed firefighting flow-rate can be derived by resorting to scientific calculation, and matching water-flow against known rates of heat release (MW) in compartment fires. It can also be calculated using formulae derived from the empirical experience of several hundred real fires, matching stated fire loads in established floor space against water flows needed to suppress fires during their growth or decay stages (the latter generally being a defensive application). In the author's own sixteen-year research project he has used both methods and eventually combined them to produce a tactical flow-rate formula of proven reliability.

Sardqvist reports that the minimum water application rate for direct extinguishing, based on experiments using wooden fuels, is 0.02 kg/sq m per second. If you consider a compartment of 100 sq m (10×10 m) (actually 1,076 sq ft but say 1,000 sq ft), then this equates to 120 liters/min (26 gallons/min) as the theoretical minimum (critical) flow-rate for such an area and fuel load. This estimate is well researched and based on a number of fire tests undertaken in various scientific research establishments around the world. Whilst these tests generally refer to wood crib or pallet stack tests, these have often been high-intensity fires burning in large compartments with heat release rates as high as 15 MW.

However, the CFR is likely to be much higher for 'real' fires where fire loading increases beyond simple 'wooden' fuels. The true CFR in an apartment fire could be said to be (author's estimate) at least double that estimated by Sardqvist for ordinary wooden fuels, and 0.04 kg/sq m per second might be a more reliable estimate. This equates to a minimum firefighting flow-rate of 240 liters/min (50 gallons/min critical flow-rate) when operating in the direct attack mode against a 100 sq m real fire burning in over-ventilated conditions. Interestingly, Stolp (1976) suggested the critical flow-rate for a 100 sq m (1,000 sq ft) compartment fire was around 200 liters/min (53 gallons/min).

Don't forget this is the **critical flow-rate,** which means that whilst it may eventually suppress the fire, there is every likelihood that it will take some time to achieve. As an example, in one 15 MW fire test the control criteria were established as a period of six minutes from the start of fire suppression to the time when loss of mass in the fuel (wood cribs) reached a point where data demonstrated such loss had ended. At this point, flows of 113 liters/min (30 gallons/min) had been unsuccessful in achieving control of a 100 sq m (1,000 sq ft) fire within the six-minute criteria set. However, much higher flow-rates were successful in achieving suppression earlier.

During the author's firefighting flow-rate research of 120 working fires in London in 1989–90, it was noted that where flow-rates were bordering on the critical rate of flow as described by Sardqvist and Stolp (above), the control of 50% of these fires was only achieved whilst in the decay stage of fire 'growth'. That is, the vast majority of the fuel-load had burned away and the energy release from each fire was in decline. Although this enabled a lower flow-rate to suppress the fire, such a tactical approach could not be termed 'successful', for where firefighters are forced into this situation, they may face greater dangers including those of structural collapse.

'Real fires' – Needed flow-rates

Sardqvist's research (1998) into actual flows used at 307 selected fires in non-residential buildings in London (UK) suggested that most working fires were extinguished with a maximum 600 liters/min (160 gallons/min) flow-rate, and that 75% of fires did not increase in size following fire brigade arrival. His studies also revealed that only a very small percentage of structural fires (in the study) exceeded 100 sq m.

It should be noted here that the author believes Sardqvist's final conclusions on flow-rate were substantially overestimated (by around 36%) due to a reliance on SRDB (Home Office Scientific Research and Development Branch) nozzle flow figures in his research. These SRDB codes were never meant to represent actual practical fire-ground flow-rate factors. The codes were used to assess a fire stream's performance and flow-rate at very high nozzle pressures whilst each nozzle was mounted in a fixed gantry. Such nozzle pressures would never have been realistically achieved in practice by a crew of firefighters advancing hose-lines on the interior of a fire-involved building. It was also the case that the UK Fire Service (and London Fire Brigade) utilized very low pumping pressures to supply the 12.5 mm, 19 mm and 25 mm nozzles in use at that time (see UK *Manual of Firemanship* for the period). These actual fire-ground nozzle pressures were far lower than the SRDB codes used for the research and therefore actual flow-rates would correspondingly reduce.

NFA Formula (USA)	0.16 gallons/min per sq ft of fire	Plus a secondary back-up line
Dunn (FDNY)	0.12 gallons/min per sq ft of fire	Plus a secondary back-up line
Grimwood (London FB)	0.10 to 0.15 gallons/min per sq ft of fire	Plus a secondary back-up line
Sardqvist (Sweden)	0.3 gallons/min per sq ft of fire	Unrelated to number of lines

Fig. 9.1 – Needed flow-rate estimates provided by several authorities that have undertaken research into the flow-rates used to suppress actual fires in structures.

Tactical flow-rate (author)

In December 2004, New Zealand Fire Engineer Cliff Barnett turned to the author's earlier practical work and fire-ground formulae to update his own world-renowned efficiency factors, used by the Society of Fire Protection Engineers (NZ), for predicting firefighting flow in designed engineering-based applications. The resulting document *SFPE (NZ) TP 2004–1* offers the most accurate fire-fighting flow-rate requirements for use by both firefighters and design engineers to date.

The author's original research into needed flow-rate occurred at 120 working (greater alarm) fires in London during the late 1980s[2] and provided a metric formula for firefighters (known as the *Tactical [Metric] Flow-rate)*, to be used as a 'rule of thumb' method to estimate the needed flow requirements at structure fires.

- **Area (sq m) of fire involvement/4 = Minimum liters/min**
- **Area (sq m) of fire involvement/6* = Minimum liters/min**

*Where walls, floors or ceilings are breached by fire, or where wind is creating a high-intensity forced draft fire, the flow rate is increased by 50%.

It should also be noted that the author recommended this formula only be applied to areas of fire involvement measuring between 50 sq m (500 sq ft) and 600 sq m (6,500 sq ft).

Where, for example, a fire has involved 25% of a 300 sq m (3,250 sq ft) single-story building, with fire showing through the roof[3], the formula would suggest:

$75 \times 6 = 450$ liters/min minimum flow-rate on the primary attack line, ***with a secondary back-up line laid in support to secure attack team safety.*** The secondary back-up hose-line must be at least equal in size and flow-rate to the primary attack hose-line. Therefore, we will be laying in a minimum flow capability of 900 liters/min on such a fire.

Now compare that formula, as derived from the author's real fire research in London, with the National Fire Academy (NFA) formula, which is actually a means tested formula also based on real fire demands in the USA.

The structure would convert to 3,250 sq ft with 25% involvement at 800 sq ft. $800/3 = 266$ gallons/min (which converts back to 1,000 liters/min)

We can see the NFA formula is very similar to the author's metric formula **where A × 6 is used,** as this converts to a needed flow-rate of:

- 900 liters/min (author) (Attack hose-line and back-up support line).
- 1,000 liters/min (NFA) (Attack hose-line and back-up support line).

Briefly, it is important to understand comparisons with other established flow-rate formulae and discuss their relevance to the above example.

2. Grimwood, P., (1992), *Fog Attack*, FMJ/DMG Publications, Redhill, Surrey, UK
3. The author would advise that, depending on structural design and integrity, such a building might not be suited to an offensive interior attack. For example, where a steel roof truss is being heated to such an extent, the likelihood of any reliable structural integrity cannot be assured.

NFA Flow-rate	133 gallons/min (500 liters/min)
Tactical (Metric) Flow-rate	120 gallons/min (450 liters/min)
IOWA Flow-rate V/100	64 gallons/min (242 liters/min)
Sardqvist	200 gallons/min (750 liters/min)

Fig. 9.2 – Comparisons of established flow-rate formulae that have been derived from 'real fire' research – 75 sq m (800 sq ft) of fire involvement.

Whilst there are many detailed engineering flow-rate formulae based on scientific theory and mathematical calculations, very few of these will align with the flows that were extrapolated from 'real world' fires, as was the case with the above four methods (see Fig. 9.2). In fact, the vast majority of these engineering flow-rate formulae will provide gross overestimates in actual needed flow-rates. However, this is the purpose of design systems (to plan for the worst-case scenario), and this may be reflected in their calculations.

The four versions of flow-rate formulae listed in Fig. 9.2 are all derived from real fire research – two programs in the USA and two based on data supplied by London Fire Brigade's Division of Investigation. It is also worth noting that this 75 sq m area represents an average-sized house in the UK. Try to picture in your mind this area of fire involvement at 75 sq m (800 sq ft) which, at full involvement in a two-story house, will easily be controlled by a single hose-line flowing 450 liters/min (120 gallons/min). However, a secondary back-up hose-line should always be laid in support where the primary line is going interior. Where the same 75 sq m area of fire involvement exists in the larger 300 sq m (3,250 sq ft) structure, then two hose-lines may be critical and the secondary support line may be needed to assist the attack (remember – these are minimum flow-rate estimates).

When comparing flow-rate formulae in this way, it is essential that the method used to suppress fires in each specific research project is also considered. As an example, the **IOWA** research is based solely on the use of water-fog directed in from a position exterior to the fire compartment. This method of attack is termed 'indirect' extinguishing and entails fog streams being directed in through windows or doorways leading to the fire compartment, where the fog stream is swirled around the room so that water droplets evaporate on hot surfaces. The effect is one of mass vaporization and the dominant mechanism of extinction is smothering, or displacement of the oxygen, with some cooling effect also. The method of attack is therefore aimed at the gas-phase fire although much surface cooling obviously occurs.

In contrast, the National Fire Academy formula was originally derived from direct attack methods, which were the dominant form of attack, where straight streams were aimed at suppressing the fuel-phase fire. In early NFA courses circa 1979–84, the academy used the ISO fire-flow formula as well as a modification of the Iowa formula. When the *Preparing for Incident Command* (PIC) course was re-written, the new NFA formula was developed under peer review of some students, who were experienced fire-ground commanders. They produced the fire-ground method of **Area (sq ft)/3 = gallons/min.** It was stated that the commanders believed an aggressive interior attack on a fire had an upper limit of 1,000 gallons/min (3,780 liters/min) and after this, the fire should generally be fought defensively.

The two other methods of estimating needed fire-flows were both based on real fire research, undertaken in London against a large number of working fires (Grimwood 1989–90: 120 fires and Sardqvist 1994–97: 307 fires). During these periods London firefighters would predominantly use the direct attack methods to control fires but would also resort to some use of fog patterns to gain knock-back against flaming combustion and provide protection to nozzle operators.

Therefore, it is important to appreciate how difficult it is to compare some flow formulae where the mechanisms of extinction are different.

NFPA 1710 flow-rate requirements
In the USA the *NFPA Standard 1710* addresses the minimum flow-rates that should (shall) be provided on the first response as follows:

- Establishment of an uninterrupted water supply of a minimum 1,480 liters/min (400 gallons/min) for 30 minutes. Supply line(s) shall be maintained by an operator who shall ensure uninterrupted water flow application.
- Establishment of an effective water flow application rate of 1,110 liters/min (300 gallons/min) from two hand-lines, each of which shall have a minimum of 370 liters/min (100 gallons/min). Attack and back-up lines shall be operated by a minimum of two personnel each to effectively and safely maintain the line (plus at least one additional support firefighter to assist advance of each individual hose-line).

The key points here are:

- Uninterrupted water supply
- Minimum 1,480 liters/min (400 gallons/min) supply feed in
- Designated operator (pump operator) responsible
- Minimum 30 minute supply to be available/provided
- Available application rate of at least 1,110 liters/min (300 gallons/min)
- Two hand-lines staffed and flowed to a minimum 100 gallons/min
- Ideally these flows should be 570 liters/min (150 gallons/min)
- At least three firefighters assigned to each hose-line, where advancing the interior is the objective
- Back-up support or exposure protection hose-lines included

The key recommendations suggest minimum flows on the attack hose-lines should be 100 gallons/min (378 liters/min) minimum but ideally 150 gallons/min (570 liters/min).

Ideal interior attack hose-line
There have been countless research projects undertaken by authorities all over the world as to what is the optimal sized attack hose-line, and flow-rate, to be advanced on the interior of a structure. The relevant factors of course are:

- Physiological limitations of firefighters
- Involved fire load
- Structural or floor layout
- Available water supply

In relation to the physiological limitations of firefighters, there have also been several research projects that have reviewed heart rates, blood pressures, and VO2 Max etc. of firefighters advancing various sized hose-lines across floor spaces. Other studies have researched further, into the physiological demands of climbing up stair-shafts even prior to laying and advancing hose-lines.

The research in general suggests that the physiological limitations on firefighters are as follows:

- Heart rates may exceed 180 bpm without fire conditions
- Heart rates may exceed 200 bpm where heavy fire conditions or high heat levels are encountered
- Nozzle reaction forces (back pressure) must be controlled
- Smaller hose-lines are lighter and easier to advance

So we must address the issue of interior attack hose-lines from the following perspective:

- What is the minimum safe flow-rate?
- What is the maximum nozzle reaction?
- What is the optimum-sized line to achieve the above two factors?

Minimum safe flow-rate

The minimum safe flow-rate for an interior attack hose-line is 100 gallons/min (380 liters/min). This is noted as the *NFPA 1710* recommendation for minimum attack line flow-rate. It is also calculated in the author's flow-rate research where moderate (average) fire loads, in commonly encountered compartment or room sizes, will provide a normally maximum level (not wind assisted) of fire intensity to the point where a single hose-line flowing 100 gallons/min (380 liters/min) should meet adequate control criteria (pre-decay stage). A flow-rate of this size will perform adequately against a moderate fire load in an area up to 60 sq m (650 sq ft).

However, be aware that once structural elements become involved where they are predominantly of timber-frame or other combustible materials, they will add greatly to the involved fire load above normal room and contents. Therefore, the needed flow-rate will also increase dramatically!

As in all cases, the secondary back-up support hose-line, of at least equal flow as the primary attack line, will doubly ensure any fire of normally sizeable limits will be safely controlled. However, it should be clear that this secondary hose-line is a tactical deployment for the safety of crews and not a flow requirement. It is critical in all situations that firefighters are equipped with the minimum rate of flow available at the nozzle to deal with a worst-case scenario, where sudden full involvement of the compartment fire load may occur (flashover).

There is an argument put forth by some CFBT instructors that low flow-rates may be equally effective from the perspective of applying a greater surface area of water (thus a greater cooling capacity) through finely divided water droplets contained in fog patterns. In theory this is of course true and when fog patterns are used against extensive amounts of flaming combustion inside Fire Development Simulators (FDS) (ISO containers) the knock-back effect is dramatic. However, remember that these simulators offer little in comparison to a real compartment fire! A 1.5 or 2.8 MW simulation of pure flaming combustion is not the same as a real compartment fire fuel load of 3–15 MW, where heavy penetration into the fuel sources may also be needed to achieve any real cooling effect.

Maximum nozzle reaction
Nozzle reaction is the result of a tremendous amount of backward-force created by the flowing jet of water and the increase in velocity as the water leaves the nozzle. Whilst firefighters may demonstrate a number of innovative ways of taking the nozzle reaction out of the equation whilst stationary, inevitably where any interior advance is to be made of the hose-line the nozzle reaction will present a fiercely opposing force to any such advance.

Again, several studies have demonstrated the practical limitations of nozzle reaction in advancing nozzles, and the author's work is reproduced here as a guide to such limitations.

By evaluating the maximum flow capability for a hose-line that could be effectively directed and safely handled whilst *advancing and working* inside a fire-involved structure, it was observed (during the author's published research in London Fire Brigade 1989) that there was a maximum nozzle reaction force that could be handled by one, two and three firefighters as follows:

- One firefighter – 266 N (60 lbf)
- Two firefighters – 333 N (75 lbf)
- Three firefighters – 422 N (95 lbf)

These were interesting findings and from these figures one is able to establish optimum baseline flows for interior firefighting operations.

However, the change to modern combination fog/straight-stream or automatic nozzles brought a demand for higher nozzle pressures to achieve similar flows, and with that comes an increased reaction force. A baseline flow of 600 liters/min (160 gallons/min) being discharged from a combination/automatic type nozzle operating at 7 bars (100 psi) nozzle pressure (NP) will produce a reaction force of 356 N (80 lbf) which may cause a two-man team to struggle with any workable advance of the line.

There are combination/automatic nozzles available that have been adjusted to provide rated flows at lower nozzle pressures, but be sure to test these yourself as manufacturer's 'rated' flows are sometimes unachievable! Top US branded nozzles must meet the stringent demands of NFPA standards and low-pressure combination nozzles are able to achieve their rated flow-rates at factory-set nozzle pressures of just 5 bars. This would enable a flow of 600 liters/min (160 gallons/min) to be achieved with a reaction force of just 303 N (68 lbf), which is more easily handled and advanced by a two-man team.

The firefighter is able to calculate the amount of nozzle reaction (NR) by resorting to various formulae:

European formulae:

- NR (newtons) $= 1.57 \times P \times d^2/10$
  (European smooth-bore)
- NR (newtons) $= 0.22563 \times \text{liters/min} \times \sqrt{P}$
  (European combination fog/jet or automatic nozzles)

These are metric formulae where P = nozzle pressure and nozzle diameter

US Formulae:

- NR (lbf) $= 1.57 \times d^2 \times P$
  (US smooth-bore)
- NR (lbf) $= 0.0505 \times$ gallons/min $\times \sqrt{P}$
  (US Combination fog/straight or automatic Nozzles)

Where P = nozzle pressure and d = nozzle diameter

Optimum sized hose-line for interior attack

Interestingly, similar research has been carried out by other fire departments, notably San Francisco, Los Angeles and Chicago, who proposed that a safe and practical baseline flow for a workable firefighting hand-line would be around 550 liters/min (150 gallons/min). More recently (1996), the City of St. Petersburg in Florida, USA have established that, for their purposes, the ideal baseline flow is around 600 liters/min (160 gallons/min) using a 22 mm ($\frac{7}{8}$ inch) nozzle with a 50 lbs psi nozzle pressure on a 45 mm ($1\frac{3}{4}$ inch) hose-line. This set-up will create an acceptable reaction force of 266 N (60 lbf) and offers a hose-line that is easily advanced and maneuvred for interior position.

Now the FDNY, and many other fire departments, have established a basic rule of thumb for sizing primary interior attack hose-lines as follows:

- Residential structures – 45 mm ($1\frac{3}{4}$ inch) ($\frac{15}{16}$ inch nozzle)
- Commercial structures – 65 mm ($2\frac{1}{2}$ inch) ($\frac{11}{8}$ inch nozzle)
- High-rise buildings – 65 mm ($2\frac{1}{2}$ inch) ($\frac{11}{8}$ inch nozzle)

Effective fire-ground flows from $1\frac{3}{4}$ inch hose potentially range from 150 to 190 gallons/min (570–700 liters/min. The City of New York Fire Department (FDNY) considers 180 gallons/min (680 liters/min) the ideal flow from $1\frac{3}{4}$ inch lines in terms of fire extinguishment capability and handling characteristics. Some members of the fire service suggest that actual fire-ground flows from $1\frac{3}{4}$ inch hose are somewhat less than the 150 gallons/min minimum given above. The main reason for this is a widespread underestimation of the friction loss in $1\frac{3}{4}$ inch hose at flows of 150 gallons/min (570 liters/min) or more.

In Europe there was some interesting research undertaken by the Building Disaster Assessment Group (BDAG) in the UK that examined various aspects associated with flow-rate, hose sizes and physiological demands on firefighters, specifically in high-rise situations.

Their conclusions suggested that 51 mm (2 inch) attack hose-lines appeared as the most viable option, especially in high-rise fires, for applying optimal firefighting flow-rates. They did not compare the 51 mm option with 65 mm ($1\frac{1}{2}$ inch) attack lines but the 51 mm hose-line did appear more suitable than the 70 mm ($2\frac{3}{4}$ inch) option, in relation to maneuvrability and physiological demands.

CFBT and firefighting flow-rates

So how does all this talk of needed flow-rates fit in with CFBT concepts? Well, as we will see, the ideal range of water droplets in a fog pattern offers a greater surface area, and therefore cooling capacity, when compared to water applied in straight stream pattern. There is no argument that the most efficient way to cool fire gases and suppress flaming combustion is to optimize the application of water by breaking the droplets down into the ideal size.

However, what is equally important is the fact that to overcome flaming combustion, or to cool gases effectively, there has to be a sufficient content of water in a fire stream, whether this is applied in either a fog pattern or a straight stream. The performance of any particular water flow-rate is dependent on its ability to take the heat out of the fire. Where the fire's energy release (intensity) is too great, or too rapid, for the available flow-rate at the nozzle(s), the fire will continue to grow and develop and may force firefighters off the fire floor or out of the building!

Whilst the CFBT training facilities offer some quite severe fire conditions, they are not truly representative of a real room or structure fire progressing past the flashover stage and into a stage of steady state burning. Therefore the flow-rates that are effective in training may not be adequate for the fire-ground, if fires are to be suppressed during their growth stages as opposed to the decay stages.

As an example already referenced earlier in this chapter, firefighters attempting to enter an apartment fire progressing towards full involvement presents a situation where flow-rate is critical. Initially, the fire had spread from a bedroom into a corridor, involving at least one other room in the apartment at the time firefighters were trying to advance the line and gain entry into the apartment to rescue their trapped colleagues. Although the fire was wind assisted and directed at the nozzle crew, the author is of the opinion that the flow-rate at the nozzle, estimated to be around 230 liters/min (60 gallons/min), was contributory to their inability to advance on the fire. The two firefighters on the nozzle remained pinned to the floor for a few short seconds, in a lobby area just outside the entrance to the apartment, with flames torching over their heads. Is it the case that twice the flow-rate at the nozzle would have enabled this crew to advance in on the developing fire instead of retreating?

There were arguments (in court) that a second crew of firefighters was eventually able to advance a hand-line in and extinguish the fire, despite the continuing wind-fed forced-draft fire. Therefore flow-rate was not an issue. However, as this second crew entered the flat, crawling in on their stomachs, they described a very intense heat – this heat was not from a developing fire but rather a fire that had entered the 'decay' stage. The walls and ceilings had retained much heat during the free-burn period as the fire burned for several minutes without water being applied. However, the firefighters described clearly a scene where the energy release was reducing – they observed 'spot fires' throughout the rooms as they advanced in. One firefighter described the scene where everything had burned completely to ash. Under these circumstances of a decay stage fire it would be easier to advance a hose-line, as the forced-draft fire would have mostly subsided.

If we are to advance CFBT concepts internationally, another area we must consider is national requirements on firefighting flow-rate. There are some countries where a minimum flow-rate on the primary (indeed any) hand-line used for interior firefighting is pre-determined by national standards. We have already discussed the *NFPA Standard 1710* (above) that recommends a minimum flow of 100 gallons/min (380 liters/min) (actual target flow-rates are at least 50% higher). In France there is a national requirement for a minimum of 500 liters/min (130 gallons/min) to be provided to each interior attack nozzle. Therefore, the consideration of flow-rate is a major issue in relation to CFBT.

9.5 WATER DROPLETS AND COOLING THEORY

Water has been known as an extinguishing agent as long as man has known fire. With the exception of helium and hydrogen, water possesses the greatest specific heat capacity of all naturally occurring substances and has the greatest *latent heat of vaporization* of all liquids. It is estimated theoretically that a single gram of liquid water can extinguish a 50-liter flame volume by reducing its temperature below a critical value – equivalent to an 'application rate' of 0.02 liters per cubic meter.

It has also been suggested that the quantity of water required to achieve control of a structure fire is between 10–18 gallons (38–68 liters) per 1,000 cubic ft of fire (28 cubic meters). Again, in the UK it is further estimated that the majority of 'typical' compartment fires are extinguished using between 16–95 gallons, which is less water than one engine carries!

Specific heat

Specific heat is the amount of heat required to raise 1 gram (g) of a substance by 1 degree Celsius (°C). Specific heat is expressed in Joules (J). The specific heat capacity of water varies slightly from 0 °C to 100 °C, but at 18 °C it is 4.183 kJ/kg °C.

18 °C is selected here because it is the typical temperature of water when it comes from an underground water main.

- **Example 1**

  Determine how much heat will be absorbed in raising 10 kg of water from 18 °C to 100 °C

  $= 4.183 \text{ kJ/kg°C} \times 10 \text{ kg} \times (100 °C - 18 °C) = 3,430 \text{ kJ}$
  Specific heat capacity is expressed in J/kgK or J/kg°C

Latent heat of vaporization

The latent heat of vaporization is the amount of heat required to change a liquid into a vapour without a change in temperature. For water, this is 2,257 kJ/kg. Water does not boil immediately upon reaching its boiling temperature (100 °C at sea level). Once boiling point is reached, the water must absorb additional heat energy to convert the water into a vapour. This is the latent heat of vaporization. Of the unique properties of water, this one is the most valuable as a fire protection tool.

- **Example 2**

  Determine how much heat will be absorbed if 1 kg of water at an initial temperature of 18 °C is perfectly converted to steam at 100 °C

  $= 4.183 \text{ kJ/kg} \times (1 \text{ kg}) \times (100 °C - 18 °C) + 2,257 \text{ kJ/kg} \times (1 \text{ kg})$
  $= 343 \text{ kJ} + 2,257 \text{ kJ}$
  $= 2,600 \text{ kJ}$
  $= \underline{2.6 \text{ MJ}}$

Combined specific heat and latent heat

The final effect of water upon a fire is a combination of specific heat and latent heat of vaporization. We have to compute the total amount of heat absorbed by a unit of water when raised from its initial temperature in a water main to the temperature of the fire gases. The total heat absorbed occurs in three stages:

(a) Specific heat multiplied by the mass of water and the increase in temperature to reach boiling temperature at 100 °C
(b) Plus the product of latent heat of vaporization at 100 °C multiplied by the weight of water
(c) Plus the specific heat of steam multiplied by the mass of steam and the increase in temperature from 100 °C to the temperature of the fire gas.

• **Example 3**
Determine how much heat will be absorbed if 1 kg of water at 18 °C is perfectly converted to water vapour at 300 °C

$$= 4.183 \text{ kJ/kg} \times (1 \text{ kg}) \times (100\,°C - 18\,°C) + 2{,}257 \text{ kJ/kg} \times (1 \text{ kg})$$
$$+ 4.090 \text{ kJ/kg} \times (1 \text{ kg}) \times (300\,°C - 100\,°C)$$
$$= 343 \text{ kJ} + 2{,}257 \text{ kJ} + 818 \text{ kJ}$$
$$= \underline{3.4 \text{ MJ}}$$

The information in Fig. 9.3 below indicates that 1 kg of water, converted to steam as in Example 3 above, would be an insufficient amount to absorb the heat released by 1 kg of any of the fuels listed. The result however is different when water is applied to a fire in typical firefighting rates of kilograms per second, that is, liters per second.

Substance	MJ/kg
Wood	16
Polyurethane	23
Coal	29
Rubber Tyres	32
Petrol	45

Fig. 9.3 – Net heat of combustion values for selected common fuels.

In Example 2 above we determined that 1 kg of water when boiled at 100 °C from an initial temperature of 18 °C can absorb 2.6 MJ. Put another way, for each MJ of fuel in the fire load a firefighter theoretically needs 0.38 kg of water as steam at 100 °C to absorb the heat output of each MJ in the fuel.

As a further example, each kg/s of water vapour at 300 °C fed into a fire is theoretically capable of absorbing 3.4 MW of fire intensity.

From this it will be apparent that 5 kg of water, as water vapour at 300 °C, has the theoretical capacity to absorb $5 \times 3.4 = 17$ MJ. This is more than enough to absorb the heat generated by 1 kg of wood or 16 MJ when consumed in a fire. It will also be apparent that 14 kg of water has the capacity to absorb the heat generated by 1 kg of burning petrol.

Efficiency in fires
Water can never be applied at 100% efficiency for various reasons, and most building fires do not retain 100% of the heat energy in the room where the fire is occurring. The net result is that both the energy absorption of the water and the energy production of the fire need to be modified by calculated efficiency factors.

These can be expressed as:

- Heat absorption efficiency of a fire hose;
- Heat production efficiency of a compartment fire.

Heat absorption efficiency of a fire stream

The heat absorption described so far illustrates perfect conditions for the absorption of heat by the water. A tactical water application directly into the fire rarely approaches 100% efficiency in most cases. Unlike a laboratory test, there will always be inefficiencies and variables in the application of water to a compartment fire. Water may also be used to cool down fire gases and hot surfaces to enable a firefighter to approach closer to the actual fire source itself to complete suppression. Parts of the fire may have to be extinguished first to enable the firefighter to reposition and carry out the extinction of other parts of the fire. In some situations, as little as 20% of the water flow may actually reach the burning fuel surface.

There have been several attempts to estimate reliable **efficiency factors** for firefighting streams, often based on extrapolated data from theoretical computer models. However in general, the most accurate of all these efficiency factors are those that result following painstaking research covering many hundreds of real fires. Previous research has indicated that to overwhelm a fire, the efficiency of water as a cooling medium is about one third, or 0.32. Thus it was proposed that the effective cooling capacity of a flow of 1 liter is 0.84 MW, or a standard 10 liter fire hose is 8.4 MW, demonstrating a practical cooling capability with 33% efficiency. However, more recent research based on extensive real fire data suggests a 33% factor may be somewhat underestimated. A figure of three quarters (75% efficient) appears more reliable for a fog pattern and one half (50% efficient) for a solid-bore stream. The cooling power of each kg (liter) of water per second applied to a fire increases with temperature. Therefore the selection of an effective cooling power of only 0.84 MW (100 °C) may be seen as somewhat conservative. At 400 °C the cooling power can be seen to be closer to 1 MW and at 600 °C it is close to 1.2 MW.

In combining Cliff Barnett's SFPE NZ engineering research with the author's fire-flow calculations based on real fire data, the updated efficiency factors are inserted into Barnett's flow-rate calculations as follows:

- **Example 4**

  Find the total heat energy absorbed (Qs) by a 7 kg/s jet nozzle if the water is initially at 18 °C, assuming that perfect steam conversion is accomplished at 100 °C

  Qs = 7 kg/s × 2.6 MJ/kg × 1.00 = <u>18.2 MW</u>

- **Example 5**

  If the efficiency of a fog nozzle delivery at 7 kg/s is only 75%, find the total heat energy absorbed.

  Qs = 7 kg/s × 2.6 MJ/kg × 0.75 = <u>13.6 MW</u>

- **Example 6**

  If the efficiency of a jet nozzle delivery at 7 kg/s is only 50%, find the total heat energy absorbed.

  Qs = 7 kg/s × 2.6 MJ/kg × 0.50 = <u>9.1 MW</u>

- **Example 7**
  An office fire burning at 100% efficiency would have an average heat release rate of approximately 0.25 MW for each square meter of area. Determining the amount of heat released for this fire in a space measuring 6 m × 6 m, we find:

  6 m × 6 m × 0.25 MW/sq m = 9.0 <u>MW</u>

  If the foregoing is true, one hose-line delivering 7 kg/s in a fog pattern at 75% efficiency or a solid-bore jet stream at 50% efficiency could both deliver enough water flow to control and extinguish this fire burning at 100% efficiency (see Examples 5 and 6 above).

Complex computer models have been developed to provide theoretical water flow estimations and are formatted to take into account additional factors, such as firefighting team intervention times; the effect of automatic suppression systems that may have operated, correcting HRR as necessary; ventilation parameters directly affecting HRR; thermal radiation and specific boundary cooling demands, thereby balancing total water requirements for a range of fires in a structural setting.

Heat production efficiency of a compartment fire
Combustion, or burning, involves causing chemical reactions that generate heat to take place between oxygen (generally supplied as air) and the combustible material (generally hydrogen or carbon or hydrocarbon compounds of these elements). Combustion of hydrocarbon fuel is brought about by the combustion of the hydrogen (H) and carbon (C) in the fuel with the oxygen (O) contained in the air (and/or in the fuel). Depending on ventilation parameters and other factors, the burning efficiency of an enclosed fuel load (within a compartment with limited openings) is never able to achieve 100%. Where compartmental ventilation openings are limited, a fire will take longer to consume any particular fuel load than it would if it were burning in the open air.

Combining efficiencies of fire streams with compartment fire burning rates
The alteration of firefighting stream (cooling) efficiency factors by Barnett, in line with the author's flow-rate research, and coupled with the burning efficiency of a compartment fire (taken as 50%), led to an updated approach by Barnett in *TP 2004/1*:

- **Example 8**
  If the efficiency of a jet nozzle at 7 kg/s is 50%, as in Example 6, but the efficiency of the fire is only 50%, find the total energy that can be absorbed by the water flow.

  Qs = 7 kg/s × (0.50 × 2.6 MJ/kg)/0.50 = <u>18.2 MW</u>

  Or, by re-arranging the equation, the amount of water required will be:

  F = (0.50 × 18.2 MW)/(0.50 × 2.6 MJ/kg) = <u>7 kg/s</u>
  F = firefighting water flow in kg/s (liters/second)
  Qs = heat absorption capacity of fire stream

In practical terms it must be pointed out that a firefighter's physiological barriers are relative to compartment size where, for example, a 1 MW fire enclosed within a 40 sq m compartment may present similar barriers to the firefighter as a 16 MW fire in a larger 300 sq m compartment.

The reliability of this method is somewhat dependent on the accuracy of the heat release rate data and cooling efficiency value used, which, in this case, is based on real fire data obtained from structural enclosures. This method considers not only the heat absorbing properties of water from a scientific viewpoint but also the efficiency of firefighting streams when used to control actual enclosure fires (exhibiting post flashover conditions and demonstrating similar HRR to common compartment fires). It should be noted that such theoretical data should only be used as a guideline for estimating absolute minimum flow-rates under the most ideal of circumstances.

Some important aspects of water droplet theory

- As water droplets evaporate in the heated fire gases existing in the overhead there will be two main effects. The first is **expansion** as the water turns to vapour. The amount of expansion will depend on how hot the overhead is and what size the droplet is. Another effect may be some **contraction** in the fire gases as they are cooled.
- The optimum application of water droplets will occur where any expansion of water vapour is immediately countered by a greater contraction in the cooling fire gases. This will ensure thermal balance is maintained, thermal inversion is avoided and visibility at the floor is maintained where smoke is not forced downwards.
- Where any expansion of water to vapour is greater than the cooling effect of contraction in the fire gases, then hot gases, steam and smoke will be driven down onto the firefighters crouching at the floor.
- The effect of droplet velocity also has an effect, as fast moving droplets will generally absorb a greater amount of heat than slower moving droplets. This is why low-flow/high-pressure systems can be so effective when comparing flow-rate performance.
- Where water-fog is applied into a room on constant flow, there will be a large flow of air drawn into the room behind the stream, due to the negative pressure created at this point. This is an undesirable effect that may 'push' fire gases (or even fire) into other areas, upset thermal balance and increase fire intensity (energy release).
- By applying water-fog in brief 'bursts' or nozzle 'pulses', this negative effect of air moving in behind the stream is controlled and easily avoided.
- One droplet of 1 mm diameter has far less surface area exposed to heat in the overhead gas-phase than if the same droplet is broken down into ten droplets each of 0.1 mm in diameter. Theoretically, this increased surface area will enable a greater amount of heat to be absorbed.
- Water droplets below 0.2 mm are not generally effective in a firefighting stream as they are unable to penetrate the fire plume and are normally carried away on convection currents before they are able to achieve any great cooling effect.
- Water droplets above 0.6 mm are generally too heavy and too large to completely evaporate in the fire gases and therefore may pass right through the gases to strike walls, ceilings and hot surfaces. The result is a greater **expansion ratio** as the area of evaporation may be superheated above 400 °C.

- A firefighting fog spray will consist of a wide range of droplet sizes right across the spectrum from extremely small to very large. The make up of the droplet range will depend upon nozzle **design** and nozzle **pressure**. Higher nozzle pressures will generally lead to smaller droplets, and vice versa. Where a nozzle is designed to function at a nozzle pressure (NP) of 7 bars (for example) the most effective range of droplets will normally be produced.
- In some instances, droplet theory suggests that some smaller droplets may follow close behind larger droplets, enabling them to penetrate further into the hot gas layers than they would normally be able to on their own.
- These theoretical effects associated with thermal balance; thermal inversions; ascending and descending smoke layers; air in-flows etc. can all be practically demonstrated in a Fire Development Simulator (FDS).

Mechanisms of fire extinction

- **Fuel-phase fire** – Cooling of the combustible solid fuel surface, which reduces the rate of pyrolysis and thus the supply rate of fuel to the flame zone. This reduces the rate of heat release by the fire; consequently the thermal feedback from the flame is also reduced and this augments the primary cooling effect of the suppression agent. The application of a water spray to the fuel bed is typical of this method although a straight-stream, or smooth-bore attack, may be equally effective, if not more so where penetration into the fuel source is needed.
- **Gaseous-phase fire** – Cooling of the flame zone directly; this reduces the concentration of free radicals (in particular the chain-branching initiators of the combustion reaction). Some proportion of the heat of reaction is taken up by heating an inert substance (such as water) and therefore less thermal energy is available to continue the chemical break-up of compounds in the vicinity of the reaction zone. One function of the latest water mist technology (for example) is to act in this manner: the fine droplets providing a very large surface area per unit mass of spray, in order to increase the rate of heat transfer. (Note: there are also other dominant mechanisms of fire extinction upon which a water-mist system will rely, such as oxygen depletion.)
- **Flame inerting** – Inerting the air feeding the flame by reducing the oxygen partial pressure through the addition of an inert gas (e.g. N_2, CO_2, vapor). This is equivalent to the removal of the oxidizer supply to the flame by the production of water vapor, and is the dominant mechanism by which the Layman/Royer/Nelson concepts of indirect water fog attack achieve suppression. In a discussion of fixed-system water-mist fire extinction mechanisms, Mawhinney added to the above three mechanisms some further effects associated with decreasing thermal radiation, dilution of the flammable vapor/air mixture, and chemical inhibition as playing a part in fire suppression.

Some interesting questions:

- When does a firefighting stream become a spray?
- When does a spray become a mist, or a fog?

These are valid questions and several references have attempted to provide the answers. It is of particular relevance to manufacturers of Water Mist Fire Suppression Systems (WMFSS) who are engaged in supplying fixed firefighting installations as a replacement for Halon gas fixed-protection systems. Herterich identified a need for consistent terminology when discussing firefighting sprays, especially when considering the characteristic size of the droplets. Grant and Drysdale adapted a 'spectrum of droplet diameters' to demonstrate the broad range of possibilities. The size ranging from 100–1,000 microns (0.1–1.0 mm) was of most interest in firefighting terms and this conformed, on the chart, to a droplet size equal to light rain or 'drizzle'.

The cut-off between 'sprays' and 'mists' remains somewhat arbitrary however. For example the US National Fire Protection Association (NFPA) have suggested a practical definition of 'water mist' as a spray in which 99% of the water volume is contained in droplets less than 1,000 microns (1.0 mm) in diameter, compared with conventional sprinkler systems where 99% of volume diameter may be in the order of 5,000 microns (5.0 mm). Some regard this NFPA definition of a 'mist' as being too 'loose' in relation to WMFSS, and an alternative definition was advanced suggesting a 'mist' should comprise 99% of volume diameter equal to or below 500 microns (0.5 mm). It is worth noting that most WMFSS produce droplets in the range 50–200 microns, and it is generally accepted that droplet sizes less than 20 microns are necessary for a spray to have true 'gas-like' attributes.

Modern firefighting nozzles produce sprays through pressure atomizing effects and the result is termed a 'polydisperse' spray – that is, it comprises a wide range of droplet sizes, ranging from coarse to very fine. There are several methods of measuring droplet sizes within a spray but the results often conflict, depending on the method used. It has been suggested that there is an optimum droplet size in terms of fire suppression, but this has never been achieved, as the objectives are variable. In terms of theory it is fairly straightforward in ascertaining the optimum size, but in real situations a firefighting spray has to contend with several hindering factors when injected into a hostile mass of super-heated fire gases. The smaller the droplet the better its cooling capacity, but if the droplets are too small then it is likely that interaction with the buoyant fire plume may prevent droplets reaching the source of the fire.

This loss of water to the surroundings is only particularly relevant where final extinction of the fire source with a spray is the objective. In terms of gas-phase cooling, this effect is not so prevalent and droplet sizing within the spray can be reduced. The ideal firefighting nozzle will produce a spray with droplets small enough to suspend in air for at least four seconds, optimizing 3D water-fog applications during gas-phase cooling. However, such a nozzle will also be versatile enough to move from spray to main stream and back again with ease to enable direct hits at the fire source. With this in mind it has been generally accepted that a water spray with a mean droplet size of around 300 microns (0.3 mm) is ideal for gas-phase cooling using the 3D firefighting applications.

Temperature inversions
Temperature inversions are where the temperature gradient between the overhead and the floor is reversed. When this occurs, the temperature at floor level (where firefighters are located) may become hotter than at the ceiling. This can be an extremely uncomfortable experience for the firefighter and is one to be avoided at all costs.

There are several possible causes here, for example:

1. The **expansion of water vapour** caused through the application of water-fog may force hot gases, which are not in the cooling zone, to move away from the expanding vapour. These hot gases may move across the ceiling to a wall, where they will then move downwards and once reaching the floor, move back into the compartment.

2. **A burst from a straight stream** may pass right through the hot gas layers before striking the ceiling, where it will break up into a large amount of water droplets. The temperature at the ceiling of a fire-involved room may be anything up to 800 °C (1,500 °F) post-flashover (or more). Where evaporation of water droplets occurs at this temperature the expansion ratio of water to steam can be excessive (5,000:1) (Fig. 9.3). This may force high temperatures existing lower down in the room (say one meter from the floor) to be pushed down onto the crouching firefighters.

3. **Water droplets that are too large** (greater than 600 microns or 0.6 mm) may also pass straight through the gas layers and strike the ceiling, where such evaporation may again force heat at lower levels, and super-heated water vapour, down towards the floor.

4. **Over-zealous use of water** may also cause excessive evaporation, which can cause similar effects in forcing heat and expanding water vapour down to the floor.

5. The **air-track is running fast** and the fire is gaining momentum. This may occur where a ventilation-controlled fire is provided with an additional supply of air and the level of fire growth is producing an energy release that is beyond the quantity of water available at the nozzle. As water is applied into the gases, the air-track fire will simply overpower any cooling effect and conditions will become increasingly uncomfortable.

Temperature °F	Temperature °C	Expansion Ratio
212	100	1,600:1
392	200	2,060:1
572	300	2,520:1
752	400	2,980:1
932	500	3,440:1
1,112	600	3,900:1
1,472	800	4,900:1
1,832	1,000	5,900:1

Fig. 9.4 – Typical expansion ratios of water to vapour at various temperatures in a compartment fire range from 1,600:1 at 100 °C to 5900:1 at 1,000 °C

Temperature inversions can be avoided with careful nozzle technique in applying the right amount of water into the gases. Where bursts from a straight stream, large water droplets, or over-zealous applications are used, high peaks of cooling will be seen on a temperature graph. These sudden and sharp drops in temperature may initially appear effective but they actually reflect temperatures that are bordering on, or surpassing, inversion limits. Where correct amounts of water-fog are applied into the gases in a controlled fashion, a more comfortable environment is created as the compartment becomes slowly fogged and cooled. On occasions, particularly where the fire is shielded (in an adjacent area or compartment) a fire's heat (energy) release may overpower the available water in the hose stream. If the MW are greater than the kg/sq m/s of available water, the actual flow-rate is deficient.

The optimum droplet size for gas-phase cooling was further addressed in a report, jointly funded by the Finnish and Swedish Fire Research Boards, where it was shown droplets below 200 microns and those above 600 microns created excessive amounts of undesirable water vapour, whilst those in the range of 400 microns (0.4 mm) optimized the effect of gas-phase cooling. The reasons for this were mainly due to the effects of fire 'plume' interaction, where smaller droplets were used, necessitating additional amounts of water in application to achieve an effective cooling rate, and an increased amount of water reaching hot surfaces in the case of the larger droplets (large droplets are heavier and have less 'residence' time in the gases).

This point was also noted in a series of tests in the USA where wall temperatures within the fire-involved compartment were greatly reduced in proportion to an increase in droplet diameter, again resulting in greater evaporation and cooling outside of the fire gases, where, during the first two minutes of application:

- Sprays measuring 330 micron droplets decreased wall temp by 57 °C
- Sprays measuring 667 micron droplets decreased wall temp by 124 °C
- Sprays measuring 779 micron droplets decreased wall temp by 195 °C

This again demonstrates that sprays producing larger droplets will reach a greater surface area (especially walls and ceiling), which in turn creates excessive amounts of steam and less contraction of the gases. Gas-phase cooling is only effective where the droplets evaporate in the fire gases, avoiding contact with hot surfaces as much as possible.

A study by the Fairfax County Fire Department in 1985 compared the cooling capabilities of smooth-bore streams against combination nozzle streams in both straight and wider fog patterns. Using protected thermocouples, they noted the combination nozzle's 'fog' pattern was three times more effective in cooling the overhead than a smooth-bore. Perhaps somewhat surprisingly the straight stream from the combination nozzle was also twice as effective as the smooth-bore in cooling the flaming overhead. The firefighters involved in the tests were convinced they would rather have the flexibility of a combination nozzle at the outset for any interior firefighting operation.

In 1994 the US Navy's Naval Research Laboratory (NRL) initiated a study on board the Navy's full-scale fire test ship to determine the benefits and drawbacks of using the three-dimensional approach in comparison to a more traditional straight stream attack to extinguish a growing Class 'A' fire within the confines of a 73 cubic meter compartment. The fuel load comprised of wood cribs and particle-board panels initiated by n-Heptane pool fires. To provide further realism, obstructions were placed between the fire sources and the entry point to the fire compartment. This forced the attack teams to advance well into the compartment before a direct hit at the base of the flames was achieved. A 38 mm hose-line was used with a flow of 360 liters/min for both the water-fog and straight stream attacks. When utilizing the fog pattern the water was 'pulsed' in short bursts from a 60 degree cone applied upward at a 45 degree angle into the flaming overhead. After the gaseous combustion was extinguished, the firefighters advanced to the seat of the fire to complete extinguishment using a straight stream. Thermocouples at various levels recorded temperatures throughout the tests and total water usage was noted. It became clear that the three-dimensional application of water-fog was far more effective in controlling the environmental conditions – the thermal balance remained undisrupted and steam production was minimal. In comparison, the straight stream attacks created excessive steam, disrupting the thermal balance and causing burns to nozzle operators, sometimes forcing them to retreat from the compartment. The reductions of compartmental temperatures were also more rapid with the pulsing tactics utilizing a fog-pattern. The US Navy report concluded that:

The three-dimensional fog attack strategy is the best method to maintain a safe and effective approach to a fire involved compartment when direct access to the seat of a fire cannot be immediately gained.

Steam expansion versus gas contraction theory

To explain the theory[4], a 60 degree fog-cone applied at a 45 degree angle to the floor into an average room (say 50 cubic m) will contain about 16 cubic m of water droplets. A 1 second burst from a 100 liters/min flow hose-line will place approximately 1.6 liters of water into the cone.

For the purposes of this explanation* let us suggest a single unit of air heated at 538 °C weighs 0.45 kg and occupies a volume of 1 cubic m. This single 'unit' of air is capable of evaporating 0.1 kg (0.1 liter) of water, which as steam (generated at this, a typical fire temperature in a compartment bordering on flashover) will occupy 0.37 cubic m.

It should be noted that a 60 degree fog-cone, when applied, would occupy the space of 16 'units' of air at 538 °C. This means that 1.6 kg (16×0.1 kg), or 1.6 liters of water can be evaporated – i.e. the exact amount that is discharged into the cone during a single 1 second burst. This amount is evaporated in the gases before it reaches the walls and ceiling, maximizing the cooling effect in the overhead. It may be seen that too much water will pass through the gases to evaporate into undesirable amounts of steam as it reaches the hot surfaces within the compartment.

Now, by resorting to Charles Law calculations, we are able to observe how the gases have been effectively cooled, causing them to contract. Each 'unit' of air within the cone has now been cooled to about 100 °C and occupies a volume of only 0.45 cubic m. This causes a reduction of total air volume (within the confines of the cone's space) from 16 cubic m to 7.2 cubic m. However, to this we must add the 5.92 cubic m of water vapour (16×0.37) as generated at 538 °C within the gases. The dramatic effect has created a negative pressure within the compartment by reducing overall volume from 50 cubic m to 47.1 cubic m with a single burst of

4. Grimwood, P., (1992), *Fog Attack*, FMJ/DMG International Publications, Redhill, Surrey UK

fog! Any air inflow that may have taken place at the nozzle will be minimal (around 0.9 cubic m) and the negative pressure is maintained.

The above calculation was subsequently amended (2006) by French Fire Engineer Frank Gaviot Blanc and presented in the following format – www.flashover.fr

Temperature in Gases °C	Flow rate liters/min	Spray volume cubic m	Water volume liters/s	Efficiency 100 %	Useful water At 100% eff	Volume of vapor cubic m	Gas Contract cubic m	Difference of volume cubic m
200	100	16	1.67	100.00	1.67	3.59	12.62	−9.03
300	100	16	1.67	100.00	1.67	4.35	10.42	−6.06
538	100	16	1.67	100.00	1.67	6.16	7.36	−1.20
600	100	16	1.67	100.00	1.67	6.63	6.84	−0.21
700	100	16	1.67	100.00	1.67	7.39	6.14	+1.25
800	100	16	1.67	100.00	1.67	8.15	5.56	+2.59
200	500	16	8.33	100.00	8.33	17.97	12.62	+5.35
300	500	16	8.33	100.00	8.33	21.76	10.42	+11.35
538	500	16	8.33	100.00	8.33	30.80	7.36	+23.44
600	500	16	8.33	100.00	8.33	33.15	6.84	+26.31
700	500	16	8.33	100.00	8.33	36.95	6.14	+30.81
800	500	16	8.33	100.00	8.33	40.75	5.56	+35.18

Temperature in Gases °C	Flow rate liters/min	Spray volume cubic m	Water volume liters/s	Efficiency 74%	Useful water At 100% eff	Volume of vapor cubic m	Gas Contract cubic m	Difference of volume cubic m
200	100	16	1.67	74.00	1.23	2.66	12.62	−9.96
300	100	16	1.67	74.00	1.23	3.22	10.42	−7.20
538	100	16	1.67	74.00	1.23	4.56	7.36	−2.80
600	100	16	1.67	74.00	1.23	4.91	6.84	−1.93
700	100	16	1.67	74.00	1.23	5.47	6.14	−0.67
800	100	16	1.67	74.00	1.23	6.03	5.56	+0.47
200	500	16	8.33	74.00	6.17	13.29	12.62	+0.68
300	500	16	8.33	74.00	6.17	16.10	10.42	+5.69
538	500	16	8.33	74.00	6.17	22.79	7.36	+15.43
600	500	16	8.33	74.00	6.17	24.53	6.84	+17.70
700	500	16	8.33	74.00	6.17	27.34	6.14	+21.21
800	500	16	8.33	74.00	6.17	30.15	5.56	+24.59

Fig. 9.5 – The expansion ratio of water to vapor may be countered by the contraction in the gas layers as they cool, as seen above in column nine where the emboldened positive figures represent unsuccessful attempts to counter water to vapor expansion where the expansion has exceeded the gas contraction in these examples. However, the negative figures above these in column nine demonstrate an effective reduction in total air volume. Note the differences in efficiency in column five, where the 100% (theoretical) efficiency is compared to Barnett's 75% efficiency factor for a water-fog application. Courtesy of Frank Gaviot Blanc (France).

In these nine columns above it becomes clear that a successful application of a bursting fog pattern (small amount of fog droplets), into the super-heated fire gases existing in the overhead, can be successful in preventing thermal inversions (steam dropping to the floor) as well as raising the smoke layer (water to vapor expansion ratio being countered by contraction in the cooling gases). This concept applies to all flow-rates (100 and 500 liters/min is used above) but effective nozzle design, nozzle pressure, and operator skills are all relevant factors.

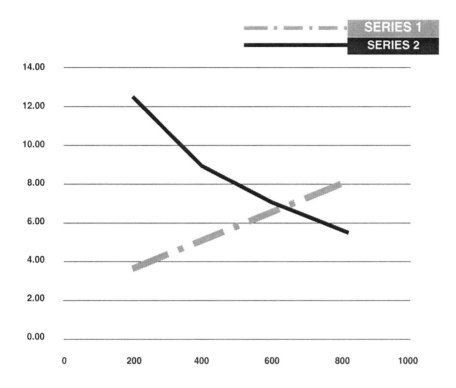

Fig. 9.6 – At 100 liters/min and 100% efficiency it is seen that effective thermal balance is maintained to ceiling temperatures up to 600 °C where, as the lines cross, the thermal balance is broken down by excessive water to vapor expansion. (Gas contraction line runs from top left of chart whilst water to vapor expansion line runs from bottom left of chart). Temperature °Celsius runs along the bottom axis and volume in cubic meters runs down the left-hand side axis.

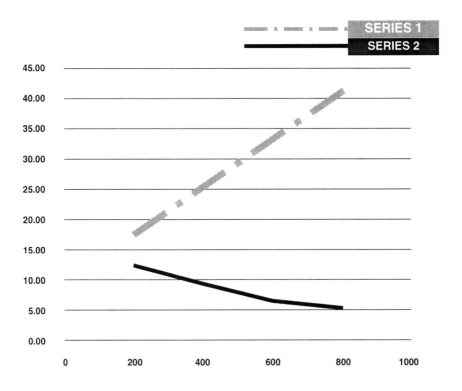

Fig. 9.7 – At 500 liters/min and 100% efficiency it is seen that the thermal balance is broken down by excessive water to vapor expansion, where the lines of water to vapor expansion (top) and gas contraction (bottom) have already crossed.

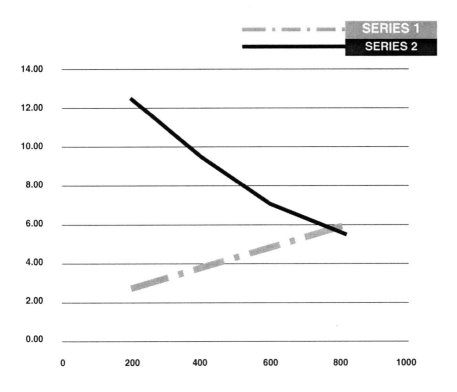

Fig. 9.8 – At 100 liters/min and 74% efficiency it is seen that effective thermal balance is maintained to ceiling temperatures around 750°C where, as the lines cross, the thermal balance is broken down by excessive water to vapor expansion. (Gas contraction line runs from top left of chart whilst water to vapor expansion line runs from bottom left of chart). Temperature °Celsius runs along the bottom axis and volume in cubic meters runs down the left-hand side axis.

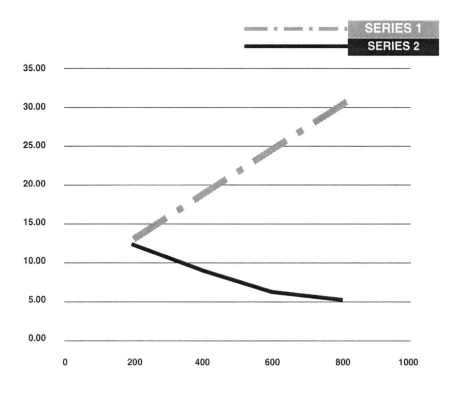

Fig. 9.9 – At 500 liters/min and 74% efficiency it is seen that the thermal balance is broken down by excessive water to vapor expansion, where the lines of water to vapor expansion (top) and gas contraction (bottom) have already crossed.

The above four diagrams demonstrate the following points:

- Flow-rates above 100 liters/min (30 gallons/min) are unlikely to allow thermal balance to be maintained, even when using short bursts at the nozzle.
- The more practical firefighting flow-rate (for direct attack) of 500 liters/min (130 gallons/min) will break down the thermal balance if applied into the heated gas layers.

However, if we are able to think about application technique, and utilize a nozzle that allows us to simply 'crack' the nozzle partially open when working at around 7 bars (100 psi) NP, then we will not be applying 500 liters/min (130 gallons/min) but about one fifth of this amount.

Some nozzle designs will not allow the nozzle operator to crack the flow control handle with such ease or, in doing so, will produce a range of very large droplets that are outside the optimum size for effective gas cooling or flame suppression. However, one type of nozzle that will allow the operator to 'crack' the flow control to the first 'indent' is that which operates using a 'slide' valve mechanism as opposed to a 'ball' valve. The slide valve will normally enable much greater control of the flow control and will not affect stream or droplet quality whilst 'pulsing,' 'bursting,' or 'sweeping' with the nozzle.

Another option is a flow-selector ring on the nozzle that might flow around 115 liters/min (30 gallons/min) at one setting but increase to 475 liters/min (125 gallons/min) when turned.

Using high-flow nozzles in CFBT FDS units?

If we are to maintain an adequate and safe flow-rate for an interior attack hose-line that can deal with both gaseous and fuel phase fire development involving realistic fire loads bordering on flashover, we need a flow-rate between 100 gallons/min (380 liters/min) and 150 gallons/min (570 liters/min). A 45 mm ($1\frac{3}{4}$ inch) line is generally optimal for such a purpose in its ability to be manageable (easily advanced) by firefighters, whilst working at reasonable pressures, which prevent excessive nozzle reaction or kinking.

The 51 mm (2 inch) or 65 mm ($2\frac{1}{2}$ inch) interior attack lines are also good options, particularly in commercial/industrial fires or on high-rise fire floors. These lines offer less frictional loss and higher flow-rates where compartment fire loading is beyond moderate, where exterior winds are fanning the fire, or where structural members have become involved.

But do such flow-rates enhance or hinder the CFBT training concepts? It is clear they are needed in 'real' fires and we should train in ISO containers as if we are facing a 'real' fire, right? However, it is a fact that most instructors and CFBT programs prefer to use 100 liters/min (30 gallons/min) maximum flow-rate in the FDS units for the reasons discussed earlier in this chapter.

In the USA, CFBT lead instructor Battalion Chief Ed Hartin[5] has taught the Rosander/Giselsson methods of fire control for many years and his firefighters regularly enter a training simulator (FDS) equipped with a 500 liters/min (130 gallons/min) flow-rate on a sliding valve nozzle. They have adapted their nozzle techniques to manage these high flows in the simulators and, as stated, the

5. Hartin, E., Battalion Chief (Training Division), Gresham Fire District, Oregon USA

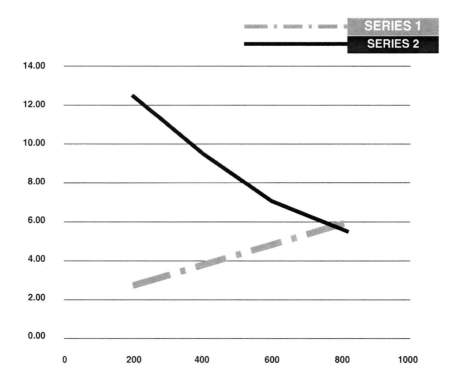

Fig. 9.10 – A 500 liters/min nozzle, using a slide-valve design incorporated in the nozzle head, is just cracked open to the first indent whilst at 7 bars (100 psi) nozzle pressure. The actual flow-rate applied into the overhead is reduced to around 20% of a fully open nozzle and is able to maintain thermal balance up to 750 °C ceiling temperature, where water to vapor expansion is countered by the cooling gas contraction. However, the higher flow-rate of 500 liters/min (130 gallons/min) remains available at the nozzle for aggressive interior attack aimed at the fuel base (direct attack). Based on the 74% efficiency ratio of a fog pattern suggested by Barnett (at 100% in the fog pattern the limit of thermal balance is encountered around 600 °C at the ceiling). (Gas contraction line runs from top left of chart whilst water to vapor expansion line runs from bottom left of chart). Temperature °Celsius runs along the bottom axis and volume in cubic meters runs down the left-hand side axis.

act of 'cracking' the flow control to a 20% opening (first indent) ensures the water applications are controlled and not excessive.

The origins of the Rosander/Giselsson techniques were optimized and taught in Swedish and Finnish FDS units during the 1980s using a nozzle (TA Fogfighter®) that flowed between 100 and 350 liters/min, with a flow-control function that followed closely the operating principles of a 'slide-valve' nozzle. In the USA the slide-valve design is common to the Task Force Tips (TFT) range of nozzles.

9.6 NOZZLE TECHNIQUES

There are several nozzle techniques that may be introduced through CFBT, aimed at dealing with the gaseous-phase fire. These may be described as follows:

- **'Pulsing' or 'spotting'** (also known as 'hole-punching'), is where very brief (half a second) bursts of water droplets (water-fog) are directed into the overhead gas layers to cool super-heated gases and smoke. Sometimes, the objective is to 'punch' very small holes in the smoke layer with water droplets. These applications are generally made to cool the gases without upsetting thermal balance, keeping the smoke off the floor. Where tongues of flames near the ceiling are indicating the onset of 'rollover', the application of a series of brief 'pulses' may be enough to offset any further development in the overhead. However, this effect will likely only last for a few seconds and further action may be needed – either more pulses or a longer 'burst'/'sweep' of a narrow fog pattern. The likelihood of 'pulsing' being an effective tactic depends on the stage of fire development, the extent of the involved fire load, the geometry of the compartment, the location of the fire, and the amount of heat or flaming combustion existing in the overhead.
- **'Bursts' or 'sweeps' are** longer bursts of the nozzle, around two to four seconds, or similar moving sweeps of the nozzle, which may apply a larger number of droplets into the overhead to deal with gaseous-phase flaming combustion, or cool pre-igniting gases in a larger area. This effect is more likely to disrupt the thermal balance, create excessive water vapour, and drive smoke to the lower regions of the room. However, it is still possible to maintain some control of the conditions and where there is adequate flow-rate in the fog pattern, a good amount of gas-fire may be knocked back. A type of fire where this particular application may be very successful is on a stair-shaft. As the nozzle is advanced up the stairs, any excessive water vapour from the brief pulses or bursts of water droplets will be carried upwards on the air-track, providing even more cooling effect. The knock-back is pretty dramatic and a rapid ascent is often achieved. In this case it is essential that a second hose-line follow the first line to ensure that the potential for any re-ignitions occurring behind the rapid advancement of the primary attack line is negated.
- **Over-pressure/under-pressure refers to** using the air-track or gravity current, where air flows in below the NPP (neutral pressure plane) and heads towards the fire, whilst gaseous-phase flaming combustion heads for an opening (which may be behind the advancing firefighters). The objective here is to apply a short burst of water droplets into the overhead and then direct the nozzle down to apply a second short burst below the NPP, aimed

at channeling the droplets into the in-flowing air-stream that is heading for the fire. Where this process is applied in reverse, an initial burst of fog aimed straight at the fire (below the NPP), preceding any fogging of the overhead, may cause a 'water to vapour expansion' that will increase flaming in the overhead and send flames in your direction.

- **The 'shark.'** A more recent application being taught in Sweden (e.g. Gothenburg) is called the 'shark'. The application promotes the opposite of the 'over-pressure/under-pressure' method described above. It is mentioned here purely because it is being taught and is in use in Sweden in 2007. Using a TA Fogfighter® nozzle, the flow goes from 100 liters/min to 300 liters/min in one sweeping nozzle movement that is intended to drive any gaseous combustion over and behind, away from the nozzle operator.
  1. Set the nozzle to straight stream.
  2. Open the nozzle, but not completely, and attack the heart of the fire.
  3. Increase the opening of the nozzle (flow), and start turning the nozzle tip in order to go to fog pattern. During this, move the nozzle to the ceiling, all in one sweeping motion.
  4. As the nozzle is directed to the ceiling, open the nozzle so you are at maximum flow with a wide-fog protection pattern and then close the nozzle.

The author has not used this nozzle application in real fires, outside a training scenario, and cannot attest to its viability. There is no doubt that the thermal layer will disrupt as water to vapour expansion may be excessive and fire gases are moved around the compartment.

- **Pencilling.** The term 'pencilling' refers to brief bursts from a straight stream or narrowed fog pattern, often applied in a lobbing fashion, to direct small 'slugs' of water onto burning surfaces in a controlled attempt to suppress flaming. This application is often useful in maintaining visibility as fuel sources are approached and then torn or cut open to reveal the base of the fire, in a controlled manner, or removed from the compartment via an adjacent window. This term (pencilling) is used in a different way in the USA, where straight streams are applied in short bursts into the overhead. This has the same effect as narrow fog patterns in cooling the overhead, although the hazards of thermal inversions are far more likely when using straight streams[6], compared to bursting water droplets in fog patterns.
- **Painting.** This term refers to the cooling of wall and ceiling linings, using a narrow fog or straight stream, in an attempt to draw heat out of the solid materials that may radiate heat back into the gases. The effect known as **'hot-wall'** occurs where compartment boundaries either retain heat or bounce it back into the compartment. This has the effect of radiating vast amounts of heat back into the fire gases and may lead to auto-ignitions of the gas layer. To avoid this situation, where surfaces and compartment boundaries are acting as heat sinks or insulators, a cooling stream may be 'painted' across the surfaces to draw the heat out.

---

6. Grimwood, P., Hartin, E., McDonough, J. & Raffel, S., (2005), *3D Firefighting*, Oklahoma State University. Refer to US Navy Tests p.54 and MAFS tests p.63.

- **Straight-stream direct attack.** When a fire remains unshielded (in view) and is accessible with a straight stream, the direct attack has the capability to penetrate a fuel-phase fire (surface and sub-surface burning) and achieve a great amount of cooling of the fuel base. This approach of a constant flow straight-stream attack on the fire is, traditionally, the most common form of fire attack. In situations where a high flow-rate is required to overcome a very intense release of energy (heat release) then the direct attack is sometimes the only method that will achieve effective suppression and rapid knock-down of the flame front.

- **Indirect attack.** This approach was very popular in the 1950s as a primary means of achieving rapid fire knock-down. The application of a water-fog pattern, rotated or 'swirled' around in a fire-involved compartment, relied on a mass expansion of water to vapor as the fine water droplets came in contact with hot surfaces. The primary mechanism of fire extinction is oxygen displacement and the technique is generally applied from an exterior position, often through a window. As the techniques generally relied on small amounts of water, in comparison to other methods, it was a popular method of attack where water resources were limited. Even to this day, this method is sometimes used to deal with compartments demonstrating the warning signs of backdraft.

9.7 ONE-SEVEN® COMPRESSED AIR FOAM

In the book *3D Firefighting*[7] we explained how CAFS (Compressed Air Foam Systems) had been subjected to some stringent research testing, against both shielded and unshielded fires, in several countries. The consensus of the research demonstrated, in broad terms, that a CAFS was superior to water in most test scenarios. However, Class A foams (without air compressors) also proved to be more effective than water and possibly more economical than CAFS in their potential to deal with room and contents and structure fires in the fuel-phase (direct attack).

The book also described a series of tests undertaken by the East Sussex Fire and Rescue Authority in the UK where a CAFS had been designed to support the 'pulsing' and 'bursting' nozzle concepts, which few (if any) systems had managed to achieve. If CAFS was to be used for interior attack hose-lines in the UK, then it was important that there was a system control function that allowed 'CAFS' to be used effectively against the gaseous-phase in a structure fire. The only system to manage this to date was the system produced by Gimaex-Schmitz, called the One-Seven® CFS.

The UK research tests compared the One-Seven® system with plain water in two different FDS units:

- **Attack 2 Scenario** is a training evolution that teaches firefighters to enter and advance (and retreat under control) a hose-line along a corridor demonstrating severe rollover conditions in progress.

- **Tactical Unit.** In this situation a tactical training unit was used to simulate an entry into a severe basement fire.

---

7. Grimwood, P., Hartin, E., McDonough, J. & Raffel, S., (2005), *3D Firefighting*, Oklahoma State University, Refer to p.390 onwards (UK Fire Brigade Experience)

In both tests the temperatures at several heights in the FDS were data-logged and there was a clear division between the plain water applications when compared to the results obtained using One-Seven® CFS. The foam system demonstrated impressive cooling ability when applied in very brief bursts into the overhead. There was remarkable knock-back of the gaseous fire and thermal balance was easily maintained. The water vapour produced by the foam system appeared far less obvious than when water was used, and the nozzle operators were able to advance with much less work of the nozzle when using foam as compared to water. Previous research had raised an issue with the excessive amounts of water-vapor produced by CAFS when applied in constant-flow direct attacks, but this was certainly not the case when using the One-Seven® CFS system, where the structure reportedly reduces this hazard.

In the tactical unit the firefighters were unable to gain entry to the basement fire when using water, but found the One-Seven® foam system provided a far more comfortable environment in which to descend, where they were then able to enter the basement compartment and achieve full extinction of the fire. In conclusion, the foam was more effective on the visible fire and the ceiling temperatures, and far easier to work with than water. This particular foam system is now in operational use in the UK (East Sussex) and is also under extensive field trials in the city of Phoenix, USA. The Phoenix Fire Department previously sent several instructors to East Sussex Fire and Rescue in the UK to learn how to operate the system, and to observe the interior firefighting tactics being used to apply foam into the gas layers to prevent flashover.

As this book goes to print there is a major European-funded research project underway in France, entitled PROMESIS, that is testing and comparing several different firefighting systems, including high-pressure and low-pressure water systems and various CAFS against *hot gas layers*. The results of this research (along with details of past CAFS research) can be located at the author's website[8] and at Euro-firefighter.com. As with any new system, make sure you evaluate thoroughly the potentials for failure, as well as the benefits which may be achieved.

9.8 WATER-FOGGING SYSTEMS

In the book *3D Firefighting*, the authors explored the tactical solutions offered by various water-fogging and piercing nozzles. There is an age-old concept of adapting lightweight 4WD mini-pumpers, or even motorbikes, fitted with small water tanks and a water-fogging system. These rapid-response fast attack units are often seen to excel in inner-city areas for attending small rubbish or trash fires, or grass and brush fires in difficult access areas.

The idea of taking this equipment into a structure fire is limited to a primary attack made under the most stringent of protocols. Firstly, it is never safe to open a door, behind which you suspect there is a fire, unless you are equipped with a minimum 'safe' flow-rate, full PPE and SCBA, and are backed up with a partner and a crew of firefighters on-scene. From this position you should initiate a correct *door entry procedure*.

8. http://www.firetactics.com/CAFS.htm

However, such equipment may be used to great effect to supplement interior attack lines, for example in attic fires or void fires. In these cases a piercing nozzle can be very effective although it may also, in some situations, 'push' fire further along voids and attics. The use of thermal image cameras must always support such tactics where viable.

There is one type of portable fog 'canon' that may be carried into a structure by a member on the primary response. The **IFEX**® gun uses 360 psi of air pressure to discharge around 0.05 gallons in each 20-millisecond burst of water-fog. With a stream velocity of 250 mph shooting 200-micron bursts of water droplets some 45 ft from the nozzle, these portable units are capable of achieving rapid (but temporary) knock-back of a fire demonstrating an energy release of up to approximately 2 MW. This particular option suffers from a time delay of three to four seconds during each recharge of the compressed air used to fire out the fog bursts, during which there is no water discharge available.

Where such equipment is used, again there must be clear and strict protocols documented, detailing the tactical limitations:

- Only for use on exterior fires where access is difficult or water supplies are extremely limited.
- Can be used as a primary suppression agent where a team is sent into a large structure to locate a reported small fire that is not obvious from the exterior.
- In doing so, firefighters **should not open doors** behind which they suspect a fire is located, unless as part of a 'safe' reconnaissance action (read on).
- Where fire is suspected behind a door, feel the door for heat using the back of your hand, starting at the midway section and moving upwards.
- Try to get a look through a fire-resisting glass panel if this exists above or to the side of the door.
- Observe the edges of the door for signs of smoke.
- If the door is warm or hot it should not be opened until the minimum flow-rate is available and firefighters are equipped with full PPE/SCBA and are being committed as part of a firefighting team.
- If the door is cool but issues smoke on opening, read the smoke conditions (density, volume, velocity, color) and follow basic SCBA (BA control) and fire attack protocols, or close the door. It may just be an incipient-stage fire so read what the smoke is telling you.
- Where occupants are 'known' to be trapped beyond this door, follow the **'Snatch Rescue'** protocol (see Chapter Three).
- Firefighters should not be committed into any known working fire situation without the minimum flow-rate, full PPE/SCBA, or without a partner.

However, there may well be occasions where the 'investigation' team stumble on a small fire of a manageable size, deep inside a large structure, during their efforts to locate the fire. Where the fire is visible, easily approachable and accessible, and is still within the capability of the equipment available, then an attempt to suppress such a fire is logical before it is allowed to spread out of control, **providing no one is placed at unnecessary risk**. It may be far safer to take such a fire early than have to retreat and then re-enter into a smoke-laden compartment with a fast developing fire within.

9.9 FIRE DEVELOPMENT SIMULATOR – DEMONSTRATOR

As of 2007, the most common containers used in international commerce (based on ISO Standard 668) were made of steel, and are 20 ft (6 m) and 40 ft (12 m), with some 48 ft units. The typical exterior container height is 8 ft 6 inches (2.4 m). So-called high-cube containers are 9 ft 6 inches (2.9 m). The standard width of containers used in international commerce is 8 ft (96 inches or 2.4 m). So-called domestic containers, used only for land transport (rail or road) are 53 ft long and 102 inches wide, 6 inches wider than standard ISO containers. These domestic containers are built to lighter standards, as they are not designed for exposure to the elements atop a ship at sea. A movement is underway in Europe for a new container width of 102 inches (8.5 ft or 2.6 m); these containers would be classified as belonging to *ISO Standard 02*.

These units may then be further adapted for CFBT with interior and exterior operating roof and wall vents; solid or fire-proof material baffles that hang from the ceiling to form reservoirs to collect smoke; interior insulation; protection of the burn chamber; protection of the flooring system, and some additional doors to provide escape routes etc.

The most basic FDS unit that is used to impart introductory CFBT training is termed the 'Demonstrator'. Its design varies but it usually consists of a 20 ft (6 m) container that sometimes has a raised burn chamber at one end. The raising of the burn chamber serves to protect firefighters to a greater extent from frontal radiant heat. Where a raised chamber is not provided, firefighter positions will normally be located further back in the container, away from the fire end.

The unit will ideally have a smoke reservoir where flammable smoke and combustion products from the burn chamber collect at ceiling level. This reservoir is formed by a 1-meter (3 ft) baffle that hangs from the ceiling about halfway along the unit. This is generally the area where the majority of ignitions of the gas layers occur in the overhead and where nozzle techniques are practiced.

The training objectives of a 'Demonstrator' unit are:

- For students to occupy the unit prior to ignition occurring;
- To watch an enclosure fire grow and develop from an incipient stage;
- To observe practical fire behavior and compare with learned theory;
- To observe the process of pyrolyzation;
- To observe the formation and effects of an 'air-track';
- To gain an understanding of air-track management;
- To understand basic concepts such as NPP and over/under-pressure;
- To experience radiant heat under controlled and safe conditions;
- To experience limited visibility in a confined compartment fire;
- To experience the basic concepts associated with vertical ventilation;
- To observe convection currents;
- To observe various warning signs of flashover;
- To observe various ignitions of the fire gases, including auto-ignition;
- To practice various nozzle techniques used to prevent, control and suppress gaseous-phase combustion;
- To practice basic communication procedures under live fire conditions where viable.

A CFBT instructor will learn to run and manage a training evolution inside a FDS Demonstrator in order to: ensure the safety of students at all times; guide students through the learning process; highlight all learning points; ensure prompt and safe crew rotations (explain and practice how this will occur prior to actually taking them into the burn); ensure correct nozzle techniques, and, most importantly, to ensure that each student gains the same experience as all others. Such a process begins with pre-planning, detailed briefings and an understanding of all of the training objectives, before the evolution begins.

9.10 FIRE DEVELOPMENT SIMULATOR – ATTACK UNIT

The 'attack' FDS unit is normally a longer 40 inch (12 m) container that is used to teach firefighters to enter (using door entry procedure) and advance (and retreat under control) a hose-line along a corridor demonstrating severe rollover conditions in progress. This evolution offers a natural progression and a more severe test than the basic Demonstrator unit.

The fire is ignited whilst the students are outside and CFBT instructors will monitor and 'set' the conditions inside the unit, using roof and door vents to enable a fire in the end burn chamber to develop a flame front at the ceiling and a reasonable smoke layer close to the floor.

The training objectives of an 'attack' unit are:

- To practice door entry procedure to a fire-involved compartment;
- To read smoke conditions during the door entry process;
- To observe and control the formation and velocity of the air-track;
- To learn how to influence the height of the smoke layer;
- To practice over and under-pressure suppression techniques;
- To advance a primary attack hose-line, in two and three person teams, a few feet into the unit to a point where any gaseous combustion in the overhead is knocked back by the nozzle operator;
- The line is then retreated under control, back out of the compartment, which is closed down awaiting entry of the next team;
- When students become more experienced and confident, to allow them to advance further into the FDS unit and along towards the burn chamber, to deal with gaseous combustion in the overhead and pencil the fuel-phase fire up ahead. This is an evolution that relies heavily on effective risk assessment and adequate safety measures, as well as the ability of students.

Some 'attack' units are designed and constructed into an L-shape or sometimes a T-shape to provide greater flexibility in the range of 'attack' training options that may be delivered.

9.11 FIRE DEVELOPMENT SIMULATOR – WINDOW UNIT

The 'window' unit is a shorter ISO container of around 16 ft (5 m) in length. They are called 'window' units because sometimes they are fitted with a closed see-through window for exterior observation of the internal fire gas ignitions. They may also be fitted with 'stable' doors at the side and ends that allow either, or both, top and bottom openings.

Straight Stream into the Overhead

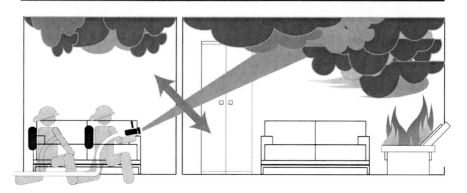

3D Water-Fog – Pulsing Nozzle

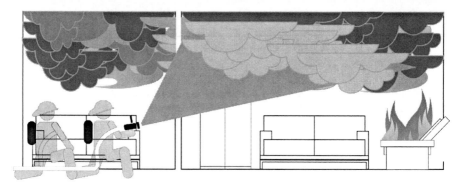

The training objectives of a 'window' unit is:

- To observe, from an exterior position, the effect of providing sudden cross ventilation to an under-ventilated fire that has progressed into a very hot smoldering state.
- The resulting effect is likened to that of 'backdraft', where a vast quantity of gaseous combustion emerges from the vent opening, sometimes with explosive force.

These window units are very dramatic in operation and leave a long-lasting effect with students. However, nothing more than simple observation of the smoke conditions and fireball effect is involved, and, as a learning tool, the effects seen in a window unit may be better provided by a video of the unit in action. There is a time and cost element to the running of such a system and, in economical terms, there may be better training options available.

9.12 FIRE DEVELOPMENT SIMULATOR – BACKDRAFT UNIT

The interior 'backdraft' unit is generally of German or Dutch design and incorporates an adaptation of the raised unit of a 20 ft (6 m) 'Demonstrator' unit, where the burn chamber becomes an internal room, served by an internal door. The process involves repeated 'backdraft' simulations to an occupied container where students will observe the opening and closing of the door to an under-ventilated fire. In some countries, the risk assessment will not allow firefighters to occupy the interior of such a FDS unit. However, the units do have an outstanding safety record and offer a training solution that can provide a useful hands-on experience to firefighters.

The training objectives of an 'interior backdraft' unit are:

- To demonstrate, at close quarters, the inappropriate opening of a door behind which an under-ventilated fire has reached a very hot stage of smolder;
- To observe the effects of pulsing smoke around the edges of a door;
- To observe the accumulation of a dangerous smoke layer in the compartment adjacent to the fire compartment;
- To teach fire suppression techniques that may quench such ignitions of the gases in adjacent compartments on opening the door to the fire compartment itself.

9.13 FIRE DEVELOPMENT SIMULATOR – TACTICAL UNITS

The Tactical Training units used for CFBT are constructed in a wide range of multi-compartment, multi-level, systems. These systems provide an opportunity for more advanced training that combines many learning objectives of the single FDS units with that of tactical deployment.

(Some) of the training objectives of a 'Tactical' unit are:

- To combine the skills learned in Demonstrator and Attack units;
- To provide a wider range of more realistic structural fire scenarios;
- To observe how 'air-tracks' and smoke/fire gases behave in varied geometrical layouts;

- To provide more opportunities to develop and practice tactical ventilation concepts;
- To provide opportunities for students to improve their tactical deployment skills;
- To ensure safe and effective deployment and placement of primary and secondary attack hose-lines, cover hose-lines and support (back-up) hose-lines;
- To establish every available means of securing team safety within the confines of a fire-involved structure;
- To utilize safe and effective deployments of search and rescue teams working behind, ahead of and above attack hose-line positions;
- To utilize effective interior search patterns, including the use of thermal image cameras (TICs);
- To implement safe and effective 'snatch rescue' procedures for interior search and rescue operations.

9.14 FIRE DEVELOPMENT SIMULATOR – GAS-FIRED UNITS

The gas-fired training units that are often found in purpose-built burn buildings may also be retrofitted in ISO FDS facilities, or similar systems. This type of compartment fire training facility burns gas to simulate flame patterns within the training facility. These facilities allow a quick turnaround between each training event, with consistently repeatable training scenarios. The system also allows precise control of the flaming combustion and temperatures, which allows continuous and consistent replication.

It must be noted that this type of system cannot truly simulate backdraft or flashover conditions, nor can it stand to represent fire behavior from any true perspective. The gas-fired simulations are turned on and off at the flick of a switch and do not allow for natural development based on pyrolysis and the production of smoke and gas layers. Therefore, any element of 'realism' is traded for repeatability and convenience. The systems are however useful to some extent in practicing nozzle techniques.

When designing or constructing gas-fired systems, particular consideration should be given to the following safety features:

- The period allowed for main flame ignition, how this is determined and monitored initially, and what emergency action is required in the event of this period being exceeded.
- The potential for explosion from delayed ignition and measures to prevent it.
- Precautions to prevent a flash-fire or explosion in the event of delayed ignition and the consequences for personnel in the vicinity.
- Gas fuel requirements, supply pressures, storage, etc.
- Gas monitoring and siting of thermocouples, their integrity and control applications.
- The monitoring and integrity of exhaust ventilation rates, and their use in an emergency.
- The quality and integrity of pipe work, joints and associated gas fittings, etc., which will be subjected to rapid heating and cooling.
- The layout of operator controls to ensure that they are safe, accessible and easy to use.

Technical standards

There are numerous technical standards that relate to the design, installation, use and maintenance of the hardware systems required to operate this type of system including:

- Electrical standards
- Gas storage and supply standards
- Pressurized systems standards
- Computer control systems

9.15 FIRE DEVELOPMENT SIMULATOR – LOADING UNITS

The loading of FDS units is an important part of any CFBT program. The use of various fuels must be closely risk assessed, documented and monitored. Where *NFPA 1403* applies, careful attention must be paid to the types of fuel used, as each particular fuel type will present different burning characteristics.

The main fuels used to load FDS units are wood chipboards to line the walls and wooden pallets as fuel bases. Here are some common examples:

- Medium density fibreboard (MDF)
- Oriented strand board (OSB)
- Wood particleboard (chipboard)

Wood particleboard (chipboard as it has been known in the UK) is an engineered wood-based sheet material in which wood chips are bonded together with a synthetic resin adhesive. (Note: the European and international standard term is wood particleboard and the use of this term is encouraged).

Wood chips comprise the bulk of wood particleboard and are prepared in a mechanical chipper generally from coniferous softwoods, principally spruce, though pine, fir and hardwoods, such as birch, are sometimes used. These chips are generally bound together with synthetic resin systems such as urea-formaldehyde (UF) or melamine urea-formaldehyde (MUF), though phenol-formaldehyde (PF) and polymeric methylene di-isocyanate (PMDI) are used by a few manufacturers.

The binding system employed depends on the end use intended and the grade of the product. The most common resin employed is urea-formaldehyde, but this is only suitable for use in dry conditions: the other three resin systems confer a measure of moisture resistance to the composite. Typical constituents of a particleboard are of the order (by mass) of 83–88% wood chips, 6–8% formaldehyde-based resin or 2–3% PMDI, 5–7% water, and 1–2% paraffin wax solids.

Particleboard has smooth, sanded surfaces. In order to achieve this smooth surface, the panel density is increased at the faces by the use of smaller wood particles with a larger percentage of resin binder compared to the core of the panel. A 2,400 × 1,200 × 19 mm panel will weigh approximately 36 kg. Veneered chipboard is widely used for self-assembly furniture, work surfaces, wall linings and partitions. High-density chipboard is often used as a basis for the carcasses of kitchen furniture, worktops, and flooring – this is hard-wearing, rigid and heavy. Other grades available are standard, flame-retardant, flooring, and moisture-resistant. Thickness normally ranges from 12–25 mm ($\frac{1}{2}$–1 inch) but for CFBT, ideally 12 mm ($\frac{1}{2}$ inch) or 19 mm ($\frac{3}{4}$ inch) is used.

Fibre boards: Types of fibreboard are differentiated by the size and type of wood fibres used, the method of drying, what type of bonding agent is used and the method by which it is pressed into shape.

Medium density fibreboard (MDF) is manufactured by a dry process at a lower temperature than for example hardboard, another type of fibreboard. The effect of this is that the natural glues and resins contained within the wood are rendered ineffective. MDF therefore uses manufactured bonding agents and resins. Varying density boards with differing finishes are used for various end uses.

Oriented strand board (OSB) is manufactured from waterproof heat-cured adhesives and rectangular shaped wood strands that are arranged in cross-oriented layers, similar to plywood. This results in a structural engineered wood panel that shares many of the strength and performance characteristics of plywood. Produced in huge, continuous mats, OSB is a solid panel product of consistent quality with no laps, gaps or voids.

It can be seen in Fig. 9.11 how different fuels are likely to affect CFBT training evolutions in terms of temperatures, radiant heat, speed of fire development, and duration of training burn. Where storage of boards is likely to allow moisture content to dry out, or where moisture (damp) is allowed to increase, this too will affect how CFBT evolutions will progress. These factors must all be built into the documented risk assessment for training. Suddenly changing from 12 mm to 19 mm boards (for example) is not acceptable without pre-planning, practice burns and a document introducing the change.

Similarly, all fire loads must be pre-planned and documented as part of the risk assessment for each individual FDS unit. Another influencing factor is that of wind and weather, for both will have an impact on how well, or how poorly, a CFBT training burn will progress. As an example, a gusting wind entering an open end of a FDS unit will likely increase the air-track and burning rate of the fuels, creating unpredictable conditions. To prevent this, many installations are designed and located with natural windbreaks.

How much fuel?

This depends on the training objectives and the type of simulator in use. Fuel type, fuel dimensions, thickness, and calculable ventilation profiles will also affect such a decision. In the author's experience, whilst training at many different locations in different parts of the world using a range of fuels and in different designs of FDS, it is important to establish and document local protocols based on how individual units are likely to operate.

It is generally the case that it takes three to four burns in each FDS unit before optimum training conditions are realized. Depending on the age and condition of FDS units, there may also be natural leakage paths that affect the quality of the CFBT training burns. It is important to work with each unit to define how venting arrangements will affect fire development, in line with fuel load.

As a general guide, both corner fires (three to four boards) and entire end fires (five to six boards) present good burning characteristics in Demonstrator units. The use of boards on all three walls in a Demonstrator burn chamber enables the effects of radiant heat and pyrolysis to be more easily demonstrated and observed by students, and allows for a longer duration burn, although the use of a corner burn is perhaps more economical. All boards should be tightly butted together to prevent fire creeping up behind them and causing an early depletion and breakdown of the fuel.

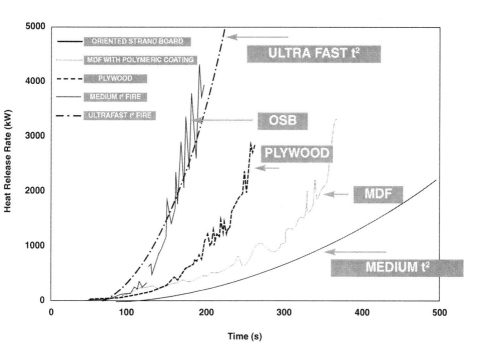

Fig. 9.11 – Typical heat release rates are compared for various lining boards and are aligned with t-squared (t^2) fire gradients to demonstrate how hot and how fast the various fuels are likely to burn. The burning characteristics for particleboard are comparable to MDF. However, note that OSB is likely to burn somewhat hotter and the heat release rate (HRR) is aligned with an 'ultra-fast' growth rate. In effect, CFBT training burns using OSB must be carefully risk assessed and pre-planned so that instructors are aware of the different burning characteristics likely during a training evolution. Note: different widths (12 mm or 19 mm etc.) of boards will also affect the duration of each training burn.

Cribbing fires. The best way to begin the fire is by use of cribbing – small pieces or strips of 2–3 inch long particleboard and wood, which are placed in a metal burn container or drum, to form a fire that will develop and burn consistently to heat up the boards. Where a central fire is used in a fully loaded (three wall) end of the burn chamber, more fuel is needed in the base fire, than if placed in a corner. The wood strips can be surrounded by a small amount of kindling of a type such as smaller pieces of wood, paper, dry straw etc., to allow the fire to grow. The cribbing fires should also be backed by pieces of board tightly butted together if corner based. An ignition source acceptable to local or national guidelines, such as a flare, can be used to ignite the kindling and base fire.

Two instructors wearing full PPE/SCBA should accompany each other at all times when lighting this base fire. Under no circumstances should it be necessary, or acceptable, to use flammable liquids to assist the initial ignition of the base fire, although the author acknowledges that in some countries this is an accepted practice.

Demonstrator FDS	Six boards (three end walls and three in the ceiling); or four boards (two end walls and two in the ceiling).
Attack FDS	Six Boards (three end walls and three in the ceiling); or nine or twelve boards for more advanced training – **to be risk assessed according to ventilation profiles**[9].
Window FDS	Twelve to fourteen wood pallets; or eighteen boards; or a mix of both, and additional off-cuts as available.
Interior Backdraft FDS	Twelve to fourteen wood pallets.

Fig. 9.12 – These guidelines for loading FDS units with 12 mm ($\frac{1}{2}$ inch) particleboard are only offered in very broad terms, but are based on normal practice in most areas.

Demonstrator	$5 \times \frac{3}{4}$ inch 8×4 ft OSB
Attack 1	$8 \times \frac{3}{4}$ inch 8×4 ft OSB
Attack 2	$8 \times \frac{3}{4}$ inch 8×4 ft OSB

Fig. 9.13 – The Fire Service College (UK) uses OSB as a fuel, and loads its FDS units as shown (2007).

Demonstrator	OSB $\frac{1}{2}$ inch 7 ft $\times$ 4 ft – six boards
Attack container	OSB $\frac{1}{2}$ inch 7 ft $\times$ 4 ft – twelve boards
Window container	OSB $\frac{1}{2}$ inch 7 ft $\times$ 4 ft – eighteen boards
Multi compartment per fire	OSB $\frac{1}{2}$ inch 7 ft $\times$ 4 ft – three boards

Fig. 9.14 – Devon Fire Service (UK) also uses OSB as a fuel source and loads its FDS units as shown (2007)[10].

9. Depending on fuel type and size, the use of twelve boards (or more) in an Attack FDS may need careful risk assessment in line with ventilation profiles, to ensure safe conditions are maintained.
10. Chubb, J. and Reilly, E., (2007), *Dublin Fire Brigade report into UK CFBT*

When loading FDS units be sure to follow local manual handling guidance and ensure that **protective respirators and PPE** are worn at all times by personnel entering the simulators, even where the structure is cold. There will always be hazardous particles floating inside and around these units and full protection must be used.

9.16 FIRE DEVELOPMENT SIMULATOR – SAFE OPERATION

The safe operation of FDS units relies on following clearly defined and documented protocols (a SOP for CFBT) based on local or national guidelines for the safe operation of live fire training. More specifically, the CFBT simulators have been risk assessed under the original Swedish model that offers some additional guidance.

- A risk-assessed documented SOP for all types of CFBT training.
- Training delivered by qualified, and competent CFBT instructors.
- Use of safe and well maintained equipment.
- Recognize the risks and implement the necessary control measures.
- Ensure efficient instructor:student ratios at all times.
- Record and log individual instructor and student times inside FDS.
- Log books for instructors (instructor welfare).
- Maximum number of burns per day/week for students/instructors.
- Monitor and record (where possible) interior conditions for each burn.
- Swedish Code of Practice.
- Attack FDS Units (venting arrangements – Attack 1/Attack 2).
- Hydration and food.
- Safety briefings.
- *NFPA 1403.*

The importance of ensuring FDS units are safe for training cannot be emphasized enough. Where there is no general scope for inspecting, recording, reporting, repairing and replacing various components written into the SOP (Standard Operating Procedure) for CFBT, then it is most likely that FDS units will gradually deteriorate and problems will arise, possibly leading to injury.

Operating features such as all doors and closing devices, ventilation hatches and operating devices, flooring, structural integrity, thermocouples and monitoring systems (where fitted), all require regular inspection and a reporting system should account for such inspections, reporting of faults and repairs etc.

Wherever FDS training evolutions occur, instructor:student ratios are an important issue as without the recommended minimum number of instructors on-scene, then safety may again be compromised. Where FDS units are used for live burns, instructors should implement the following Risk Control Measures:

- Ensure that all firefighters are familiarized with the unit;
- Know where the **escape routes** are located;
- Understand the procedure to be followed for **emergency evacuation;**
- Deliver **safety briefings** both before and after training evolutions;
- Always operate with an exterior **safety hose-line** (separate supply source if available) located at an entry point near to the fire;
- This hose-line must be crewed by at least one instructor;
- The second safety line can be sited and crewed on the interior of a Demonstrator FDS unit where it can also be used to control the fire during operator rotations of the attack nozzle;

Demonstrator	Six to ten students	Minimum two instructors
Attack 1*	Two students interior	Minimum two instructors
Attack 2**	Two to six students interior	Minimum three instructors
Tactical unit	Six to eight students interior	Minimum three instructors

Fig. 9.15 – Instructor:student ratios will vary. The above are guides as to what may be considered as safe minimums.

Swedish Code of Practice[11]

It can take five to seven days for instructors to acclimatize to the excessive heat conditions experienced at instructor locations in FDS units. Where instructors have just returned to work following an extended period away, they should gently break themselves in, taking shorter periods at hot locations in the FDS units. Inner core body temperatures of all instructors and students will provide a starting point for safe practice when undertaking training in FDS units.

- A normal body core temperature is 37 °C.
- Core temperatures of 39.3 °C have been recorded following 25 minutes in an FDS unit.
- Above 39 °C the body begins to lose its efficiency.
- A core temperature of 43 °C may prove fatal.

The following additional Risk Control Measures are also recommended:

- No one should be allowed to take part in training where they register an inner core body temperature (aural) above 38 °C.
- A maximum compartment temperature of 320 °C (600 °F) at 2.2 m (h).
- A maximum compartment temperature of 250 °C (480 °F) at 1.2 m (h).
- A maximum heat flux of 5 kW/sq m at firefighter locations.
- A maximum skin temperature of 55 °C under PPE.
- Decontaminate on exit and dress-down.
- Rehydrate and rehabilitate firefighters.

Attack 1*

The compartment fire is allowed to develop to the extent where there is heavy gaseous combustion occurring at the ceiling. Students are then brought in to practice branch techniques using correct door entry procedure. The neutral plane, fire frontage and steam egress is managed by an outside instructor at the vent location. A dual control is available to interior instructor if required.

Attack 2**

This exercise reflects the elements of Attack 1. Instructors set the conditions to increase the level of difficulty for students. The top vent is closed to extend and lower the flaming combustion. This presents an exercise demanding a precise aggressive attack from students. Students are encouraged to advance to boards and perform pencilling techniques[12].

11. Graveling, R.A., Stewart, A., Cowie, H.A., Tesh, K.M., and George, J.P.K. (2001), *Physiological and Environmental Aspects of Firefighter Training*, ODPM UK
12. Chubb, J. and Reilly, E., (2007), *Dublin Fire Brigade report into UK CFBT*

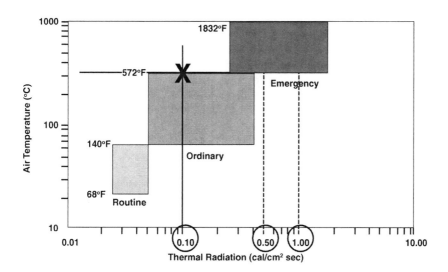

Fig. 9.16 – The range of thermal conditions faced by firefighters at 'real' fires demonstrates routine, ordinary, and emergency conditions. In terms of FDS training the cross denotes the approximate maximum limits that instructors and students should be exposed to (these are 320 °C [600 °F] at 2.2 m [7 ft] height and 250 °C [480 °F] at 1.2 m [4 ft] height – helmet tip of a crouching firefighter), and 5 kW/sq m (0.10 cal/sq cm/s). Note also that 0.5 cal/sq cm/s = 20 kW/sq m which represents the scientific definition of 'flashover' conditions.

Example safety brief

1 Introduce yourself.

2 Welcome to the training centre.

3 If the fire alarm should sound you will leave the building and gather in the car park opposite for role call.

4 Toilets are located.

5 Remember to maintain your fluid, salt and sugar levels at all times, both before and after working in CFBT units.

6 Before travelling up to the CFBT training units, a few safety points need to be brought to your attention.

7 Keep away from units: they may be hot.

8 Do not enter any units unless you are wearing respiratory protection.

9 Medical aid is located at.

10 Smoking is not allowed; smoking items are to be left back at the training centre.

11 Full PPE is to be worn at all times when inside the 'training' zone (also think decontamination).

12 Long-sleeved garments are to be worn for all exercises.

13 Do not breathe smoke coming from any of the units.

14 SCBA only to be stored and serviced in designated areas.

15 Students must be fit and well and able to take an active part in CFBT training.

16 All injuries must be reported at the time of occurrence or before taking part in CFBT training.

17 Ensure that students are aware of heat stress symptoms.

18 Indicate the dangers of consuming alcohol prior to course attendance (i.e. heat stress and dehydration).

19 Ensure that fluids are taken at start of the course and regularly throughout the day.

20 Undergarments, approved gloves and flash hoods are to be worn and be correctly fitted.

21 Long-sleeved garments are to be worn for all exercises.

22 Jewellery is to be removed.

23 Ensure all PPE is in good condition and that gloves and flash hoods are dry.

24 Check pockets for flammable items, e.g. cigarette lighters!

25 CODE RED indicates a real emergency, as does the activation of a DSU/PASS. Staff will evacuate students to the SCBA entry control (ECP)/Medic point.

26 Ensure students are conversant with all operational equipment being used.

27 Rig SCBA in 'safe air', ECO or buddy to check teams are correctly dressed.

28 Fires are 'REAL' – Crews must listen to instructors and 'STAY LOW'.

29 Anyone finishing the exercise early must report to ECO and remain at the entry control point until completion of the exercise.

30 Personnel must follow the directions of the instructors, but if suffering any discomfort they are permitted to withdraw under the direction of an instructor.

31 On completion of the training you may close down your SCBA set under the supervision of ECO – in safe air! **COLLECT YOUR BA TALLY FIRST!**

32 Obtain refreshments/fluid intake at the end of the exercise.

33 Advise instructors of any injuries, serious discomfort or heat-induced illnesses during or immediately after the exercise.

34 Rehabilitate, cool down and rest after CFBT live burns.

35 No eating is allowed on-site.

36 When drinking, ensure that hands and face are washed first.

37 **Reinforce hydration.**

NFPA 1403 and **CFBT**

Ed Hartin is a Battalion Chief in Oregon, USA and heads the CFBT training program delivered by the Gresham Fire and Rescue Services. The Gresham CFBT program is one of the very few in the USA that closely follows European safety codes of practice and training standards. Chief Hartin writes:

In the United States, National Fire Protection Association 1403 Standard on Live Fire Training *is the standard that identifies requirements for training firefighters under live fire conditions. In examining the impact of this standard on CFBT, it is important to understand the difference between regulations and standards as well as how occupational safety and health regulations are administered. A consensus standard does not carry the weight of law and is not directly enforceable (unless adopted by an enforcement agency directly or by reference). Regulations are administrative rules adopted by government agencies within a scope defined by legislative action. Occupational safety and health regulations are enforced by either the Federal Occupational Safety and Health Administration (OSHA) or state occupational safety and health agencies.*

Federal OSHA does not have regulations related to live fire training, but many states do. Some have adopted NFPA 1403, *others have adopted portions of this standard (and not others), and others have developed their own rules. However, state or federal occupational safety and health regulators generally default to industry standards if there are no rules under general regulatory provisions requiring a safe workplace.*

The following discussion examines key issues related to the application of NFPA to CFBT, but is not intended as a comprehensive review of the provisions of NFPA 1403. *A more detailed examination is provided in Chapter 10 of* 3D Firefighting: Training, Techniques, and Tactics[13]. *However, the standard has undergone inor revisions since the publication of this text and readers are encouraged to review the current standard.*

Understanding the application of NFPA 1403 *to CFBT requires an understanding of how this standard has evolved, its intent, and the letter of the standard as applicable to this type of training.* NFPA 1403 *was initially developed in 1986 following the tragic deaths of two firefighters during a drill to develop skill working in an acquired structure filled with smoke. However, the smoke was created by burning ordinary combustibles and combustible liquids. After extended operations, fire spread to the combustible interior finish and trapped the firefighters. In many of the subsequent revisions, errors by instructors causing the deaths of firefighters participating in training have resulted in specific prohibitions (particularly related to fuel, location of fires, and use of humans as simulated victims). When* NFPA 1403 *is applied literally to CFBT in a single compartment cell, the provisions seem to be a bit excessive (particularly in the areas of water supply, hose-line flow rate, and staffing).*

The minimum water supply specified by the standard is 7,569 liters (2,000 gallons), hose-lines must be capable of flowing 360 liters/min (95 gallons/min), and attack and back-up lines supplied by independent sources. These provisions are reflective of firefighting practices in the United States which favor vigorous use of tactical ventilation and high flow hose-lines as standard practice. Note that it does

13. Grimwood, P., Hartin, E., McDonough, J. & Raffel, S., (2005), *3D Firefighting*, Oklahoma State University

not specify that the hose-line has to flow 360 liters/min, simply that it must be capable of doing so. Use of a variable flow nozzle (with a maximum flow rate of at least 360 liters/min) meets this requirement. The volume of water required far exceeds that necessary for the typical CFBT session. Use of a continuous water supply such as a hydrant exceeds this requirement. If tank water must be used, there is no reasonable work around, but this requirement is easily addressed using a portable water tank and/or water tender. Attack and back-up lines must be supplied from separate sources. The standard does provide that a single source can be used, provided that it has sufficient flow and backup power and/or pumps to ensure an uninterrupted supply in the event of a power failure or malfunction.

NFPA 1403 requires assignment of an instructor in charge (incident commander), safety officer, and sufficient instructional personnel to maintain a student:instructor ratio of 5:1. This also requires close reading of the standard and consideration of intent. The standard is equally applicable to single compartment cells (where a limited number of participants can enter and work inside) and complex acquired structures. The typical staffing for a CFBT session of three instructors can meet the requirements of this standard if instructors have clearly defined roles, effective operating positions, and the number of participants inside the cell is not excessive. An additional consideration is provision of a Rapid Intervention Team (RIT) outside the cell (not specifically mentioned in the standard, providing a team of at least two firefighters outside the hazard area ready to respond to emergencies is required in the United States by Federal and state respiratory protection regulations).

The provisions of NFPA 1403 are intended to provide a safe training environment when working with live fire. They are by no means the only way to do so (as evidenced by safe and effective training conducted throughout the world using other systems of work). However, where applicable, this standard can be simply and effectively applied to CFBT with good result. Hopefully subsequent revisions of this standard will address some of the differences in purpose built live fire training props and structures used in CFBT.

9.17 PPE AND FIREFIGHTER HEAT STRESS/THERMAL INJURY

The thermal performance of firefighters' protective clothing has been a point of interest and discussion for several decades. However, little detailed scientific information is available on the technical issues. Much of these discussions are based on fire service field experience, and many of these studies are difficult to reproduce. Very little has been done to develop methods for predicting the thermal performance of protective clothing throughout the *range* of fire environments normally faced by the fire fighter.

Firefighters can be burned by radiant heat energy that is produced by a fire or by a combination of radiant energy and localized flame contact exposure as replicated by the Thermal Protective Performance (TPP) test. Some injuries also occur as a result of compressing the protective garment against the skin, either by touching a hot object or by placing tension on the garment fabric until it becomes compressed against the skin. In addition to these mechanisms, moisture in protective clothing can significantly change the garment's protective performance. Garments that are wet may exhibit significantly higher heat transfer rates than garments that are dry. Burn injuries that result from the heating and evaporation of moisture trapped within one's protective clothing is also significant. These injuries are generally referred to as scald or steam burns. Moisture may also help to store heat energy in protective clothing[14].

Thermal Protective Performance testing measures the amount of heat transfer through a firefighter's clothing composite (the combination of all layers) when exposed to a combination of convective heat and thermal radiation. The NFPA standard TPP test method measures heat flow through the garment while exposed to an 84 kW/ sq m (**2 cal/sq cm/s**) thermal environment (see Fig. 9.16). This level of flux is chosen in order to replicate a flash fire or mid-range post-flashover exposure (a 'fireball'). A minimum TPP rating of 35 is required according to the NFPA standard. At this level of protection a firefighter would have approximately 17.5 seconds (theoretically) to escape from a flashover exposure before sustaining *second degree* burns. It is important to recognize that TPP measurements do not imply a *certain* protection time, because the testing only simulates one condition amongst an unlimited set of clothing exposure conditions. Research by Krasny[15] suggests that firefighters wearing TPP 35 garments are likely to receive **serious** burn injuries in less than 10 seconds when exposed to a flashover fire environment. Other data indicate that a firefighter can survive flashover conditions of 816 °C (1,500 °F) for up to 15 seconds depending on the conditions[16].

A Conductive and Compressive Heat Resistance (CCHR) test is used for evaluating the garment shoulder and knee areas of structural and proximity fire-fighting protective garments. In general, testing of existing and potential reinforcements with this method has shown that thicker materials provide higher CCHR ratings; however, several other observations have been made[17]:

- The requirement in the 2007 edition of *NFPA 1971* has been raised to 25. At this level, all coat shoulder areas must be reinforced with at least one layer, and trouser knees must have several layers of additional reinforcement.
- Heavy dense reinforcement materials such as coated fabrics generally offer lower CCHR ratings compared to standard textile materials. Leather also provides relatively low CCHR ratings when compared to composite reinforcements of similar weight and thickness.

Total Heat Loss (THL) is used to measure how well garments allow body heat to escape. The test assesses the loss of heat, both by the evaporation of sweat and the conduction of heat through the garment layers. As clothing is made more insulative to high heat exposures, there is a trade-off with how well the heat build-up in the firefighter's body (that can lead to heat stress) is alleviated. Garments that include non-breathable moisture barriers or very heavy thermal barriers prevent or limit the transmission of sweat

moisture, which carries much of the heat away from the body. If this heat is kept inside the ensemble, the firefighter's core temperature can rise to dangerous levels if

14. NISTIR 6299, *A Heat Transfer Model for Firefighter's Protective Clothing*
15. Krasny, J.F., Rockett, J.A. and Huang, D. (1988), "Protecting Fire Fighters Exposed in Foom Fires: Comparison of Results of Bench Scale Test for Thermal Protection and Conditions During Room Flashover, *Fire Technology*, National Fire Protection Association, Quincy, MA
16. Kerber, S. and Walton, W., (2006), *NIST Report NISTIR 7342*, Building and Fire Research Laboratory
17. Total Fire Group 2007 Reference Guide

other efforts are not undertaken (i.e. limiting time on-scene, rotating firefighters, providing rehabilitation at the scene).

Thus a total heat loss test has been included in several NFPA standards to provide a balance between thermal insulation for protection and evaporative cooling insulation for stress reduction.

The Total Heat Loss (THL) requirement in *NFPA 1971* provides a tool for examining the trade-off between thermal insulation (from heat) and the stress-related aspects of clothing materials. In general, as the material composite thickness increases, higher levels of thermal insulation (measured using TPP testing) are obtained. At the same time, thicker composites typically create more stress on the firefighter. With the advent of THL testing, organizations can now choose to optimize the selection of their composites by balancing composite Total Heat Loss values with thermal protective performance values (while still meeting the minimum performance for both areas of performance).

In late 1998, the International Association of Firefighters sponsored the study entitled, *Field Evaluation of Protective Clothing Effects on Firefighter Physiology: Predictive Capability of the Total Heat Loss Test*. This study demonstrated that the THL was the test that best predicted changes in firefighter core and skin temperatures as related to work stress. The study is significant because it was the first firefighter field test with real time monitoring and simulated fire-ground activity. It also provided a basis for reducing stress in firefighter protective clothing by specifying a minimum THL value for the garment composite. While NFPA originally adopted a lower requirement of 130 W/sq m in the 2000 edition of *NFPA 1971*, the IAFF recommended value of 205 W/sq m was finally adopted for the new 2007 edition of *NFPA Standard 1971*.

The **European Committee on Standardization**, or CEN establishes standards in Europe. Membership in CEN is made up of the individual countries in Europe, although voting is based on the size of the country's population. CEN has prepared standards on the major elements of the firefighting protective ensemble, including:

1. Protective clothing for firefighters (*EN 469*)
2. Helmets for firefighters (*EN 443*)
3. Gloves for firefighters (*EN 659*)
4. Footwear for firefighters (*EN 345, Part 2*)
5. Hoods for firefighters (*EN 13911*)

Unlike NFPA, the various CEN standards have been developed by different committees or work groups. Consequently, the types of requirements and levels of protection are not consistent between ensemble elements. While many of the same kinds of tests are performed on each ensemble element, there are substantial differences in the way that these tests are conducted that make it nearly impossible to compare results from NFPA tests to those from CEN tests.

Garment requirements in *EN 469* – For protective garments for structural firefighting, there are significant differences between *EN 469:2006* and *NFPA 1971, 2007 Edition:*

- No moisture barrier is required, but optional tests are provided.
- There are no requirements for trim other than that it not interfere with the function of the clothing.

- Substantially lower levels of thermal insulation are required (two levels are provided). Testing is performed in two ways for flame transfer and radiant heat transfer. Performance is based on temperature rise with no relationship to predicted burn injury.
- Flame resistance testing is performed on the composite with examination of after-flame and after-glow, but no char length measurement is made.
- Heat resistance testing is performed in an oven at 355 °F (180 °C) instead of 500 °F (260 °C) as required in *NFPA 1971*. This permits the use of materials that melt, such as nylon.
- The thermal shrinkage requirement is more severe for *EN 469* (<5%) than for NFPA (<10%), though testing is performed at a lower temperature.
- Cleaning shrinkage is limited to 3% by *EN 469* while *NFPA 1971* allows 5%.
- A liquid runoff test is used for assessing chemical penetration using a different battery of chemicals.
- Water penetration and breathability tests are provided at two levels.
- No wristlet performance requirements are specified.
- Trim requirements are less extensive.

Helmet requirements in *EN 443* – *EN 443* has fewer requirements than *NFPA 1971* for helmets. For example, *EN 443* compliant helmets are not required to have chinstraps, neck-guards, face-shields or ear covers. The majority of requirements parallel *NFPA 1971* but use different test methods:

- Impact and penetration testing are conducted with a different mass and after different types of preconditioning.
- A different electrical insulation test is used.
- Strap elongation and breaking strength are measured in *EN 433*, while the entire retention system is evaluated in *NFPA 1971*.

International standard garment requirements in *ISO 11613* – Due to an impasse between Europe and North America, *ISO 11613* for structural firefighting contains two parts, or separate sets, of requirements. One part reflects requirements consistent with *EN 469* (2006), while the other part is based on the 1991 edition of *NFPA 1971*.

Because each part is based on a different existing standard, the test methods used for similar performance determinations are often very different and make product comparisons extremely difficult. Jurisdictions may choose either one part or the other; however, requirements are not to be mixed between the two parts. Jurisdictions are instructed to choose the appropriate set of requirements based on a risk assessment of their activity.

Since *ISO 11613* was based on earlier versions of both *EN 469* and *NFPA 1971*, clothing that is compliant with this standard may not conform to either the proposed 2006 edition of *EN 469* or the 2007 edition of *NFPA 1971*. Part A (based on the CEN standard) includes no criteria or requirements for a moisture barrier. Part B contains fewer requirements on the moisture barrier, no overall shower test, no Total Heat Loss test and no conductive and compressive heat resistance test (unlike the 2007 edition of *NFPA 1971*). *ISO 11613* is being extensively revised for expected release in 2008. The new edition will address all parts of the ensemble. Since there are fewer required components, there are fewer overall required tests.

Likelihood of burn injuries in CFBT

The potential for students or instructors receiving burn injuries whilst occupying CFBT FDS units is minor. Where systems are monitored (temperature thermo-couples to exterior data logger); where fuel loads are risk assessed; where flow-rates are sufficient; where PPE is adequate according to the standards and is worn correctly, with all skin covered; and where CFBT instructors are qualified and competent in managing the training evolutions in FDS units, there should be no burn injuries.

The types of injuries that may occur are normally those encountered during fire service training, such as minor knee and ankle injuries. Where a CFBT evolution does not follow the Swedish code of practice then students may occasionally receive minor skin reddening (first degree burns) on their shoulders or arms, particularly if long-sleeved under-garments are not worn or where there is no air gap between clothing and skin, perhaps where SCBA straps cause compression. Such injuries are not painful but may be slightly sore and will disappear within a few hours of leaving the units.

Where moisture is trapped within PPE (sweat or water permeating through gloves) then this may cause minor first or even second-degree burns, where the student is working too close to the fire. Remember the limitations of temperature and radiated heat flux, which can only be assured by strict monitoring or by following clearly defined protocols in terms of acceptable firefighter locations. Some units have lines painted on the floor to denote areas that should not be crossed, ensuring that firefighters are always working in safe locations and are not too close to the fire whilst occupying FDS units. Students should be encouraged to regularly 'shake' their upper torso whilst crouching in a position where they might feel hot. This will relieve compression and assist in creating air gaps between PPE and inner garments. They should also be encouraged to inform instructors if they are feeling too hot and be allowed to leave the training unit immediately where discomfort persists. In doing so they should be accompanied.

Heat stress

The physiological and psychological effects on firefighters taking part in hard work, whilst wearing full structural PPE and SCBA, are well researched. There have been several studies that assess to what limits a firefighter may function effectively, whilst working the interior of a structure fire or training burn. Elevated internal body core temperatures can increase mental and cognitive impairments, such as increasing decision-making time and decreasing memory functions.

Dehydration and thermal strain, along with excessive VO2 Max, increased heart rates (to 200 bpm) and high blood pressure during firefighting operations, have all demonstrated through research that firefighters may work to extreme limits inside a fire-involved structure.

There are three different types of heat-related injuries[18]:

- Heat cramps
- Heat exhaustion
- Heat stroke

18. Baird, C., (2006), Gresham Fire and Emergency Services, Oregon USA

Heat cramps are the least damaging and are characterized by muscular-skeletal cramping resulting from loss of water and electrolytes while the body is attempting to cool itself through sweating. Treatment includes rest, along with fluid and electrolyte replacement. Gentle stretching of the affected muscle group can help relieve the pain.

Heat exhaustion results from significant fluid loss from profuse sweating. Signs and symptoms include:

- Weakness
- Nausea
- Rapid heart rate
- Hypotension
- Pale, diaphoretic skin
- Poor skin turgor

Skin turgor is an abnormality in the skin's ability to change shape and return to normal (elasticity). Skin turgor is the skin's degree of resistance to deformation and is determined by various factors, such as the amount of fluids in the body (hydration) and age.

Treatment for heat exhaustion begins with moving the patient out of the sun and to a cool area. The patient should be placed supine, with legs elevated, if tolerated. Loosen their clothing and provide active cooling through increased air movement.

Heat stroke is the most severe form of heat-related illness and marks the point at which the body is no longer able to cool itself adequately. Typical symptoms of heat stress include excessive facial skin reddening; heavy perspiration; headaches; cramps; weakness; dizziness; fainting, and loss of concentration to the point where the victim is almost in a daze with a staring look of confusion in the eyes.

Proper rehabilitation, hand and forearm cooling, monitoring of firefighters, and effective use of the three crew rotation system, are all useful methods to combat heat stress.

In an ODPM report 5/2003[19] the authors provided useful information based upon the research of the physical state of CFBT instructors undertaking training evolutions in FDS units. To gauge whether any performance decrement might have occurred, the instructors were asked whether they thought they could perform a rescue at the end of the training exercise. Although all the instructors believed they would have had no problems performing a rescue after live fire training exercises conducted in modified containers, this was not the case for those conducted in the fire buildings ('Hot Fire' exercises). After three (out of twenty exercises involving twelve different instructors) of these exercises, the instructor doubted his ability to perform a rescue, and one instructor was sure he would not be capable. As the key function of the instructors is to act as safety officers during the training exercises, and hence be responsible for rescuing a collapsed trainee firefighter, these findings were cause for concern.

In an extension of this study, the authors measured the energy demand of rescuing a 50 kg dummy wearing SCBA from a fire building. Even though the dummy was considerably lighter than the average firefighter, there was no heat exposure and the instructors were assisted, the simulated rescues required heart rates of 160 bpm and

19. Elgin, C. and Tipton, M., (2003), Department of Sport and Exercise Science, University of Portsmouth for Office of Deputy Prime Minister (UK)

an average energy expenditure of 47 kcal. If no heat was dissipated, this would result in an increase in deep body temperature of 0.6 °C. Given the highest deep body temperature at the end of a Hot Fire exercise was 40.6 °C and heart rates up to 194 bpm were observed, it was concluded that the ability to perform a rescue at the end of an exercise may be severely compromised. However, it should be emphasized that these conclusions were based on the very worst-case scenarios of highest body core temperatures and heart rates.

Conclusions from the study:

• The physiological responses of the instructors observed during the Hot Fire training exercises were within the range of those reported previously.
• The rescue tasks devised were representative of the worst-case scenario that a single instructor may face according to the responses to a questionnaire sent out to all the brigades in the UK.
• These rescue tasks were very demanding and approached the physiological limits of the majority of current instructors.
• Despite the arduous nature of the rescue tasks, the instructors monitored were capable of performing a rescue task after acting as a safety officer in a live fire training exercise.
• Evidence from this study showing a sweat loss of 1.5 liters during the Hot Fire exercise and rescue task confirms the importance for all firefighters involved in Hot Fire training exercises to be fully hydrated at all times.
• It is likely that in less favourable situations (higher deep body temperatures, greater levels of dehydration, less fit or experienced instructor, or a casualty heavier than 85 kg) a rescue may not be possible, or attempts to continue to do so may result in a heart attack in the rescuer.
• It should also be considered whether it is acceptable to expect less fit firefighter instructors to undertake such a strenuous task in combination with heat exposure.

9.18 CFBT HEALTH AND SAFETY

The health and safety of all personnel during operations and training is of the utmost importance. There are always technical issues evolving that affect how we should operate in hazardous areas and it is useful to look at research and knowledge from a global perspective because the questions we are asking ourselves today may have already been answered yesterday!

One such question that offers extensive and useful in-depth answers concerns the *CFBT Exposure Study*[20] undertaken on behalf of the New South Wales Fire Brigades (NSWFB) in Australia in 2007. The main objectives of this study were to determine, under current standard work practices including the use of personal protective equipment, the extent of **firefighters' exposure to particleboard combustion products** and to assess the adequacy of existing control measures during CFBT and associated work activities.

The study investigated exposure of firefighters/instructors during CFBT activities to polycyclic aromatic hydrocarbons, volatile organic compounds, inorganic compounds, formaldehyde and hydrogen cyanide, as well as systemic uptake of PAH as

---

20. Aust, N., Forssman, B. and Redfern, N., (2007), *Report 10-5289-R1D3/2007 CFBT Exposure Study*, Heggies PTY Ltd NSW, Australia

measured by its urinary metabolite. The CFBT exposure study was initiated by the Health and Safety Branch of the NSWFB due to concerns about the current control measures for airborne hazardous substances present in particleboard wood-smoke and the potential exposure of NSWFB personnel to chemicals in particleboard wood-smoke during CFBT activities.

The conclusions of the research suggested that:

- It is unlikely that, with the correct use of SCBA, there would be significant inhalation exposure to any of the chemicals of concern measured in this study, even though notable concentrations of PAH and formaldehyde were measured in the personal breathing zones of firefighters participating in CFBT activities.
- Biomonitoring demonstrated small but statistically insignificant increases in urinary metabolites of PAH, suggesting some uptake may have occurred.
- Testing of structural firefighter's ensemble and duty wear suggests that a small amount of naphthalene is present on new structural firefighter's ensemble, and may also penetrate this to reach duty wear. Cross contamination of garments is also a potential source of garment concentrations.
- With the correct use of PPE, the risk of short-term or long-term health effects from these low concentrations of naphthalene is negligible.
- With the correct use of PPE, the risk of short-term or long-term health effects (including cancer) from formaldehyde exposure during CFBT activities is negligible.
- There was negligible exposure to volatile organic compounds, inorganic compounds, and hydrogen cyanide during CFBT activities.
- There is little to no foreseeable risk of exposure to chemicals of concern for individuals located in the vicinity of CFBT activities – but not participating – as long as they are not in direct contact with the smoke.
- There may be a small amount of dermal exposure to naphthalene despite the use of structural firefighter's ensemble, however the concentrations at which the firefighters may be exposed are likely to be well below the concentration at which health effects would occur in the majority of the population.

Based on this present study the following recommendations were made to further limit risk of exposure of NSWFB personnel to hazardous substances present in the particleboard wood-smoke:

- A review of current NSWFB procedural hygiene requirements i.e. disrobing, hand washing, eating areas, washing of clothing and location of observers.
- CFBT burn activities limited to one burn per day for trainers.
- Health surveillance and monitoring of NSWFB trainers.

CFBT instructor welfare (example SOG)

- Benchmark medical on joining CFBT faculty.
- Annual medicals thereafter or on request of instructor.
- Monthly self-assessment log forwarded to occupational health unit.
- Currency maintenance requirement at least one course in six months.
- One hot wear allowed per day for two consecutive days, rest day mandatory – can rotate instructors from hot to cool (exterior) wears.
- Cool vests provided and mandatory.

- Instructors take and record core temperature before and after each exercise.
- Instructors should not load containers.
- Sauna provided to assist and maintain acclimatization.
- The use of skin barrier creams are considered optional, but not essential, as added personal protection.

9.19 SMALL-SCALE DEMONSTRATION UNITS

There are several different types of small-scale demonstration units that can be used to introduce and enhance the learning experience of FDS units. Many of the different forms of fire phenomena may be demonstrated on a smaller scale, prior to experiencing the same events in the larger FDS units. This approach to learning creates a useful link as it reinforces how much more intense the various events may be when scaled up. This method of training firefighters is also something that can often be taken back to the station and repeated over and again.

The 'doll's house' or small-scale fire box

There are various designs of this basic small-scale fire simulator, some are made of steel and are lined with fuel for hundreds of repeatable demonstrations. Another more common form is where these little simulators are made up using off-cuts of the particleboard used as the primary fuel source in the FDS units. A small fire is then started in a corner of the 'doll's house' using a very small amount of wood cribbing and an entire class of students can observe the fire growth and development through the opening.

All the basic teaching points may be demonstrated in what is a very cheap and economical way to impart knowledge or deliver a basic introduction to fire behavior:

- Air-rack
- Over-pressure/under-pressure
- Pyrolysis
- Smoke accumulation
- Neutral Pressure Plain (NPP)
- Rollover
- Flashover
- Backdraft
- Smoke burns
- Auto-ignition

Constructing dolls' houses (mini FDS unit)

- Based on 12.5 mm width ($\frac{1}{2}$ inch) particleboard.
- Each unit will require six sheets to construct.
- Sides, front and back sections should actually measure 400 mm × 362 mm.
- Sides, front and back sections should actually measure 16 inches × 15 inches.
- Floor and roof sections should measure 400 mm × 400 mm
- Floor base and roof sections should measure 16 inches × 16 inches.
- Cut a door opening 180 mm wide × 250 mm high, in the front face.
- Cut a door opening 7 inches wide × 10 inches high, in the front face.
- Glue or nail all sections tightly together.

- Construct a sheet larger than the door opening and fit a handle to one side – this will be used to close off the door to starve the fire of air, allowing a build-up of heat and a smoldering fire before sliding it away to allow a sudden in-rush of air and a backdraft (may take several attempts!).

A useful video demonstrating the correct use of the doll's house, along with several ways to ignite the gases, and other small-scale demonstrator FDS units is available on DVD in the book *3D Firefighting*[21].

Important: As the generation of toxic gases will occur during this demonstration, it should be controlled by one or two instructors (or students) wearing full PPE and SCBA with observers far enough away from the smoke that they do not breathe the products of combustion.

Giselsson Bang Box[22]

The Bang Box was originally developed as a classroom (laboratory) tool for illustrating the **range of flammability** and the **potential of energy inside this range**. A certain number of droplets of volatile liquid are used: normally eight drops are equal to the lower limit, ten droplets to the ideal mixture and approximately seventeen drops are equal to the upper limit of flammability. The stirring is switched off before igniting and the subsequent explosion makes the cork lid jump up to 2 meters or more. The results of a test series are marked on a scale. To show the maximum power from a few droplets of flammable liquid, the stirring unit is left running when igniting the gases. The mechanical distribution of the flame will increase the effect the same way that is designed for in modern car engines.

Pure ethanol can be used as an alternative flammable liquid. About 4 cl should be poured into the chamber. If the temperature is approximately 20 °C it will become an ideal gas mixture in the box after stirring. Repeating this experiment without adding new fuel but slowly raising the temperature will produce a decrease in energy as the gas-mixture is more and more over-carburated. Finally it is over the upper limit. Diluting the alcohol with about 3 cl of water at approximately the same temperature again gives explosive gases. A TWA800 experiment can be made with diesel or jet-fuel if the cylinder is pre-heated to about 70 °C.

The 'Aquarium' FGI (fire gas ignition) small-scale demonstrator

Another classroom tool used to demonstrate the presence of flammability limits of gas/air mixtures is termed by Giselsson as the Aquarium. The tank, measuring approximately 0.5 m wide, 0.7 m long and 0.5 m high is constructed from a metal frame holding 6 mm-thick laminated glass on three sides. The fourth side has an opening which can be sealed with a sliding cover. The top of the tank has some opening flaps, which act as pressure relief vents.

A solid base contains a mixing fan, propane supply and supports electrodes for a spark igniter. The gas supply and ignition spark are controlled remotely by the instructor using a small handheld unit. While gas is being supplied, a display at the front of the tank records time. This display is calibrated to show the times taken to reach the lower and upper flammability limits in the tank.

21. Grimwood, P., Hartin, E., McDonough, J. & Raffel, S., (2005), 3D Firefighting, Oklahoma State University
22. http://www.uclan.ac.uk/facs/destech/builtenv/facilities/firelab/equipment/MFbangbox.htm

Four demonstrations of FGI may be conducted:

- The tank is filled with propane with the spark operating. An explosion occurs when the lower flammable limit of the mixture is reached.
- A much more violent explosion is achieved by filling the tank to an approximately stoichiometric mixture before operating the spark.
- A 'smoke explosion' is simulated by placing a compartment wall around the spark so that lean ignition inside the compartment ignites a much richer mixture outside.
- A rich mixture is created and the spark started. The vent on the side of the tank is opened and air is allowed to enter. As the gas/air mixture drops into the flammable range, a simulated fire gas ignition (FGI) occurs.

9.20 READING THE FIRE – B-SAHF (BE SAFE)

There are various mnemonics used in firefighting texts that serve as a prompts for firefighters to consider or address various tactical issues at fires. One of the best I have ever seen is that devised by Shan Raffel (Queensland Fire and Rescue in Australia) and further adapted by Battalion Chief Ed Hartin of Gresham Fire District in Oregon USA. The mnemonic **B-SAHF** (Be Safe) represents the following points in relation to 'reading the fire':

B – Building
S – Smoke
A – Air-track
H – Heat
F – Flame

Reading the smoke

- Smoke volume (stage of fire development)
- Smoke density (visibility)
- Smoke velocity (pressure)
- Smoke content (rich or lean)
- Smoke color (stage of fire development)
- Smoke movement (hazards)
- Smoke interface (NPP) (stage of fire development)
- Smoke location (not a true guide to the **fire's** location)
- Smoke variations (from different openings)

Dangerous smoke movements (rapid fire warning signs)

- Pulsing movements of smoke at a doorway or window;
- Turbulence at the smoke interface (repeated rise and fall of the smoke layer – NPP);
- Sudden lowering of the smoke interface (NPP);
- Sudden movements of air towards the fire and smoke away from the fire;

Reading the fire

- Flame volume
- Flame color
- Flame velocity

Why read smoke conditions?

- To assist us in sizing the fire;
- To assist us in locating the fire;
- To assist us in anticipating hostile fire events or rapid fire phenomena;
- To assist us in locating tactical venting points.

Smoke is fuel

- Smoke has trigger points
- Flash point
- Fire point
- Ignition temperature
- Limits of flammability

Limits of flammability

- Fuel lean fire gases (below the LFL)
- Flammable limits (between the LFL and the UFL)
- Stoichiometric point (Ideal mixture of gas/air)
- Fuel rich fire gases (above the UFL)

Air-track

- The point to point air-track is from **inlet** to **fire** to **outlet**;
- The air inlet may also serve as the outlet (may be the same window);
- The inlet and outlet may be the entry door;
- There may be radiant heat from the overhead, **between fire and outlet**;
- There may be more than one inlet/outlet;
- As further openings occur, the direction of the air-track may change;
- Any such change may be to the advantage or disadvantage of occupants or firefighters working the interior;
- We can sometimes take actions that will reverse the direction of the air-track to our advantage;
- We may sometimes take actions that will alter the direction of an existing air-track to our disadvantage!;
- Air-tracks are greatly influenced by exterior wind and interior building pressures such as stair-shaft stack effects in tall buildings;
- The potential for an 'auto-ignition' of superheated fire gases within a compartment are far greater in locations adjacent to vent inlets/outlets.

Fire-fighting strategy based on conditions

- RECEO (Rescue-Exposure-Confine-Extinguish-Overhaul)
- Locate and size the fire
- Evaluate resources on-scene
- Recognize and assess the risk
- Select a safe system of work
- Manage the risk
- Monitor the risk

9.21 STABILIZING THE ENVIRONMENT

It's a fact that at least a quarter of structural fires get worse from the moment we arrive on-scene. This is due, in part, to our actions or non-actions during the vital first few seconds of arrival. What do we do, or fail to do?

- We create openings;
- We open doors;
- We open windows;
- We cut holes to remove smoke, heat and gases;
- Often, all this is done before we get water on the fire or into the gas layers.

So what is 'stabilizing' the environment and why should it be our primary task (after immediate rescue)?

1. **Controlling the air-track;**
2. **Cooling the gases in the overhead;**
3. **Venting with a clear directive and an objective.**

From the very first moment you arrive you should be looking for the B-SAHF indicators that will inform you at what stage the fire has developed and how much of a hold on the structure the fire has. Why make an opening where an opening is not pre-existing? There must be a good reason to make any opening in a fire-involved structure, even where this is the entry doorway.

Reasons to open doors

- To gain access;
- To locate the fire;
- To undertake a search for suspected occupants;
- To undertake a search for known occupants;
- To advance the primary attack hose-line.

Reasons to open windows and cut roofs

- To ventilate rooms, compartments and spaces;
- To release flaming combustion;
- To relieve conditions inside the structure;
- To provide an air-flow to any occupants that might remain within.

So how can we address our first-in response tactics differently, in a way that might lead to better control and stabilization of the fire's development?

Controlling the air-track

In any situation where we arrive at a door, we should check it for heat layering by applying the back of a hand to feel the door at different heights. Where there are signs of heat layering (hotter at the top compared to lower down) then we know the fire is beyond its incipient stage and SCBA is required. We know, or at least should suspect, this before we even open the door. At this stage we must carry out an effective door control procedure that should prevent the fire from worsening to any great extent. If we are working ahead of the hose-line at this point we need to prioritize and justify our actions. The only reasons to enter ahead of the hose-line in this situation are:

- A reasonably suspected life risk may exist;
- A known life risk definitely exists;
- It is necessary to locate the fire.

It is logical to take a quick look to see if the fire's location is immediately obvious for we may even be able to close the fire down, by closing an interior door, which might slow the spread of fire. In taking this action we must carefully risk assess our situation. Such an action should only be undertaken if following these protocols:

- The reconnaissance team should consist of at least three firefighters.
- Two firefighters enter as a team and remain together.
- One firefighter remains at the door to observe conditions and control the door opening.
- All should have a communication link with each other.
- A portable fire suppression device should be taken in support (IFEX?).
- If at any stage an **'air-track'** forms with heavy smoke moving out on the over-pressure, then the reconnaissance team should return to the outside immediately and the entry door closed.
- In any 'recon' situation into smoke and heat, penetration into the structure should be no further than ten to twenty paces without a hose-line, depending on conditions. At this point, if the fire (or approximate area) has not been located then the 'recon' team should return to the exterior and the door(s) should be fully closed.

Where there are viable reasons to **suspect** there may be occupants inside a structure, then it is again a case of risk assessing the situation. The reasons to 'suspect' should be heavily weighted in the favour of a strong likelihood, rather than a slight possibility, where fire conditions are deteriorating. Searching ahead of the primary attack hose-line being laid is a strategy fraught with danger and this is a tactical approach that should be carefully assessed in the risk versus gain conundrum.

- A 'search' team should consist of at least three firefighters.
- Two firefighters enter as a team and remain together.
- One firefighter remains at the door to observe conditions and control the door opening.
- All should have a communication link with each other.
- A portable fire suppression device should be taken in support (IFEX?).
- If at any stage an **'air-track'** forms with heavy smoke moving out on the over-pressure, then the search team should return to the outside immediately and the entry door closed.
- In any interior search situation into smoke and heat, for **suspected** life risk, penetration into the structure should be no further than ten to twenty paces without a hose-line, depending on conditions. At this point, if the fire (or approximate area) has not been located then the search team should return to the exterior and the door(s) should be fully closed.
- In any interior search situation into smoke and heat, for **known** life risk, penetration into the structure without a hose-line is also dependent on conditions. However, a much greater exposure to risk is acceptable under such circumstances.

In all of the above situations, where entering a structure to locate a fire or search for suspected or known occupants, you will see we have implemented some Risk Control Measures in an effort to stabilize interior conditions and reduce unnecessary risk to firefighters. The function of the door control firefighter serves two critical needs:

- Controlling the 'air-track';
- Observing the conditions.

Stabilization or 'mayhem'

In situations where firefighters arrive on-scene and proceed to break out all the windows in a structure without any purpose, objective or tactical reason, then they are more likely to create mayhem rather than stabilize interior conditions. It is certain that a serious working fire existing in an under-ventilated state *(it may have been developing high heat conditions with a continual process of recycling fire development inside the structure, leading to heavy pyrolyzation and a large build-up of very hot smoke and fire gases)* will unleash large amounts of intense gaseous flaming combustion where opened up in such a way. If there were any occupants remaining inside the structure it is most likely that they are no longer viable rescues.

Cooling the gases in the overhead

Another way we might attempt to 'stabilize' conditions is by directing water applications into the gaseous overhead to reduce compartment temperatures and radiant heat transmission. We can do this in several ways but our main objective is to take the maximum amount of heat out of the overhead with the least amount of water. In doing this we are able to maintain the thermal balance, keep the smoke layer (NPP) high, maintain whatever visibility we might have near the floor, and avoid temperature inversions where the expansion of water to vapour drives hot gases and condensing steam down at us.

We can apply water into the **overhead** in several ways:

1. Bursts from a straight stream pattern (1–3 seconds);
2. Bursts from a fog pattern (1–3 seconds);
3. 'Pulses' from a fog pattern (0.5 second);
4. Sweeps of a narrow-fog or straight stream in an attempt to draw heat out from the ceiling and walls (**painting**);
5. Note: **pencilling** is not a technique that is generally directed into the overhead, but is rather aimed at cooling hot or burning surfaces of the fuel-phase (base) fire (i.e. couches, beds, furniture, contents etc.). This term was transferred from Sweden to the USA but misinterpreted by instructors during the early years of using flashover training 'cans'.

Effective gas-cooling applications into the overhead demand:

- Adequate nozzle pressure to provide an effective droplet range;
- Small amounts of water to prevent vapour hitting the floor;
- Nozzles capable of enhancing nozzle actions;
- Nozzle operators who are trained in gas-phase cooling techniques.

Venting with a clear directive and an objective
A third and final way of stabilizing the environment is through careful and controlled tactical ventilation of a fire-involved structure/compartment. Such an action must be directed (pre-assigned or ordered), be precise (location), coordinated (timing), and serve a purpose (objective). This strategy may not suit all situations but in some cases is critical to a successful firefighting operation.

9.22 DOOR ENTRY PROCEDURE

The main objective of carrying out a routine 'door entry procedure' is to assert some element of control over fire conditions from the outset. The very first opening we create in a fire-involved structure is generally the entry door, unless of course this is already wide open on our arrival. If this is the case, we might want to think about closing it, depending on B-SAHF conditions as they are presenting.

Door entry procedures may vary somewhat but all maintain the same objectives:

- Select the right door to enter by.
- Is this entry point the fire-side or the non-fire side?
- What direction is the wind, if any?
- Does the entry point serve logical escape routes?
- Does the entry point provide the best access to all parts of the structure, e.g. basement?
- Get a quick visual of B-SAHF indicators (see above).
- Are there adjacent windows that may provide hazard warnings?
- Feel the door with the back of a hand to ascertain thermal layering.
- If its warm or hot, or if there is fire/smoke suspected behind the door, follow the door entry procedure:
    1. Don't enter without a charged hose-line at the door;
    2. All firefighters under air (SCBA);
    3. If this is an interior door either serving or near to a staircase, are the stairway and immediate area above free of firefighters and occupants?;
    4. Position forcible entry firefighter at door;
    5. Attack team on hose-line in crouching position;
    6. Apply a brief burst of fine water-fog above the door and then;
    7. Immediately open the door to **6 inches maximum (150 mm)** *(if the door opens inwards* **use a strap** to retain the door in a partially closed position);
    8. Observe the smoke movements;
    9. Apply a brief burst (1–2 seconds) of water-fog into the overhead by placing the nozzle tip at 45 degrees to the floor, just inside the door;
    10. Immediately close the door fully and wait for 15 seconds;
    11. Repeat the process, each time feeling for heat reduction, observing the smoke movements, and comparing to the previous cycle;
    12. Where conditions are improving, or where there appears to be no fire or heavy smoke movement behind the door, make full entry and advance the hose-line.

By taking these precautions, implementing such control measures, and following this procedure on every occasion as a routine, the safety of firefighters is improved and the prognosis for any remaining occupants is greater still, as such procedures serve to maintain a stable working environment and reduce the chances of rapid fire progress during entry being made to a fire involved structure/compartment. This is a routine that should be practiced over and over, on various types of door, in line with forcible entry practice.

Think how many times you, or your firefighters, have experienced sudden fire development on opening a door, or how many times you have seen this occur in training videos? Could this procedure have prevented such uncontrolled rapid fire growth and served to stabilize the conditions on entry?

9.23 GASEOUS-PHASE COOLING

It was shown above how effective finely divided water droplets might be at cooling heated fire gases. To effectively cool a layer of super-heated fire gases that has accumulated in the form of a smoke reservoir, perhaps just hanging there in the overhead of an adjacent compartment, requires very small amounts of water. A few brief pulses, in series, will dramatically lower temperature in the gases without bringing the smoke layer to the floor.

However, where there is a fast moving 'air-track' with rapid smoke movement of high velocity gases in the overhead, in exchange for air flowing in below, then gas cooling may be less effective (*depending on the stage of fire development and the ventilation profile*).

9.24 GASEOUS-PHASE COMBUSTION

The term 'gaseous-phase' combustion mainly refers to the volume of burning gases and combustion products that exist within a compartment. Such combustion may be seen to fill an entire volume within a room or space.

The various forms of gaseous-phase combustion may exist as:

- Fire gases burning and filling the entire volume within a compartment, possibly as a result of flashover;
- Fire gases only partially burning within a compartment volume (under-ventilated);
- Fire gases burning in a substantial rolling fire-front in the overhead;
- Detached tongues of flaming existing in the overhead (rollover);
- Fire gases igniting and burning as they exit windows (auto-ignition);
- Fire gases burning as they suddenly auto-ignite inside a structure;
- Floating (also termed 'ghosting') flames of gas/air mix;
- Pre-mixed explosive burning of gases and combustion products that may result in a 'backdraft' or 'smoke explosion'.

Is it possible to have a fuel-phase fire burning on its own in a compartment? Yes certainly, this is termed a 'fuel-controlled' fire. Is it possible to have a gas-phase fire burning on its own in a compartment? Yes this is also possible. The potential for fire gases and combustion products accumulating in areas, voids, spaces and pockets in the overhead and then igniting on their own exists in every developing compartment fire. The combustion could be occurring in the overhead, hidden by the smoke layer as firefighters are extinguishing the fuel-phase fire below. This

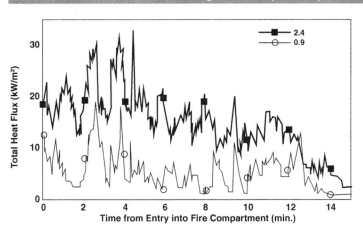

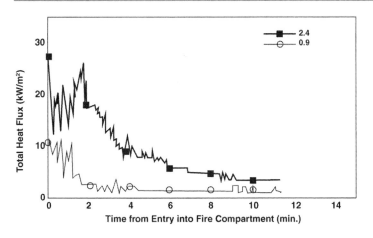

Fig. 9.17 – US Navy fire tests demonstrating the differences between bursting straight streams (top graph) and fog patterns (bottom graph) into the overhead. The severe spikes in the top graph demonstrate how bursting straight streams may cause temperature inversions, repeatedly driving super-heated steam down on nozzle operators, whereas the fog bursts (lower graph) produce a more even rate of cooling in the overhead and less steam at lower levels.

gas-phase fire may continue to burn for quite some time and spread down behind firefighters to trap them inside. We are able to simulate this event during live fire training in FDS units.

Air-track

Where fire intensity is increasing, such a dangerous development of a compartment fire occurring in this way is most likely being driven by one (or a combination of) three factors:

- The fire has found added fuel (probably spread to other items);
- The fire has found additional air/oxygen;
- Fire gases have reached dangerous limits in both temperature and range of flammability.

Where the fire has spread to additional fuels, then it is a case of getting water on the fuel-phase fire quickly before the heat energy release from the fuel-load reaches a stage where it overpowers the available flow-rate at the nozzle.

If the fire is suddenly feeding on an additional source of air-supply, then perhaps a window serving the fire compartment has broken through the heat, or maybe a vent opening has purposely been created. In this instance the smoke layer may lift slightly as the fire becomes more intense. The air-track development may affect firefighters locations in several ways.

- As flaming combustion develops through an increase in ventilation, there will be an air-inlet and a combustion/smoke outlet in the structure.
- If there is only one opening in existence then this will serve both needs.
- Where flames fill a window opening then the air-track is entering from another point.
- If a window is only flaming at the top section then the air-track (or part of it) will be entering in the lower section.
- The most likely air inlet will be via the entry doorway following the path to the fire.
- The air inlets and combustion/smoke outlets will most likely change or increase in number during a serious fire.
- Sometimes, the location of air inlets and combustion/smoke outlets will reverse.
- Wind may affect the arrangement between inlets and outlets.
- Sudden wind gusts may have devastating affects on fire development.
- Momentum and inertia forces may suddenly develop within a structure, due to the differences in internal pressures, driving the air-track with some great velocity towards the fire. This too may have some devastating effects in terms of sudden fire development.
- An exterior window venting action, in line with an entry door to the structure or compartment being closed, may create most favorable conditions at firefighter locations as the air inlet/fire outlet will now be transferred away from the entry route.
- In effect, this may redirect flaming combustion in the overhead away from firefighter locations as it heads in a new direction, towards the sole inlet/outlet opening at the window. This will reduce heat flux being radiated down to the floor where firefighters are located.

9.25 FUEL-PHASE COMBUSTION

Earlier in the chapter it was demonstrated how a fog pattern offered a 75% efficiency factor when suppressing a compartment fire compared to a 50% efficiency factor for a straight stream. These theoretical efficiency factors were derived from research into hundreds of real fires. That is to say, that for every 4 gallons/min (15.1 liters/min) applied to a fire, only 2 gallons/min (7.5 liters/min) will be effective in suppression in a direct attack mode but 3 gallons/min (11.3 liters/min) will be effective if used in a fog pattern.

However, despite the theory, a straight stream or smooth-bore stream will offer greater penetrating qualities when directed at the flaming solid fuel source. Any fire burning across the surfaces of contents or compartment linings is termed 'fuel-phase' combustion and burning may have extended down some way into the fuel base. Therefore, in most instances it is necessary to reach and penetrate the fuel source if the emission of further dangerous gases is to be prevented.

Some fuels will require greater quantities of water to achieve suppression, depending on their potential for energy release, their make-up and location. Hidden or shielded fires may be more difficult to reach and various nozzle techniques may be required, such as bouncing the stream off the ceiling or off facing walls, to reach the fire's source.

However, one situation that should be avoided is the application of a straight stream pattern, directly at the base of a serious compartment fire, where large amounts of unburned flammable gases remain within. You might see a heavy flame front issuing from a doorway and hit the fire head-on at the base of the flames, low down in the room. Where gases and unburned combustion products remain in the room, the sudden transition of water to vapour may drive these gases directly at you, if this is their only escape route. As these fuel-rich gases reach an air supply near to the door they will erupt into flame. The effect of applying a straight stream in such a way can be pretty dramatic and may drive firefighters back from any advance. Where these gases have a safe escape route to the exterior then it is simply a case of overpowering the fuel-phase fire with sufficient flow-rate.

9.26 WORKING WITH CASE HISTORIES

One of the best learning tools in the fire service is to review past incident reports in an attempt to learn from the real fire-ground experiences of others. There is generally a widespread availability of established case history reports and the database at NIOSH[23] provides an extensive range of LODD (line-of-duty death) reports that may be used for classroom presentations and to generate useful debate.

Another useful online research link to similar reports exists at the National Fire-fighter Near Miss Reporting System[24] where a database detailing a wide range of reports into non-fatal incidents can be used to search for specific reports using a keyword locator.

These reports offer an extensive range of tools to the CFBT instructor. Such reports can be downloaded and used to create Powerpoint presentations. They may also have their 'recommendations' removed to allow students to form their own

23. http://www.cdc.gov/Niosh/fire
24. http://www.firefighternearmiss.com

list of recommendations, or learning points, from both a fire behavior and a tactical perspective.

It is critical that we learn from past experiences of others in the hope that we will not repeat history, where particular events were/are preventable.

9.27 TRAINING RISK ASSESSMENT

Undertaking and completing a documented risk assessment for CFBT training should be based on local standards, codes of practice and occupational health standards. The preceding information offers a guide to the various aspects and issues that may need addressing. A training risk assessment should be detailed in its coverage and should generally include:

- Training needs analysis
- Training objectives
- Learning outcomes
- Health monitoring (students and instructors)
- Physiological and environmental limits
- Occupational safety
- System safety
- Temperature monitoring
- Training evolution safety
- Instructor:student ratios
- Water supplies and minimum/maximum flow-rates
- Site safety
- PPE safety
- FDS design aspects
- FDS fuel loading
- Ignition process
- Decontamination processes

9.28 OPERATIONAL RISK ASSESSMENT

Having implemented a CFBT training program and equipped specialist instructors with the knowledge to deliver the training throughout an entire force of firefighters, a fire authority will want to justify the financial outlay by ensuring that what has been learned on the training ground is easily transferable to the realities of the fire-ground.

In order to do this it is essential to provide a clear set of documented protocols, in the form of Standard Operating Procedures (SOPs), that will guide firefighters in the operational objectives and limitations of CFBT, whilst equipping commanders with a clear understanding of how CFBT may affect more traditional fire-ground strategy and tactics. In order to do this the SOPs need to address the following issues:

- What fire conditions will dictate which size attack hose-line is to be laid in?;
- What limitations are placed on the suppression techniques learned? For example, what is the largest floor space, or highest ceiling, in which gas-cooling applications might be effective?;
- Fire load and potential energy release effect;

- Attack line crew size;
- Snatch rescue procedure;
- Back-up hose-line;
- Venting protocols – under what circumstances?;
- Door entry procedure;
- How will crew numbers or staffing affect tactics?

Chapter 10

Compartment Fire Behavior

10.1 INTRODUCTION

Indianapolis 1992

Then conditions abruptly changed. I'd never seen anything like this. I've fought a lot of fires in different kinds of buildings, in all kinds of weather, with all kinds of combustibles. I thought I'd seen a lot. I thought I'd seen enough that I could deal with whatever happened and I could take care of my crew. But, as I said, this thing abruptly changed. To this day, I'm still amazed that this happened.

In the darkness, I could see little orange flickers around me. The heat was unbelievable. 'Unbelievable'. The heat from this flashover was like a blast furnace, and that causes you to turn into an instinct-driven animal. I've seen people in videos jump out of windows several floors up, and I thought, 'What the hell were they thinking? We could save those people.' Now I know. The pain from the heat and the feeling of being trapped is overpowering. If I was on the ninth floor, I would have jumped.

Unfortunately, John Lorenzano and Woodie Gelenius died in the fire. They were found in separate locations on the third floor. I don't know how John and I got separated. I was the last one to talk to John; I was the last one to see Woodie. Why did I get rescued and they died? I don't know. It's a thought that will always be with me.

Captain Mike Spalding on the Indianapolis Athletic Club Fire 1992

A Compartment Fire Behavior Training (CFBT) instructor must gain a detailed theoretical knowledge and a practical appreciation of fire behavior and fire dynamics. The theory can be easily learned in a classroom but the practical aspects of fire behavior can only be learned through real fire experience. This real fire experience will come in the form of live fire training burns under controlled conditions, and further experience gained out on the fire-ground will greatly enhance an instructor's ability and credibility to teach other experienced firefighters. Whilst never a complete text on the theory of fire behavior or fire dynamics, this chapter introduces fire behavior theory to a level that should enable a CFBT instructor to gain a working knowledge of compartment fire development and behavior. It will also provide the instructor with the background knowledge needed to determine burning characteristics of various fuels and how varying ventilation parameters might influence these factors.

10.2 LEARNING OBJECTIVES[1]

- Describe the basic chemical and physical processes involved in combustion;
- Explain fire phenomena using the fire triangle and tetrahedron as simple models of combustion;
- Explain basic concepts of thermal dynamics, including thermal energy, temperature and methods of heat transfer;
- Describe the combustion process for gaseous, liquid and solid fuels;
- Explain the concepts of heat of combustion and heat release rate;
- Describe the influence of the fuel/oxygen mixture on combustion;
- Explain the concept of chemical chain reaction as it relates to flaming combustion;
- Recognize characteristics of common types of combustion products;
- Be able to use common terminology related to combustion and fire dynamics.

Fire triangle	Classification of fires
Fire tetrahedron	Auto-ignition
Exothermic reaction	Flashover
Degrees Celsius [C]	Backdraft
Degrees Fahrenheit [F]	Fire gas ignition
Degrees Kelvin [K]	Flash fire
Convection	Smoke explosion
Conduction	Heat release rate (HRR)
Thermal radiation (radiant heat flux)	Heat of combustion
Combustion	Temperature

The Edexcel CFBT (BTEC) Syllabus for Fire Behavior recommends that instructors should understand the principles of combustion and compartment fire behavior relating to the following specific areas:

Combustion
Triangle of fire (interaction of heat, fuel and oxygen); propagation (conduction, convection, radiation); process (pyrolysis); chemistry; types of combustion (complete, incomplete); products (carbons and unburned pyrolysis products).

1. Hartin, E., Gresham Fire and Rescue Service CFBT Training Program, Oregon USA

Compartment fire behavior

Combustible gases; limits of flammability (lower explosive limit, upper explosive limit, ideal mixtures); ignition sources; fire gases; types of flame, e.g. colors, premixed, diffused.

It is beyond doubt that a lack of knowledge concerning the practical aspects of fire behavior and fire dynamics is one of the biggest causes of firefighter fatalities in both the training environment as well as on the fire-ground.

10.3 COMBUSTION[2]

Combustion is an oxidation reaction. Several factors need to be present before combustion can occur. The first requirements are fuel and oxygen. **Fuel** can range from a forest to home furniture, or from crude oil to gasoline. A fuel can present itself in any physical form i.e. gases, liquids or solids can burn.

What is fire? Flaming fires involve the chemical oxidation of a fuel (combustion or release of energy) with associated flame, heat, and light. The flame itself occurs within a region of gas where intense exothermic reactions are taking place.

The visible flame has little mass, and it is comprised of luminous gases that emit energy (photons) as part of the oxidation process. The color of the flame is dependent upon the energy level of the photons emitted. Lower energy levels produce colors toward the red end of the light spectrum while higher energy levels produce colors toward the blue end of the spectrum. The hottest flames are white in appearance.

The **oxygen** required usually originates from the surrounding air. The oxygen concentration in normal air varies around 21%. If the oxygen concentration is lowered, the combustion will be hindered and eventually stop. As oxygen levels drop below 14%, flaming combustion will become problematic and the fire will take on a stage of smolder as flaming combustion ceases. During this stage of the fire the **combustion is termed 'incomplete'** as the efficiency of the burn rate is reduced. This will increase the amount of smoke, fire gases and other flammable combustion products that will fill the space. In fact, although enclosed fires may develop in a plentiful supply of oxygen, most fires will become **ventilation controlled** and rarely burn with anything greater than 50% efficiency. Where an additional supply of oxygen is provided, possibly through a window breaking or a door opening (admitting air/oxygen into the fire compartment), the fire will increase in intensity and flaming combustion will return. There may even be a sudden transition to flashover or possibly a backdraft.

Another source of oxygen is the one contained in the molecule. In organic or inorganic peroxides, the oxygen present in the molecule can sustain the combustion. This effect is used in gunpowder or in fireworks.

In scientific terms one can describe a fire as being an **exothermic reaction between fuel and oxygen**. This means that the reaction produces energy, i.e. heat. Next to heat a fire generally produces light, combustion gases and soot. An endothermic reaction is one where energy (heat) is absorbed.

Fires may occur in the **fuel-phase** (at the surface of fuels) or in the **gaseous-phase** (flaming combustion). Flaming combustion may detach from the fuel surface to burn independently in fire gases mixed with air/oxygen.

---

2. Desmet, K. and Grimwood, P., (2003), *3D Tactical Firefighting*, Crisis and Emergency Management Centre CEMAC (Belgium), available at http://www.firetactics.com

Flames may exist in a **diffused** state where fuel and air mix in the region where combustion is taking place or in a **premixed** state where the fuel and air have already mixed into a flammable state before the combustion takes place. Any combustion that takes place in a premixed region is usually very intense and sometimes explosive.

In combustion, a **diffusion flame** is a flame in which the oxidizer combines with the fuel by diffusion. As a result, the flame speed is limited by the rate of diffusion. Diffusion flames tend to burn slower and to produce more soot than premixed flames because there may not be sufficient oxidizer for the reaction to go to completion, although there are some exceptions to the rule. The soot typically produced in a diffusion flame becomes incandescent from the heat of the flame and lends the flame its readily identifiable orange-yellow color. The orange flame is indicative of a diffusion flame that is short in supply of oxygen. Diffusion flames tend to have a less-localized flame front than premixed flames.

A **premixed flame** is a flame in which the oxidizer has been mixed with the fuel before it reaches the flame front. This creates a thin flame-front, as all of the reactants are readily available. If the mixture is rich, a diffusion flame will generally be found further downstream.

If the flow of the fuel–oxidizer mixture is laminar, the flame speed of premixed flames is dominated by the chemistry. If the flow-rate is below the flame speed, the flame will move upstream until the fuel is consumed or until it encounters a flame holder. If the flow-rate is equal to the flame speed, we would expect a stationary flat flame front normal to the flow direction. If the flow-rate is above the flame speed, the flame front will become conical such that the component of the velocity vector normal to the flame front is equal to the flame speed. As a result, the flame-front of most premixed flames in daily life is roughly conical.

Flame types[3]

- Laminar, premixed
- Laminar, diffusion
- Turbulent, premixed
- Turbulent, diffusion

An example of a **laminar premixed flame** is a Bunsen burner flame. Laminar means that the flow streamlines are smooth and do not bounce around significantly. Two photos taken a few seconds apart will show nearly identical images. Premixed means that the fuel and the oxidizer are mixed before the combustion zone occurs.

A **laminar diffusion flame** is a candle. The fuel comes from the wax vapor, while the oxidizer is air; they do not mix before being introduced (by *diffusion*) into the flame zone. A peak temperature of around 1,400 °C is found in a candle flame.

Most **turbulent premixed flames** are from engineered combustion systems: boilers, furnaces, etc. In such systems, the air and the fuel are premixed in some burner device. Since the flames are turbulent, two sequential photos would show a greatly different flame shape and location.

Most unwanted fires fall into the category of **turbulent diffusion flames**. Since no burner or other mechanical device exists for mixing fuel and air, the flames are diffusion type.

3. http://www.doctorfire.com

The temperature of flames with carbon particles emitting light can be assessed by their color[4]:

Red

- Just visible: 977 °F (525 °C)
- Dull: 1,290 °F (700 °C)
- Cherry, dull: 1,470 °F (800 °C)
- Cherry, full: 1,650 °F (900 °C)
- Cherry, clear: 1,830 °F (1,000 °C)

Orange

- Deep: 2,010 °F (1,100 °C)
- Clear: 2,190 °F (1,200 °C)

White

- Whitish: 2,370 °F (1,300 °C)
- Bright: 2,550 °F (1,400 °C)
- Dazzling: 2,730 °F (1,500 °C)

Thermal energy – Internal kinetic energy.

Heat – Thermal energy in transit due to temperature difference.

Temperature – The average thermal energy of molecules in a substance.

It is common for firefighters to confuse 'heat' with 'temperature' when referring to compartment firefighting. A compartment fire will burn within a level of intensity that is dictated by the amount and form of the fuel load, the amount of ventilation available, and the geometry of the fuel. This level of intensity is measurable by calculating the energy release as the fuel mass decreases. This release of energy **(heat)** is recorded in kW or MW. A typical range of energy release measurements often refers to the maximum (peak) heat release of an individual item or a total fuel load in a compartment fire or the rate of heat release. Typical heat release peaks seen in post-flashover room fires are around 0–5 MW (for an average sized residential or office space); around 15 MW for a 70 sq m (750 sq ft) five-roomed apartment or open plan office space; and around 50 MW for a heavily-involved large warehouse fire. It is worthy of note that heat release in gaseous combustion may be greater near windows or doorways where ventilation (oxygen) levels are increased.

Typical compartment fire **temperatures** are measured (a) in the flame, and (b) in the non-burning (or burning) gases at various levels from floor to ceiling. Temperatures may vary from 0–100 °C at the floor, to 100–350 °C at helmet tips of kneeling firefighters, to 350–700 °C at the ceiling of pre-flashover fires verging on flashover at the upper limits. A post flashover fire will burn with temperatures at the ceiling around 1,000 °C and a strong in-blowing wind may intensify the fire to produce temperatures of 1,200 °C.

Although the relationship between heat and temperature is invariably linked, you can still have a 1 MW fire producing 600 °C at the ceiling and a 5 MW fire producing the same.

---

4. The Stirling Company, (1905), *A Book of Steam for Engineers*

There is fairly broad agreement in the fire science community that flashover is reached when the average upper gas temperature in the fire compartment (2.3 m ceiling) exceeds about 600 °C. Prior to that point, no generalizations should be made: there will be zones of 900 °C flame temperatures, but wide spatial variations will be seen. Of interest, however, is the peak fire temperature normally associated with room fires. The peak value is governed by ventilation and fuel supply characteristics and so such values will form a wide frequency distribution. Of interest is the maximum value that is fairly regularly found. This value turns out to be around 1,200 °C, although a typical post-flashover room fire will more commonly be 900–1,000 °C[5].

	Fahrenheit	Celsius	Kelvin
Water boils	212°	100°	373°
Water freezes	32°	0°	273°
Absolute zero	−460°	−273°	0°

In a fire, the initial energy sources that cause the fire can be multiple e.g. a spark, an open flame, electricity, sunlight. The type and format of the fuel will dictate the amount of energy in the ignition source needed to initiate the combustion process. Once the reaction is started, however, it generates more than enough energy to be self-sustaining and a **chain reaction occurs**. The energy given off in excess can be seen as light and heat generated by the fire.

The energy liberated in the combustion process causes **pyrolysis** and the evaporation of the fuel. In the pyrolysis process the chemical composition of the fuel is broken down into small molecules. These molecules evaporate and react with the oxygen in the air. This process is complex and involves sublimation, melting, evaporation and decomposition with changes in state from solid fuel to liquid to vapor. Take note of a wood panel board as radiated heat causes it to emit liquid, fuel vapor and white smoke combustion products as it begins to pyrolyze.

Stoichiometric or complete combustion means that just enough oxygen molecules are present, to oxidize the fuel molecules. When hydrocarbons undergo complete combustion only water and carbon dioxide would be formed. Such conditions are, however, rare, therefore we need to note that other **combustion products** will also be formed. In the case of hydrocarbons the formation of carbon monoxide, pyrolyzates (gaseous volatiles) and soot increases with the oxygen deficiency. If other types of fuel are burned additional toxic products are formed based on their molecular composition e.g. hydrogen chloride, hydrogen cyanide, hydrogen bromide, sulfur dioxide, isocyanates.

Combining the factors that we already mentioned above one can create the fire triangle, which symbolizes all the factors needed for combustion. However next to fuel, oxygen, and energy one should also note the **mixing ratio** between oxygen and fuel. A log of wood will not sustain a fire if it is ignited with a match; an amount of wood shavings, however, will. There is a better mixture between the fuel and the air, which supports the combustion. A much larger surface of the fuel is in contact with the air, thus a greater **reaction surface** is offered.

5. http://www.doctorfire.com

A further factor in the combustion process should be added, which is called the **inhibitor**. In a combustion process a chemical chain reaction occurs, radicals of fuel react with radicals of oxygen heat and combustion products are formed. If one adds a chemical molecule (inhibitor) which reacts with those radicals without sustaining the combustion process, one can stop the fire. This principle is used in dry chemical extinguishers, which contain, for example, potassium or sodium bicarbonate, or in the now banned Halon gas extinguishers. A **catalyst** has the opposite effect to an inhibitor. A catalyst is a substance which promotes the reaction (without being altered or used in the reaction), e.g. adding metal shavings to oil rags aids their combustion.

The **ignition temperature** of a substance (solid, liquid or gaseous) is the minimum temperature to which the substance exposed to air must be heated in order to cause combustion. The lowest temperature of a liquid at which it gives off sufficient vapor to cause a flammable mixture with the air near the surface of the liquid or within the vessel used, that can be ignited by a spark or energy source, is called the **flashpoint**. Some solids such as camphor and naphthalene already change from solid to vapor at room temperature. Their flashpoint can be reached while they are still in solid state. The lowest temperature at which a substance continues to burn is usually a few degrees above its flashpoint and is called **fire point**. A specific ignition temperature for solids is difficult to determine because this depends upon multiple aspects, such as humidity (wet wood versus dry wood), composition (treated or non-treated wood), and physical form (dust or shavings or a log of wood).

The **auto-ignition temperature** is the lowest temperature at which point a solid, liquid or gas will self-ignite without an ignition source. Such conditions can occur due to external heating – a frying pan that overheats causing the oil to 'auto-ignite'. They can also occur due to chemical or biological processes – a silo fire may result because of the biological processes in humid organic material. The auto-ignition temperature of a substance exceeds its flashpoint.

When considering vapor or gas explosions, or fires, it is important to look at their **vapor or gas density** relative to air. In this way air has a coefficient of one. A substance having a relative vapor of 1.5 will be one and a half times as heavy as air, while a substance with a relative vapor density of 0.5 is half as heavy as air. Gases or vapors that are heavier than air stay low to the ground or enter lower-lying areas such as sewers or cellars.

Next to vapor pressure when handling liquids, their **volatility** is also important. Volatility refers to how readily a liquid will evaporate. The volatility of a product is closely linked to its boiling point. The higher the **boiling point of a liquid** the harder it will be for the liquid to evaporate. An amount of highly volatile fluid spilled will be of greater concern than the same amount of low volatile liquid, because of its ease to find an ignition source or because of the toxicity of the vapors. A more scientific term for volatility is the saturated **vapor pressure** of a liquid at a certain temperature; this is the pressure exerted by the vapor at that temperature. The larger the vapor pressure of a liquid, the more vapor is produced. The vapor pressure has an impact on the extent and area of the gas/air release. The vapor pressure of a liquid rises with the rise in temperature. The boiling point of a liquid is defined as the temperature at which the vapor pressure reaches 1 atmosphere. The lower the boiling point, the greater the vapor pressure at normal ambient temperatures and consequently the greater the fire risk.

In case of a gas-air or a vapor-mixture, an explosion can only occur under certain circumstances. An underground tank half full or near full with gasoline will not explode due to an above ground fire. The amount of vapor (density greater than air) present will cause a too rich mixture which will not ignite. If, however, the tank is near empty, air will already have entered the tank; otherwise the resulting vacuum would damage the tank (implosion). The amount of liquid left will dry out and gradually disperse, not generating enough vapor to reach a rich atmosphere. A spark or flame entering the tank at that point could cause an explosion. Modern underground gasoline tanks are fitted with a wire mesh flame guard at the air entry, hindering the introduction of an energy source.

Fuel burns only when the fuel:air ratio is within certain limits, known as the **flammable (explosive) limits**. In cases where fuels can form flammable mixtures with air, there is a minimum concentration of vapor in air below which propagation of flame does not occur. This is called the **lower flammable limit (LFL)**. There is also a maximum concentration above which flame will not propagate called the **upper flammable limit (UFL)**. These limits are generally expressed in terms of percentage by volume of vapor or gas in air. Where gas/air mixtures exist below the LFL they are considered too 'lean' to ignite and where such mixtures of gas/air exist above the UFL they are considered too 'rich' to burn. The flammable limits reported are usually corrected to a temperature of 32 °F (0 °C) and 1 atmosphere.

Increases in temperature and pressure result in reduced lower flammable limits, possibly below 1%, and increased upper flammable limits. Upper limits for some fuels can approach 100% at high temperatures. A decrease in temperature and pressure will have the opposite effect. Caution should be exercised when using the values for flammability limits found in the literature. The reported values are often based on a single experimental apparatus that does not necessarily account for conditions found in practice.

The range of mixtures between the lower and upper limits is called the **flammable (explosive) range**. For example, the lower limit of flammability of gasoline at ordinary temperatures and pressures is 1.4%, and the upper limit is 7.6%. All concentrations by volume falling between 1.4% and 7.6% will be in the flammable (explosive) range. All other factors being equal, the wider the flammable range, the greater the likelihood of the mixture coming into contact with an ignition source, and thus the greater the hazard of the fuel. Acetylene, with a flammable range between 2.5% and 100%, and hydrogen, with a range from 4% to 75%, are considered very dangerous and very likely to be ignited when released.

Every fuel/air mixture has an optimum ratio at which point the combustion will be most efficient. This occurs at or near the mixture known by chemists as the **stoichiometric** ratio. When the amount of air is in balance with the amount of fuel (i.e. after burning there is neither unused fuel nor unused air), the burning is referred to as stoichiometric. This condition rarely occurs in fires except in certain types of gas or gas-phase fires.

Only a few materials like ethylene oxide are able to decompose and burn when no oxygen is present.

A mixture of vapor or gas with air, within the explosive range, will ignite if the energy source presented has enough energy. The **minimal ignition energy**, which is the minimal amount of energy that is needed to set-off the explosion, can be found in literature. The minimal ignition energy of a gas or vapor/air mixture varies between 0.01 and 0.30 millijoules.

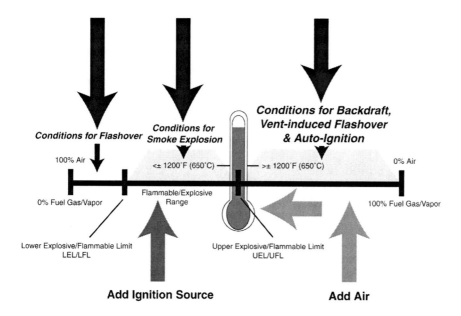

Fig. 10.1 – Limits of flammability

A rise in ambient temperature causes the explosive range to broaden, enlarging the concentration range where an explosion can occur. As well as a rise in temperature, an increase in oxygen concentration can also widen the explosive range of a substance.

The ferocity of an explosion depends on the speed of the flame front. If the flame-spread remains lower than 340 m/s the explosion is called a **deflagration**. If this speed exceeds 340 m/s – and they can reach up to 1,800 to 2,000 m/s – one calls it a **detonation**. In layman's terms the differences are defined in being faster or slower than the speed of sound, respectively **supersonic** and **subsonic**. After the ignition, the flame front passes upstream through the flammable mixture, propagated by the volume expansion of the exothermic combustion reaction. This volume expansion causes a pressure surge, which compresses the flammable mixture ahead of the flame-front.

Flammable dust from metals such as aluminum, or that of organic compounds such as sugar, milk powder, grain, plastics, pesticides, pharmaceuticals, wood-dust etc. can explode. A dust explosion is an explosive combustion of a mixture of flammable dust and air. In other words it is a combustion reaction in a mixture of finely mixed dust and air, which starts due to a local heat rise and propagates itself through the complete mixture. A dust explosion is generally considered as a deflagration.

The dust explosion range is more abstract than that of gas explosion because it is difficult to determine in real life. Next to the **concentration of dust in air** the explosion range depends on:

- **Particle size**
  The finer and more irregular the form, the more explosive the dust (greater reaction surface). In reality a dust cloud is built of a mixture of different particle sizes.

- **Moisture content**
  The larger the moisture content, the more difficult the explosion becomes. The finer and drier a dust cloud is, the more explosive the dust may become.

- **Hybrid mixtures**
  The presence of flammable volatiles in the dust, as in polystyrene granules, extracted soya beans, other seed waste, or even wood-dust containing paint or varnish, can promote an explosion. In this case the ignition energy required is less.

- **Dwell time**
  The time the dust remains in the air, and is thus explosive, depends on it density.

- **Oxygen concentration**
  The higher the oxygen concentration, the easier the combustion reaction.

- **Turbulence**
  This is a factor which can speed up the flame front but can also hinder the explosion.

- **Temperature**
  The higher the ambient temperature, the easier the ignition.

- **Inert particles**
  The presence of inert particles as water vapor or inert dust slows the reaction.

A dust explosion can cause secondary explosions; the fact that a primary limited dust explosion can cause further explosions makes dust explosions very deceptive. A small explosion in a room can cause dust – which had settled on surfaces – to swirl, allowing it to be ignited by the primary explosion. In this fashion a chain reaction can occur which can continue throughout an entire installation/compartment if sufficient dust is present.

The ignition of a dust-air mixture requires **much higher ignition energy** than a gas-air mixture (around 10 millijoules, hybrid mixtures require less). The above factors all influence the sensibility to ignition of the dust-air mixture.

The ignition temperature of common dust mixtures lies around 330–400 °C. This can easily be achieved by industrial hot surfaces. A layer of dust lying on a hot surface can start smoldering because the upper layers insulate the lower ones, causing the temperature to rise. The thicker the layer of dust, the lower the temperature required to cause smoldering. A layer of 5 mm of flour only requires a temperature of 250 °C to begin smoldering in less than 2 hours. Such a temperature is easily attained by the surface of a glow-bulb. Regular cleaning (up to 1 mm of dust can be tolerated) of an installation is therefore a must.

10.4 UNITS OF MEASURE

It is important for CFBT instructors to understand how the following units of measure may be used, in order to be able to quantify the fire loading of various types of fuel in association with expected heat release, burning rates and combustion characteristics for training burns and real fire situations.

The joule (J) is the international standard unit for **energy**. 4,186 J of heat energy are required to raise the temperature of one kilogram (1 liter) of water 1 °C. **Specific heat** is the amount of heat per unit mass required (kJ/kg) to raise the temperature by 1 °C. British thermal units **(Btu)** (no longer used in Europe or UK) refer to the energy required to increase the temperature of 1 lb of water at 59 °F by 1 °F (one Btu = 1,054.8 J or just over one kJ).

A watt (W) is the SI unit for **power** [Btu/hour].
1 watt = 1 J/s.
A 100 W light bulb transforms 100 J of energy to light and heat every second.

10.5 HEAT RELEASE AND COMBUSTION DATA

Design fire load mass	kg
Energy release	MJ
Design fire load	kg/sq m
Fire load energy density (FLED)	MJ/sq m
Maximum burning rate – MBR (Mass loss rate – MLR)	kg/s

Heat of combustion (calorific value)	MJ/kg
Peak heat release rate (HRR). Maximum fire intensity	kW/sq m
Peak heat release rate (HRR) MFI	kW or MW
Radiated heat flux	kW/sq m
HRR = MBR (kg/s) × heat of combustion (MJ/kg)	HRR
MW = kg/s × MJ/kg	HRR

- For most fuels the heat released per mass of air consumed is a constant approximately equal to 3,000 kJ/kg. Therefore, the rate of energy release of a confined fire can be approximated from the air in-flow rate.
- Given that HRR per unit of oxygen is relatively constant at 13.1 kJ/g for common fuels, for every 1 MW of heat release rate, 76 g/s of oxygen is consumed.
- Effective heat of combustion of wood in the flaming phase for a fully developed compartment fire is 10.75 MJ/kg.
- Effective heat of combustion of wood volatiles is 16.4 MJ/kg.
- For a mixture of wood and plastics, the effective heat of combustion is in the order of 16 MJ/kg (16 kJ/g).
- *HRR may also be referred to in relation to time as proportional to the rate of energy release (as in 'T-squared' fires).*

It is necessary to understand the burning characteristics of fuels used in training scenarios. As an example of using the above measurements in a CFBT training evolution we might consider the following example:

Fuel type	Veneer particleboard
Ambient heat of combustion	10 MJ/kg
Fire load energy density (FLED)	134 MJ/sq m
Design fire load mass	121 kg
Design fire load energy	1,206 MJ
Maximum burn rate	0.1 kg/s
Maximum fire intensity	1.4 MW (0.152 MW/sq m)

Fig. 10.2 – Simple laptop software may be used to calculate the above 'known burning characteristics' by entering basic input data. Such software is able to further calculate burning rate and fire intensity with variable venting parameters. For access to this software use the readers' online link code at the front of the book.

10.6 FIRE GROWTH AND DEVELOPMENT

The energy liberated during combustion can radiate back on the fuel substance, where it causes pyrolysis and evaporation of the fuel. It can also aid further pyrolysis of the products in the gas-phase. The heat liberated by the fire also causes the surrounding materials to warm up. The **heat transfer** is accomplished by three means, usually simultaneously: conduction, radiation, and convection.

Conduction is direct thermal energy transfer due to contact. The heat on a molecular level means that the kinetic energy of molecules, that is to say their movement, increases. This energy is than passed on from one molecule to the next. Materials conduct heat at varying rates. Metals are very good conductors while concrete and plastics are very poor conductors, hence good insulators. Nevertheless a fire in one sidewall of a compartment will result in the transfer of heat to the other side of the wall by conduction. If a metal beam passes through the wall, this effect will be even larger.

Radiation is electromagnetic wave transfer of heat to an object. Waves travel in all directions from the fire and may be reflected or absorbed by a surface. Absorbed heat raises the temperature of the material, causing pyrolysis, or augmenting the material's temperature beyond its ignition point causing it to ignite. The amount of radiant heat transfer (radiant heat flux) is measured in kW/sq m.

- 0.67 kW/sq m heat from the sun on a warm sunny day
- 1 kW/sq m for indefinite skin exposure
- 6.4 kW/sq m pain after brief skin exposure
- 12.5 kW/sq m sufficient pyrolysis for piloted ignition of wood
- 20 kW/sq m auto-ignition of wood

Convection is heat transfer through a liquid or gaseous medium. This transfer is caused by density difference of the hot molecules compared to the cold ones. Hot air, gases expand and rise. Convection normally determines the general direction of the fire spread. Convection causes fires to rise as heat rises.

Radiation, convection and conduction next to flame contact consist of normal **fire growth**. Burning embers carried by the wind, debris falling, breakdown of recipients containing flammable liquids or gases, or the melting of lead pipes or plastics can cause fire spread in an unforeseen manner.

Compartment boundaries – Walls and ceilings

The compartment boundaries can influence the conditions inside a fire compartment in three ways:

1. There is a large heat loss through non-insulted walls to the exterior. In this situation the wall is cold and the hot layer is losing heat.
2. The wall may be insulated and prevent heat escaping from the hot layer. In this case the heat is retained in the hot layer and radiated heat flux increases dramatically.
3. The wall itself may act to absorb large amounts of heat (i.e. brick or concrete walls etc.), and this hot wall condition will serve to increase the heat in the hot layer further still.

As radiation becomes the dominant mechanism of heat transfer in a compartment fire, the situation may be termed 'hot wall' where the walls are able to retain heat.

Where walls are not able to retain heat the heat is either lost through the wall to the exterior or bounced back into the hot layer (insulation).

In **cold-wall** situations where **convection is the dominant mechanism** of heat transfer, venting leads to heat loss from the hot layer and a reduction in compartment temperature.

In **hot-wall** situations where **radiation is the dominant mechanism** of heat transfer, venting generally leads to some heat loss from the hot layer but the radiation from hot walls may overwhelm these losses as the combustion process accelerates (thermal runaway).

Stages of growth and development

The temperature versus time plot of a normal compartment fire is shown in Chapter Two. Three different fire phases can be distinguished, namely the growth phase, the steady state phase, and the decay phase. The early stage of a fire during which fuel and oxygen are virtually unlimited is the **growth phase**. This phase is characterized by an exponentially increasing heat release rate. The middle stage of a fire is the **steady state phase**. This phase is characterized by a heat release rate which is relatively unchanging. Transition from the growth phase to the steady state phase can occur when fuel or oxygen supply begins to be limited. The final stage of a fire is the **decay phase**, which is characterized by a continuous deceleration in the heat release rate leading to fire extinguishment due to fuel or oxygen depletion.

Flashover is normally the culmination of the fire growth phase, and occurs when the ceiling temperature reaches around 500–600 °C, depending on the materials present in the compartment and the geometric arrangement. After flashover, room temperature rapidly increases to reach up to 1,000 °C.

Depending on the air inflow or the amount of oxygen present in a compartment, a fire can evolve to flashover as described above but it may also slowly die out as a result of the lack of oxygen. This lack of oxygen in a compartment is mostly due to modern heat-saving construction utilizing double or even triple glazing. In **modern buildings** a fire can smolder due to the lack of oxygen producing large amounts of carbon monoxide and pyrolysis gases. Due to the high thermal insulation of modern buildings, a major heat build-up may occur, even from a small fire. Due to the sudden opening of a door or window, the sudden intake of oxygen-enriched air can cause the combustible gases to explode in what is called a **backdraft**. This is not only a dangerous situation for intervening fire crews but it can be even more dangerous to an untrained occupant of the premises.

An enclosed building fire will behave with reasonable predictably. Fire development is not something that scientists cannot generally account for, and there are many computer models and detailed engineering calculations that will, with correct data input, demonstrate how an enclosed fire is likely to behave under varying parameters. Unfortunately, the firefighter is not in a position to apply detailed calculations on the fire-ground and he/she must make critical decisions and take pre-calculated actions, all within a few short seconds following arrival. Past experience has shown us that actions taken (or not taken) within the 'five minute' window following fire service arrival on scene, is more likely to shape the outcome of any particular fire situation than any other period during the fire-fight.

One important feature of fire growth and development (especially to firefighters), particularly in large floor areas, is associated with the heat release rates of 'T-squared fires' that are seen to grow proportionally to the square of the time period. In the 1980s, fire protection scientists and engineers introduced the concepts of 'slow', 'medium', and 'fast' T-squared fires to represent a range of expected rates of heat release for fire modelling. Basically, a slow T-squared fire reaches a burning rate of 1,000 Btu/s (1,055 kW) in 600 seconds, while a medium T-squared fire reaches that rate in 300 seconds and a fast fire in 150 seconds.

The concept of the 'ultra-fast' T-squared fire was introduced shortly after the concepts of the slow, medium, and fast fires, when it became apparent that the range of those three design fires wasn't sufficient to capture some of the more important fire challenges. The ultra-fast T-squared fire reaches the burning rate of 1,000 Btu/s (1,055 kW) in 75 seconds.

If we take a look at the Power Laws related to fire growth and development, we become aware that even average fires, of medium fire loads (offices and residential occupancies for example), existing between normal ventilation parameters within the confines of their compartmentation, are expected to double in size every 60 seconds where there are adequate amounts of air or oxygen. In areas with higher fire loads or high velocity winds feeding in, the growth rate may well develop on a faster time/area gradient (doubling in size every 30 seconds) or even ultra-fast gradient (doubling every 16 seconds).

- Slow developing fires – Double in size every 120 seconds
- **Medium developing fires – Double in size every 60 seconds**
- Fast developing fires – Double in size every 30 seconds
- Ultra-fast developing fires – Double in size every 15 seconds

Placing these guidelines into a fire-ground perspective, where the fire load is excessive and the supply of air is plentiful, a large non-compartmented area involved in fire can **double in size every 15 seconds**. If this particular area is, for example, open-plan to 20,000 sq ft (1,860 sq m) and filled with 'fast burning' upholstered furniture, we might expect a fire involving 500 sq ft (47 sq m) to double in size every 15 seconds. Within a minute of committing firefighters inside the building, this average-sized fire may have developed rapidly in area and intensity to involve over 2,000 sq ft (nearly 200 sq m) of floor space! This fire may have developed so fast that it was already beyond the control and capability of a single 150 gallons/min (567 liters/min) hose-line within 15 seconds of entry, or even two hose-lines within 30 seconds.

10.7 CLASSES OF FIRE

Fires are divided into classes depending on the materials that burn. Commonly the classes A, B, C and D are recognized.

Class A fires are fires in ordinary solid combustible materials such as bedding, mattresses, paper, and wood. Class A fires must be dealt with by cooling the fire below its ignition temperature. Most Class A fires leave embers, which are likely to rekindle if air comes in contact with them. A Class A fire should therefore not be considered extinguished until the entire mass has been cooled thoroughly. Smothering a Class A fire may not completely extinguish the fire because it doesn't reduce the temperature of the embers below the surface.

Class B fires are those that involve flammable liquids such as gasoline, kerosene, oils, paints, tar and other substances, which do not leave embers or ashes. Class B fires are best extinguished by providing a barrier between the burning substance and the oxygen. Most applied are chemical or mechanical foam.

Class C fires involve gases like natural gas, propane, butane etc. Extinguishing such a fire equals shutting of the source of the gas. Putting out the flames without being able to reach the valve creates a dangerous situation where a spark can cause an explosion.

Class D fires involving burning metals are less common. Combustible metals include sodium, potassium, lithium, titanium, zirconium, magnesium, aluminum and some alloys. Most of the lightweight metal parts in cars contain such alloys. The greatest hazard exists when they are present as shavings or when molten. Fighting such fires with water can cause a chemical reaction or it can generate explosive hydrogen gas. Special extinguishing powder based on sodium chloride or other salts are available. Extinguishment by covering with clean sand is another option.

Class E fires. There are some thoughts that electric fires aren't really considered a true fire class. Electricity doesn't burn but, for example, a short circuit can cause a fire of the insulating material around the wires, which can propagate the fire. Extinguishing electrical fires is best done by using carbon dioxide or by using a powder extinguisher. The use of water is not advised, certainly not as a direct jet on apparatus remaining live. Water spray or mist might be used but with **great caution**. Due to the air between the water droplets, a much larger resistance exists than when using a direct jet. Where possible, the electrical supply should be isolated prior to applying water in any form.

The USA classification of fires is somewhat different:

Class A – As above
Class B – Flammable and combustible liquids and gases
Class C – Electrical fires
Class D – As above

10.8 FIRE PLUMES AND DETACHED FLAMING COMBUSTION – FIRE TYPES

- **Type 1** – Fire plume under or up to the ceiling.
- **Type 2** – Fire plume bending and running along the ceiling.
- **Type 3** – Fire plume bending and detached gas-phase 'fire snakes' (rollover).
- **Type 4** – Fire plume with velocity and momentum caused by exterior wind (forced draft) or fast running 'air-track'.
- **Type 5** – Detached gas-phase fire (rich-mix hanging gas layer) burning at smoke/air interface (NPP) or within the upper regions of the layer.
- **Type 6** – Detached gas-phase fire burning in small 'pockets' and moving around within the compartment (ghosting flames or auto-ignition).
- **Type 7** – Detached gas-phase fire (pre-mixed combustion) possibly resulting in some form of rapid fire phenomena.
- **Type 8** – Flashover. Full compartment involvement.
- **Type 9** – Black fire (super-heated smoke) so hot it is verging on auto-ignition and temperatures are near flame temperature. (therefore heat transfer mechanisms may be similar to that of full flaming gas-phase combustion).

ISO container (FDS) fires

In a container fire the types of fire we encounter will depend on:

Size of the fire load;
Type of fire load (floor pallets or wall-mounted boards);
Size and location of the vent openings;
Size of the door opening;
Direction and force of exterior wind conditions.

If we have a high fire load on the floor (say 4–6 pallets) we will get a Type 1, 2, 3, or 4 fire. This will be fast running, very hot and hard to control without adequate flow-rate. The amount of ventilation will increase or decrease the speed of the fire growth, although it will likely burn with some intensity.

If we have a very small fire load (wood cribbing) at the floor, but a higher fire load (multiple boards) at the ceiling, we may experience Type 5 and 6 fires with limited ventilation, Type 2 with additional ventilation, and Type 3. which is most common in the FDS.

In a real room fire the fire is nearly always hidden in the smoke, which may be down to, or near the floor. Firefighters are most likely to encounter Type 1, 2, and 3 fires. These fires may be deeper seated in the fuel-base than in FDS units, and generally require a higher flow-rate to extinguish or control than the other fire types. They also require different nozzle techniques. For example, Type 5, 6, and 7 fires can normally be dealt with in FDS units by 'pulsing' small amounts of water droplets into the overhead, but real room fires may demand longer bursts of water, or continuous water applications at the fuel base (direct attack). Gas cooling applications can still be 'pulsed' into the smoke layer to reduce temperatures in the overhead and maintain thermal balance.

Firefighters must also cool the boundaries (walls and ceiling) aiming for a 2:1 ratio of droplets as gas to fuel phase. If we don't cool boundaries then heat absorbed into the boundaries will immediately radiate back into the gases and may cause them to auto-ignite. In real fires involving large compartments, or in situations where a room fire is shielded and difficult to access, the heat in the overhead may be extreme and cooling of the gaseous phase may become problematical. Where a shielded fuel-based fire continues to transfer heat into the overhead, cooling of the gas phase may require higher flow-rates than normal and other tactics, such as vertical venting or large calibre streams, may be more productive.

Whilst all fires may be unpredictable, FDS fires develop with known burning characteristics and within calculated venting parameters. This ensures a very controlled and safe environment in which to train. Real room and compartment fires will not burn under such controlled conditions. However, the FDS training will demonstrate to firefighters how various conditions will occur, how they may develop, and how using variable nozzle techniques and altering venting parameters may stabilize conditions.

10.9 RAPID FIRE PROGRESS (RFP)

What is 'flashover'? There are so many different forms of **flashover-related phenomena** it can become confusing for the firefighter. We have grouped the various phenomena under the single heading: Rapid fire progress (RFP). These are all events that are **known KILLERS of firefighters!** To simplify the understanding of critical issues, it is essential for firefighters to know:

- What actions might **CAUSE** an event of RFP.
- What actions might **PREVENT** an event of RFP.
- What fire behavior indicators might offer some warning of impending events.

The types of RFP discussed here are all forms of fire gas ignition (FGI):

- Auto-ignition (is actually a source of ignition);
- Smoke explosion;
- Flash fire;
- Ghosting flames;
- Backdraft;
- Progressive flashover.

What firefighting actions might lead to an event of RFP?

- Incorrect location of vent opening;
- Mistimed vent opening;
- Inappropriate vent opening;
- Inappropriate entry point/procedure for gaining access to structure;
- Creating vent openings without confining the fire or laying a charged primary attack hose-line;
- Delay in getting water on the fire or into the gas layers;
- Inadequate flow-rate at the nozzle.

The actions that can be taken by firefighters to counter or prevent RFP are:

- 3D tactical 'door entry' procedure;
- Confine the fire to room of origin (close doors);
- Get sufficient water on the fire as quickly as possible;
- Get water into the gas layers as quickly as possible;
- Tactical ventilation (under strict protocols);
- Anti-ventilation.

What typical fire behavior Indicators may signal an impending event?

- A smoke layer that is moving up and down or is very turbulent;
- A sudden lowering of the smoke layer;
- Detached flaming 'fingers' in the smoke existing at the ceiling;
- Detached ghosting flames moving around the compartment;
- 'Pulsing' smoke pushing and sucking back and forth at an opening;
- Heavy black staining or crazing (trailer cracking) of windows;
- An increase in heat in the overhead that forces you to crouch low;
- Smoke or flames being sucked back into the building on the under-pressure (fast moving air-track);
- A darkening of smoke from white to brown to grey to black;
- Smoke issuing from openings, or under roof eaves, with an appearance of being under great pressure;
- Black fire (see definition above).

The first known reference, or use of the term FLASHOVER, was made in the 10th edition of the NFPA *Handbook for Fire Protection* in 1948, where a 'flashover point' was used to describe enclosure fires reaching a stage of development where all the combustible material in the area will flash into flame.

In 1961 US Fire Investigator John Kennedy wrote about the phenomenon of flashover, noting the ability of fire to leap across rooms or down corridors at 'express train speed'. The first scientific discussion of the phenomenon appeared in UK Fire Research note 663 (December 1967) where Dr Philip H Thomas referred to the term as 'the theory of a compartment fire's growth, up to the point where it became fully developed.' Customarily, this period of growth was said to culminate in flashover, although Thomas admitted his original definition was somewhat imprecise and accepted that the term could be used to mean different things in different contexts.

Thomas's original 1967 definition of flashover

In a compartment fire there can come a stage where the total thermal radiation from the fire plume, hot gases and hot compartment boundaries causes the generation of flammable products of pyrolysis from all exposed combustible surfaces within the compartment. Given a source of ignition, this will result in the sudden and sustained transition of a growing fire to a fully developed fire ... This is called flashover.

In *RN 663* (1967) Thomas informed us that there could be more than one kind of flashover and described flashovers resulting from both ventilation and fuel-controlled scenarios.

Then in 1995, Walton and Dr Thomas further informed us through the SFPE Handbook that flashover is not a precise term and that several definitions in the literature can be found. Although there appear to be several definitions, they all allude that flashover results in full total room surface involvement with sustained flaming.

The newly crafted ***NFPA 921–2004*** definition of flashover is:

A transitional phase in the development of a compartment fire in which surfaces exposed to thermal radiation reach ignition temperature more or less simultaneously and fire spreads rapidly throughout the space resulting in full room involvement or total involvement of the compartment or enclosed area.

According to 2007 updates of *ISO 13943*[6], **flashover still remains an official term with a current ISO definition** as follows:

Flashover is a stage of fire transition to a state of total surface involvement in a fire of combustible materials within an enclosure.''

The main use of the term flashover was always meant (and is used) to describe fire safety tests that develop to full room involvement. The term rapid fire progress, used by the NFPA, better describes the wider range of gaseous-phase combustion that firefighters are likely to encounter at structure fires.

Dr Thomas himself suggests (2005) that:

'Flashover' is now a problem word of which there seem to be several definitions. The fire services seem to stick to a gas phase definition, and yet the ISO and other definitions

6. ISO 13943: Fire Safety Vocabulary

refer to fire spread and fuel surfaces. These are, to me, not alternatives but different types of flashover: the essence is 'flash' and 'over' – 'overhead' and 'over surfaces' are two varieties. ISO 13943 does refer to 'transition', but it could be 'slow' or 'fast'.

However, for firefighting purposes, the NFPA have recognized for over twenty years (in their annual reports concerning firefighter life losses) that there are several other forms of related phenomena or terms used, such as smoke explosion, flameover, backdraft, flash fire etc., and that many of these phenomena cannot be explained or directly attributed by on-scene firefighters.

Therefore, the NFPA reporting system has established the term 'rapid fire progress' to cover all situations where some form of fire phenomena led to an extreme event of combustion causing sudden transition from a small fire to a large fire, even where flaming is not sustained. They further refer to various sudden or extreme fire phenomena as falling into one of three categories:

Progressive flashover

The fire dynamics associated with normally-accepted definitions of flashover preclude such an event from occurring in high-ceiling, large volume structures. What normally occurs here is an accumulation of flammable combustion products and fire gases in the smoke at ceiling level. This smoke may be collecting in a flammable reservoir, and may be visible or hidden in a ceiling attic space.

When this layer of smoke enters the flammable range, either from the lean side or the rich side, and sufficient fire or heat energy reaches the smoke layer, a fast developing escalation of flaming combustion will spread across the ceiling. The heat flux radiating downwards will eventually ignite fuel sources at the lower level. This ignition of high level fire gases will most likely occur very quickly, possibly faster than a firefighter can run, and will be preceded by a layer of thick black smoke hitting the floor and reducing visibility to zero. The most intense areas of burning may be at walls as the gases deflect downwards with some high velocity and force, similar to a flame-thrower effect. Some have referred to this form of event as a **'progressive' flashover**. It can equally be argued that this is a form of **flash fire**, where the smoke layer is igniting – a fire gas ignition.

Forced draft wind-assisted fires

The hazard to firefighters from *forced draft fires* is immense. In 1999 the author published an article at Firetactics.com entitled *Flashover Pathways*, which investigated the effects that point to point air-tracks have on the flashover phenomenon. Some of the author's earlier articles described a phenomenon that eventually became known as *'high pressure backdraft'*[7], where exterior wind effects cause an undesirable increase of pressure inside a structure. The sudden release of such pressures can lead to devastating effects on the fire development.

Unlike a backdraft, the forced draft fire is driven by air being forced rather than sucked into the fire compartment. The outcome may be the same, although the tactical approaches may be different from a firefighting point of view.

Further research by scientists at the University of Manchester[8] in the UK have looked at how air flows into windows affect the rate of heat release inside a fire

7. White, B. (Captain), (2000), *FDNY Fire Engineering Magazine*, Penwell Publications USA
8. http://www.mace.manchester.ac.uk/

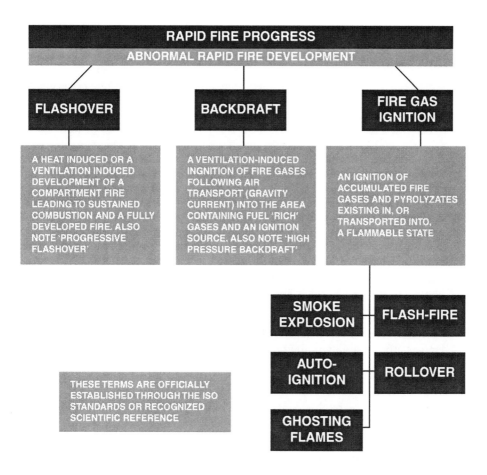

Fig. 10.3 – Rapid fire development

compartment, and how the external flaming effect may spread the fire. They have further looked at the *forced draft* effect that is created where:

- There are windows on opposite sides of the fire compartment; or
- Additional air is being fed to the fire from another source (other than windows).

Having been on the wrong end of such a situation on more than one occasion, the author can attest it's not an experience you need! The forced draft fire is also a killer of firefighters, and is most certainly a situation you need to **plan, train,** and **equip** for. This is a warning to firefighters of a fire-ground hazard that may be well known, but is little understood and rarely credited with the respect it deserves.

Such fires are renowned for creating havoc on the fire floor and, although most commonly encountered in tall buildings, this hazard may also affect your strategic approaches at ground level on a day where wind is gusting moderate to heavy.

If you look back at fire reports, there are countless incidents where wind speed and direction have played a major part in causing abnormal and 'rapid fire' development. Such fires may burn with great intensity and create excessively high temperatures, forcing firefighters to retreat from their position with great haste. Many others have not always been so lucky.

As an incident commander, make sure you account for this hazard when you position and flow the primary attack hose-line. Never underestimate the potential effects on the rate of heat release where a second opening is created (point to point air-track), where a window causes unplanned ventilation, or where an exterior wind (or interior stack effect) might initiate an event of 'rapid fire' development that overcomes the capability of the hose-line in use. If you care about your firefighters you need to make them aware of these hazards.

Smoke explosion

There are three basic requirements[9] that must be met before a smoke explosion can occur. They are:

1. A **contained smoke layer** that consists of enough unburned pyrolyzates that places the mixture within its limits of flammability. For example, the flammability limits for carbon monoxide are 12.5% and 74%, for methane the range is between 5% and 15%, (SFPE, 1995, p.3–16).
2. To ignite the flammable mixture, an **ignition source** is needed. There is a minimum amount of energy that will ignite the layer.
3. The last requirement is enough **oxygen** to support combustion.

Further information:

- A smoke explosion can involve both **cold smoke** and **hot smoke.**
- A rich-mix of super-heated fire gases in smoke may **auto-ignite.**
- *All that is needed in this case is the addition of* **air** – this is not truly reflective of a 'smoke explosion' as a **premixed** state in the gases should normally exist for a smoke explosion to occur.
- A smoke explosion usually entails structural damage caused by pressure waves, whereas the lesser event, termed flash fire, does not.

9. Sutherland, B.J., (1999), *Smoke Explosions Research Report 99/15*, University of Canterbury, Christchurch, New Zealand

What is the relevance? Well, it is relevant to **firefighting actions** because if firefighters are creating openings that allows air in to feed the fire at the time of the rapid fire development, then it is the venting action that might initiate the RFP. If however the action of firefighters was to:

1. Uncover an ignition source by disturbing debris; or
2. 'Push' a flaming ember up into a flammable smoke layer through in-appropriate use of a fog nozzle (for example), or a PPV fan (another example), then a smoke explosion or flash fire may result.

The countering actions to avoid each event are:

- A super-heated fuel-rich smoke layer needs cooling before venting.
- A heavy premix layer of smoke with a suppressed fire needs removing (tactical venting) before overhaul, or disturbing hot spots.

There is also a definite common factor that has so far been overlooked: The **white smoke warning**!

A recent incident in the UK (2004), where two firefighters lost their lives, reiterates previous warnings at Firetactics.com on 'white smoke conditions' being a classic warning sign for an impending SMOKE EXPLOSION.

A basement fire had developed slowly with average amounts of heat, to create smoky conditions throughout the four-story structure. First responders reported a small amount of whitish/greyish smoke coming from an upper floor window. As the fire progressed it became clear that pyrolysis products were forming at all levels in the structure as white smoke began to emit from all openings.

After just over an hour on-scene, the white smoke suddenly darkened as the fire became under-ventilated, whilst an ignition of the accumulated fire gases on the ground floor simultaneously occurred as fire reportedly broke through the floor at the rear of the premises. The fire gas ignition (FGI) trapped the two firefighters who had been attempting to reach the basement to tackle the fire. A venting pathway created on the top floor just a few seconds prior to the smoke explosion is also suggested as a possible catalyst in creating the inertia which led to unfavorable air movements within the structure.

It is well established and documented that white smoke conditions may well offer a warning sign of an impending smoke explosion. The possibility of such phenomena originating in compartments adjacent to, or some way from, the fire compartment itself is also known. It is part of carbonaceous CFBT training sessions for instructors to point out how pyrolysis products present themselves as white smoke early in the live burn. This is seen quite clearly as the fire heats the wood fibre boards causing white smoke to form a highly flammable gas layer in the overhead, eventually igniting as the gases mix in proportion to available air within the compartment.

John McDonough writes in *3D Firefighting*:

When the temperature is too low to support flaming combustion, or when oxygen levels drop below 15%, the fuel package breaks down (pyrolysis) without active flaming and most of the carbon remains on the material. This produces a lighter colored smoke. It is important to realize that as the fire develops, heat will be transferred to neighboring compartments, which can result in pyrolysis of the contents and an accumulating of white smoke, which contains a very high percentage of unburned fuel. As a general guide, lighter colored smoke often indicates that there is an accumulation of pyrolysis products due to increasing compartment temperatures. This is often seen in rooms or spaces adjacent to the fire compartment.

In his paper *Smoke Explosions* Sutherland reports how white/grey smoke was seen to precede experimental smoke explosions following a period of smoldering. He states on p.47 how grey smoke turns white during the transition period towards unstable conditions. '*Smoldering is seen externally as the production of thick white smoke*' (p.50). He also describes how grey smoke signalled Stage 2 and white smoke Stage 3 in the gradual progression to Stage 4 (smoke explosion).

The author has repeatedly warned since 1991 of the explosive dangers posed by wood or FIB wall and ceiling linings, and it is believed they played a major part in the accumulation of flammable pyrolysis products in this situation. These linings will emit dangerous volumes of white smoke when subjected to a slowly developing fire. Wood products contain large amounts of water vapor and sometimes formaldehyde glues that may increase the white smoke effect. Whilst not every white smoke situation is explosive by nature, any slow build-up of confined white/grey smoke conditions should always be take seriously.

Perhaps one of the most well-known fires, where the existence of white smoke led to a major FGI that killed thirty-one people, was the 1987 **King's Cross** underground railway fire in the heart of London. The author was a fire investigator working for London Fire Brigade at the scene of the fire and remembers the repeated statements made by firefighters and other witnesses.

Taken from the official London Fire Brigade report into the fire:

Wisps of white smoke were seen coming up from underneath the escalator.

They walked up the central escalator past the smoke which was white.

The two members of staff entered the upper machine room which they found full of white smoke.

There were large amounts of white smoke building up under the escalator in the machine room.

Smoke was building up in the roof area of the booking hall.

I looked up the escalator and could see a haze of smoke in the air.

At 1937 hours white smoke was seen by a police officer coming from the street entrance to the station.

The fire had involved a large amount of wood and timber paneling on the wooden escalator involved. However, subsequent reports were unable to explain why the fire had developed so rapidly into the booking hall to kill so many people within a few short seconds. Did this white smoke ignite in a fireball?

It is sometimes the case that firefighters enter compartments where a fire has smoldered for some time, producing a vast amount of highly flammable smoke that is contained in the room or space. The smoke may even be cool and appear unthreatening to firefighters, who proceed to search for the source of the fire. When they find it, they reveal the fire by cutting into the sofa, or lifting the mattress, or peeling back the layers of polystyrene plastic to get to a fire underneath. Suddenly – BANG! The ignition source has been uncovered and there is a dramatic explosion as it meets with the accumulation of premixed gas, often at its stoichiometric point. This type of explosion has lifted firefighters off their feet and thrown them 6 meters (20 ft) through the air. On occasions it has killed them in the intense developing fire that followed. These firefighter losses sometimes come in their multiples. The space fire in question can come in the form of a relatively harmless understairs cupboard that appears in an incipient or decay stage, an attic that is smoky but cool, a large warehouse floor, or a small room fire where smoke is light but fire is hidden. Beware of this situation and fully vent the smoke before uncovering the fire's source. (Also beware of an exterior wind that may enter to stir up fire embers when venting this smoke from the compartment).

Auto-ignition

The phenomenon of 'auto-ignition' is one of the most common forms of 'rapid fire' phenomena but is rarely mentioned in training texts. When a crew makes entry to a fire-involved compartment, or structure, where there is an exchange of hot outgoing smoke with cool incoming air:

1. If the smoke is above its auto-ignition temperature (AIT); and
2. There are sufficient fire gases and combustion products to create a flammable mixture; and
3. Air enters the mixture to bring it within its flammable range without cooling the smoke below the AIT; then . . .

. . . there will be an auto-ignition of the smoke. This may occur at the exterior of the exit point only; it may occur inside the entry point or window; or it may burn back into the compartment following an exterior ignition.

Another way an auto-ignition can manifest itself is within a compartment or space where a roof vent is opened adjacent to super-heated smoke. There may also be similar auto-ignitions where pockets of hot fire gases mix with air in balloon-sized flames. These flames may move around in a compartment and are termed 'ghosting flames.'

However, to be more accurate, auto-ignition is **not necessarily an event** in itself but primarily a **source of ignition**. The interior auto-ignition will likely manifest as either a 'backdraft' or as 'ghosting flames.' The exterior auto-ignition will most likely exist purely as flaming combustion at a window with rich fire gases feeding them from within.

Backdraft

The UK's 40,000 firefighters encounter around fifty backdrafts a year. On other occasions they experience various events associated with 'unknown' forms of 'rapid fire' progress around 600 times a year: once every 187 working structure fires[10]. In the USA there is an event of 'rapid fire progress' every day. The vast majority of these events are passed off without injury to firefighters. '*We opened the door and there was a flashover,*' is a typical statement made every week by firefighters to interested media reporters. However, many of these events catch firefighters during their occupation of the structure and multiples of firefighters are seriously burned every month. Many are also killed by sudden and unexpected fire development. If you want to learn how wide a problem this is becoming, go to Google Alerts[11], a free

10. National Fire Service Incident Report Forms (FDR1) from 58 UK Fire Brigades
11. http://www.google.com/alerts

web-based service, and enter keywords such as flashover, backdraft and smoke explosion, to receive daily updates on flashover-related reports. Also remember these are just those events that were reported to the media. The actual number of such 'events' occurring is probably five or ten times greater!

Several conditions are necessary in order for a backdraft to occur within a compartment. The fire must have progressed into a ventilation-controlled state with a high concentration of pyrolysis products and flammable products of combustion. Oxygen concentration in the compartment is low, generally to the point where flaming combustion is incomplete. In addition, there must be sufficient temperature to ignite the fuel when mixed with air. The energy release from a backdraft is extremely rapid and is generally transient, lasting only a short time. However, the fire often advances to a fully-developed state due to changes in ventilation resulting from the over-pressure and heat release caused by the backdraft.

Backdraft definitions[12]

Steward 1914:

These smoke explosions frequently occur in burning buildings and are commonly termed 'backdrafts' or 'hot air explosions.' Fire in the lower portion of a building will often fill the entire structure with dense smoke, before it is discovered issuing from crevices around the windows. Upon arrival of the firemen, openings are made in the building which admit free air, and the mixture of air and heated gases of combustion are ignited with a flash on every floor, sometimes with sufficient force to blow out all the windows, doors of closed rooms where smoke has penetrated, ceilings under attics etc.

The Institution of Fire Engineers (IFE) defines backdraft as:

An explosion of greater or lesser degree, caused by the inrush of fresh air from any source or cause, into a burning building, where combustion has been taking place in a shortage of air.

The NFPA definition is:

A deflagration resulting from the sudden introduction of air into a confined space containing oxygen-deficient products of incomplete combustion.

C. Fleischmann and P. Pagni define backdraft as:

If the compartment is closed, the excess pyrolyzates accumulate, ready to burn when a vent is suddenly opened, for example, as may happen when a window breaks due to the fire-induced thermal stress or a firefighter enters the compartment. Upon venting, a gravity current carries fresh air into the compartment. This air mixes with the excess pyrolyzates to produce a flammable, premixed gas, which can be ignited in many ways.

Enclosure fire dynamics – Quintiere and Karlsson:

Limited ventilation during an enclosure fire can lead to the production of large amounts of unburned gases. When an opening is suddenly introduced, the inflowing

12. Gorbett, G.E. and Professor Hopkins, R., *The Current Knowledge and Training Regarding Backdraft, Flashover, and Other Rapid Fire Progression Phenomena*, available at http://www.kennedy-fire.com

air may mix with these, creating a combustible mixture of gases in some part of the enclosure. Any ignition sources, such as a glowing ember, can ignite this flammable mixture, resulting in extremely rapid burning gases out through the opening and causing a fireball outside the enclosure. (Quintiere, 1999)

All of the various aforementioned definitions of backdraft contain one or more of the following elements:[13]

- **Ventilation-controlled fire** – Combustion cannot sustain itself without adequate oxygen. This oxygen typically comes in the form of atmospheric air. When a compartment does not have any open ventilation to re-supply the air/oxygen, the fire will begin to decay.
- **Unburned pyrolysis products** – Incomplete combustion of the fuel(s) produces heavy volumes of unburned pyrolyzates, which are suspended in the compartment.
- **Confined space or contained fire** – There must be an enclosed space or compartment such as single room or enclosure.
- **Sudden introduction of air/oxygen** – An opening is suddenly introduced into the compartment, and allows fresh air to enter into the compartment.
- **Rapid burning of pyrolysis products** – Ignition occurs of the suspended pyrolyzates and a flame front begins to progress through the compartment.
- **Fire spreads out of the compartment** – The flame front will exit the compartment via an open vent and result in a fireball and overpressure.

Components that control backdraft[14]

- Under-ventilated compartment fire;
- Production of unburned, oxygen-deficient pyrolysis products (excess pyrolyzates);
- Sudden introduction of air (i.e. window or door);
- A gravity current carries fresh air into compartment;
- Air mixes with unburned oxygen-deficient pyrolysis products, creating a
- flammable/combustible mixture interface;
- If ignition source is present at this flammable/combustible mixture interface, then ignition will occur;
- Turbulent mixing of air and unburned, oxygen-deficient pyrolysis products results from the ignition of this interface, which results in further flame spread;
- A deflagration occurs as the flame propagates through the compartment;
- Excess unburned pyrolyzates are forced through the opening by the positive pressure build-up and heat created by the propagating flame front;
- The excess pyrolyzates outside the compartment ignite once presented with fresh air and ignited by the following flame front, creating a fire ball and blast wave.

---

13. Gorbett, G.E. and Professor Hopkins, R., *The Current Knowledge and Training Regarding Backdraft, Flashover, and Other Rapid Fire Progression Phenomena*, available at http://www.kennedy-fire.com
14. Gorbett, G.E. and Professor Hopkins, R., *The Current Knowledge and Training Regarding Backdraft, Flashover, and Other Rapid Fire Progression Phenomena*, available at http://www.kennedy-fire.com

Indicators of a backdraft
The following are indicators that a backdraft may occur:

- The fire may be pulsating. Windows and doors are closed, but smoke is seeping out around them under pressure and being drawn back into the building;
- No visible flames in the room;
- Hot doors and windows;
- Whistling sounds around doors and windows. If the fire has been burning for a long time in a concealed space, a lot of unburned gases may have accumulated;
- Window glass is discolored and may be cracked from heat (Norman, 1991);
- The key indicator that has been witnessed in the past is the in and out movement of the smoke, which gives the appearance that the 'building is breathing.'

A new practical definition of backdraft[15]
The most fitting current definition for backdraft is modified from Quintiere and the Pagni/Fleischmann study.

Limited ventilation during an enclosure fire can lead to the production of large amounts of unburned pyrolysis products. When an opening is suddenly introduced, the inflowing air forms a gravity current and begins to mix with the unburned pyrolysis products, creating a combustible mixture of gases in some parts of the enclosure. Any ignition sources, such as a glowing ember, can ignite this combustible mixture, resulting in an extremely rapid burning of gases/pyrolysis products forced out through the opening and causing a fireball outside the enclosure.

Flameover

Flameover – A fire that spreads rapidly over the exposed linty surface of the cotton bales. In the cotton industry, the common term is flashover and has the same meaning.[16]

This term has since been commonly redefined (in the USA) from its original meaning, to describe rapid flame spread across the surface of highly flammable surfaces, such as wall or ceiling linings painted in varnish or lacquer. This is a practical definition that is based on firefighters' observed experiences of such 'rapid fire' spread and is not a scientific definition as such.

Standard ISO and established scientific terminology

There are still some issues of conflict with the terminology and theories brought through the translation from Swedish to English, and this has caused much confusion. In the 1980s, Swedish fire engineers had begun to redefine terms that had already been established by scientists and firefighters in the US and the UK several decades before, by using new terms, definitions and explanations for events associated with various rapid fire phenomena.

15. Gorbett, G.E. and Professor Hopkins, R., *The Current Knowledge and Training Regarding Backdraft, Flashover, and Other Rapid Fire Progression Phenomena*, available at http://www.kennedy-fire.com
16. NFPA, (1999), *NFPA230 Standard for the Fire Protection of Storage*

Fire gas ignition

From Lund report 1019 Bengtsson gives a definition of a 'brandgasexplosion' (fire gas explosion):

> *The fire gas explosion concept is not defined in any ISO standard. This concept is, however, used in many countries and those definitions that exist are largely similar. One possible definition is given below:*
>
> *'When fire gases leak into an area adjacent to a burning compartment they can become well mixed with the air in that adjacent compartment. This mixture can fill all or part of the available volume and may be within appropriate flammable limits. If the mixture is ignited this may cause a large increase in pressure. This is called a fire gas explosion.' A fire gas explosion occurs without changing the status of any opening in the compartment. In order for backdraft to occur, the ventilation conditions in the compartment must change during the development of the fire. Naturally, the boundary between the two concepts can at times be hazy.'*

The Swedish term 'brandgasexplosion' (fire gas explosion) and its associated definition do not account for the fact that **smoke explosion** has existed for many years in the English language and has been used practically by firefighters in both the UK and USA and documented by scientists from at least 1975. It is difficult to find the exact origin of the term, but is clear to see it is almost 100 years old, and was originally used to describe an ignition of combustion products under circumstances similar to backdraft. More recent scientific research has defined this term more accurately. The most detailed paper by Sutherland (1999) clearly described the phenomenon of fire gases igniting with explosive force. However, this paper also described other events where smoke (fire gases) may ignite **without explosive force**. There are references to earlier work by Croft (1980) and Wiekema (1984), who inform us that high-pressure waves associated with ignitions of the gases (in excess of 5 kPa) may be termed 'smoke explosions' and other such ignitions with minor pressure waves should be termed 'flash fires'. Then there are also auto-ignitions of the gases where they meet additional oxygen supplies at exit points etc. These cannot be termed 'explosions' but are more suited to 'ignitions' as a description of the stated event.

The main issue here is that not all ignitions of the gas layers are explosive. The author spent much time with Dr Martin Thomas, a senior fire research scientist in the UK, in defining these terms, and he was in agreement that the term 'fire gas ignition' is a more descriptive term when applied to the broader range of events that includes 'ghosting flames', 'smoke explosion' and 'flash fire', as opposed to fire gas 'explosion'. He was also of the opinion that 'smoke explosion' was the established scientific reference used for several decades in the UK, and that any alteration of pre-existing English terminology, to 'fire gas explosion', served no logical purpose.

Most importantly, it is essential to differentiate the various phenomena here so that firefighters are able to gain a wider appreciation of slow-rolling flame ignitions (more controllable) as opposed to the more dangerous and explosive situations associated with smoke explosion (take fire gas explosion).

Some authors continue to use pseudoscientific terms in their training texts and this has caused further confusion. The original terminology translated from Swedish, as it related to the term 'flashover', has since been reverted back, in order to conform with current European and North American accepted scientific and ISO definitions as follows:

- Swedish 'lean flashover' is **ROLLOVER**
- Swedish 'rich flashover' is **BACKDRAFT**
- Swedish 'delayed flashover' is **SMOKE EXPLOSION**
- Swedish 'hot rich flashover' is **AUTO-IGNITION**
- Swedish 'fire gas explosion' is **SMOKE EXPLOSION**

Black fire

A term that is commonly used by North American firefighters to describe very dark hot curling 'mushroom-shaped' smoke, that is issuing with some great velocity. This smoke is demonstrating the transition from an under-ventilated fire, between 'boiling' smoke and flaming combustion. It precedes (normally by a few seconds) a stage where smoke auto-ignites at the point of exit. It may also signal the onset of a ventilation-induced flashover. The temperature of this smoke is generally close to 500 °C (932 °F) or more.

Actions (non-actions) by firefighters

The complex nature of various events associated with rapid fire progress (RFP) prevents a practical understanding of such phenomena by firefighters. We should therefore approach the subject with a more basic training objective. From a CFBT fire instructor's standpoint, it is not necessarily the precise science behind each event that is important to the firefighter, but rather the **actions (or non actions)** that he/she might take to cause, or prevent/counter such forms of extreme fire behavior.

There are several forms of fire phenomena that may lead to sudden and extremely intense combustion (RFP) in a compartment fire. The firefighter needs a basic understanding of how and why these events may occur. For example, the simple action of opening a door can lead to an event of RFP. Furthermore, firefighters must gain a practical understanding of actions they might take to prevent such occurrences. One example might be to place just enough water into the overhead to cool the gases, but maintain the thermal balance. Inappropriate amounts of water may cause a 'thermal inversion' and drop the smoke layer to the floor, hindering any visibility.

Cold air events – Temperature and pressure differentials

For some years the author has promoted his beliefs of how the 'air-track' in a fire-involved building can be affected by internal pressures. For example, the sudden release of interior 'fire' and 'wind' pressure caused through opening a door or venting a window where an inlet also exists elsewhere in the structure, can lead to sudden and devastating fire development. Further, his research of 'rapid fire' phenomena has created a link with temperature differentials where exterior temperatures may affect the **likelihood** and **intensity** of any sudden fire phenomena.

There have been several severe events in Canada, where very cold exterior temperatures may have led to very intense backdrafts and ignitions of the fire gases. The Blaina event in Wales, UK occurred on a very cold day also. This is not to suggest that backdrafts will only occur on very cold days, but moreover that the likelihood is increased and the intensity may be far more severe.

A fire engineer in Sweden[17] familiar with this theory suggested the following as an explanation:

The ideal gas law states that $pV = nRT$, which, when simplified, means that pressure × volume = amount × temperature. For a given amount at constant pressure, this gives us Charles Law $V = kT$. This means that a gas expands by 1/273 for each degree (Kelvin) temperature rise.

For example, a 40 °K difference gives a 40/273 (14.6%) difference in volume. So on a cold day, 1 cubic meter of air contains **15% more oxygen** than on a hot day.

The additional cooling effect of the colder air as it enters the structure is negligible in relation to the higher oxygen content.

10.10 FURTHER TERMINOLOGY

Auto-ignition – The auto-ignition point is the temperature at which a flammable mixture ignites spontaneously in air. Auto-ignition temperatures (AIT) refer to near stoichiometric mixtures for which the AIT is a minimum.

Dancing flames – See ghosting flames.

Diffusion flame – Most flames in a fire are diffusion flames: the principal characteristic of a diffusion flame is that the fuel and oxidizer (air) are initially separate and combustion occurs in the zone where the gases mix.

Flammability of fire gases – Fire gases, including carbon monoxide and methane, are capable of burning in both diffusion and premixed states. The smoke given off in a fire is flammable. Particulate smoke is a product of incomplete combustion and may lead to the formation of a flammable atmosphere, which, if ignited, may lead to an explosion.

Fuel-controlled fire – Free burning of a fire that is characterized by an air supply in excess of that which is required for complete combustion of the fuel source or available pyrolyzates.

Ghosting flames – A description of flames which are not attached to the fuel source and move around an enclosure to burn where the fuel/air mixture is favourable. Such an occurrence in an under-ventilated situation is a sure sign that precedes backdraft. Also termed dancing flames.

Gravity current – Also termed gravity wave. An opposing flow of two fluids caused by a density difference. In firefighting terms this is basically referring to the under-pressure area where air enters a building or compartment and the over-pressure area where smoke, flame or hot gases leave: the mixing process between fresh air and combustible fire gases. (Also see hot layer interface below).

High velocity gases – Where the ignition and movement of super-heated fire gases are accelerated through narrow openings, corridors etc., or are deflected, the effects can be dramatic. The deep levels of burning (referred to in the UK as a *local deepening*) will cause unusual patterns of burn as if an accelerant has been used to increase fire intensity. On occasions, where high-velocity gases escape to the outside without being deflected, their flow is such that they may cross an entire street, creating a flame-thrower effect from a window or doorway.

17. Mandre, J., Swedish Fire Engineer, Author's personal communication

Hot layer interface – Often referred to as the NPP (neutral pressure plane). It is assumed that the hot smoky upper layer that forms below the ceiling and the lower cool layer that shrinks as the hot layer descends are joined at a distinct horizontal interface (computer model). This is obviously a simplification because the turbulence within a fire compartment would prevent any true formation of such an interface. Also, highly turbulent plumes and hot layers, as well as strong vent flows, may cause the destruction of a clear interface. However, a noticeable change in conditions from the upper layer to the lower has been observed in many compartment fires. The hot layer *interface* plane and *neutral* plane are not the same. The interface is the vertical elevation within the compartment, away from the vent point, at which the discontinuity between the hot and cold layer is located. The neutral plane (or point) is the vertical location *at the vent* at which the pressure difference across the vent is zero. The terms 'over-pressure' and 'under-pressure' are also used by firefighters to describe the area above the NPP (over-pressure) and the area below the NPP (under-pressure).

Premixed flame – In premixed burning, gaseous fuel and oxidizer (air) are intimately mixed prior to ignition – the flame propagation through the mixture is a deflagration (e.g. smoke explosion).

Pulsation cycle – An indication of the presence of unburned fuel vapours within a compartment with the potential for premixing and a potential explosion. A warning sign for backdraft as smoke 'pulses' intermittently in and out at a ventilation/entry point.

Pyrolysis – The second stage of ignition, during which energy causes gas molecules given off by a heated solid fuel to vibrate and break into pieces. Regardless of whether a fuel was originally a liquid or solid, the overall burning process will *gasify* the fuel. With liquids, the supply of gaseous fuel is a result of *evaporation* at the surface from the heat generated by the flames. Solids entail a significantly more complex process involving chemical decomposition (*pyrolysis*) of large polymeric molecules. Certain combustible solids such as sodium, potassium, phosphorus, and magnesium can even be oxidized directly by oxygen in the air without the need of pyrolysis.

Rapid fire progress – An NFPA definition of all types of rapid fire escalation that may occur and be linked to the above phenomena and their associates.

Regimes of burning

1. Fuel-controlled,
2. Ventilation-controlled,
3. Stoichiometric.

Step events – The heat release rate (HRR) is either controlled by the supply of fuel or the supply of air. Therefore, in principle, four transitions (steps) are possible:

1. Fuel control to new fuel control;
2. Fuel control to air control;
3. Air control to new air control;
4. Air control to fuel control.

In each of these cases the new fire is sustained.

The event defined as 'flashover' is usually related to Step 2, although it may also occur through an increase in ventilation (Step 3).

Thermal balance – The degree of thermal balance existing in a closed room during a fire's development is dependent upon fuel supply and air availability as well as other factors. The hot area over the fire (often termed the fire plume or thermal column) causes the circulation that feeds air to the fire. However, when the ceiling and upper parts of the wall linings become super-heated, circulation slows down until the entire room develops a kind of thermal balance with temperatures distributed uniformly and horizontally throughout the compartment. In vertical terms the temperatures continuously increase from bottom to top with the greatest concentration of heat at the highest level.

Transient events – These are short, possibly violent, releases of energy from the fire which are NOT sustained:

1. Adding fuel;
2. Adding air/oxygen (backdraft);
3. Adding heat (smoke explosion).

Under-ventilated fire – Unlike the *ventilation-controlled fire* an *under-ventilated fire* is not recognized as a burning regime but rather a situation where fuel-rich conditions have accumulated within a compartment. The situation may not involve a fully developed fire and may only be in a state of smoldering. The conditions may or may not present warning signs related to backdraft.

Ventilation-controlled fire – Sometimes referred to as an 'under-ventilated fire,' although this may be incorrect (*see 'under-ventilated' fire above*). Most fully developed fires that occur under confinement or within a compartment are ventilation controlled and burn under fuel-rich conditions. In these situations, the highest temperatures are normally noted at the ventilation openings. The rate of air supply is insufficient to burn all the fuel vapors within the compartment, possibly leading to much external flaming.

Chapter 11

High-rise firefighting – The basics

11.1 INTRODUCTION

The author was invited to address the Seoul, Korea Conference on high-rise fire-fighting in November 2007. His simple message was this:

*When we are faced with a serious fire at ground level, our firefighters often encounter great difficulties and exposure to some element of risk. When they are faced with that same fire, thirty stories above ground, the physiological demands are much greater and the difficulties and risks are greatly magnified. There may be long time delays between a fire commander's chosen strategies becoming tactical operations on the fire floors. There may be changing circumstances during this delay that require the strategy to be altered. There will be a great demand for effective **staffing** to accomplish even the most basic operation and then, where firefighters are working hard, the need to support them in a sustained attack on the fire will treble the resources operating on the fire floors.*

*To be effective you must have a **pre-plan** that is based on the experience of those who have fought these types of fires and learned many lessons. The pre-plan must be well understood by everybody and to achieve this requires frequent practice in such buildings. The **communication** process at a high-rise fire will inevitably break down and the pre-plan must ensure that critical tasks, such as searching stair-shafts, elevators and roofs, are documented as written assignments into the pre-plan. The* *objective is to enable firefighting teams to adapt and function in small teams with pre-assigned tasks and, on occasions, without fire command supervision.*

Above all, avoid **complacency!** *This is inevitably the firefighter's worst enemy. Approach every situation (even calls to automatic fire alarms) with care and professionalism and always try to be at least one step ahead of the fire's next move.*

There are four key words there and if additional factors could be added to summarize the critical areas that let firefighters down in high-rise fires, they are surely these:

- Staffing
- Pre-plan
- Communication (both human and technology failures)
- Complacency
- Minimum flow-rate
- Exterior wind and building air dynamics

Some interesting statistics from the USA[1]:

- Each year, an estimated 15,500 high-rise structure fires cause 60 civilian deaths, 930 injuries, and $252 million in property loss;
- High-rise fires are more injurious and cause more damage than all structure fires;
- Three-quarters of high-rise fires are in residential structures, but these cause only 25% of dollar loss;
- The leading cause of all high-rise fires is cooking (38%), but cause patterns vary by property type;
- 69% of high-rise structure fires originate on the fourth floor or below;
- 60% of high-rise structure fires occur in apartment buildings; 43% originate in the kitchen.

11.2 TELSTAR HOUSE FIRE (LONDON, ENGLAND 2003)

The Telstar House office fire in London's West End district tore through five floors of a thirteen-story high-rise after firefighters were forced off the fire floor due to inadequate flow-rate.

Telstar House is a thirteen-story steel-framed concrete office building built in the late 1960s and measuring approximately 300 ft × 50 ft. It has a ground floor entrance lobby and the upper floors are built above an open parking area. There is also a two-story retail/living unit used as a public house/restaurant built adjacent to the ground floor entrance lobby. There is a staircase at each end of the building. The staircase at the eastern (main entrance end) also has a protected elevator/firefighting lobby. No sprinklers are fitted within the structure but all areas are covered by an automated fire detection system. Each floor of the building is open plan with multiple workstations and low level partitioning. There is no HVAC system fitted in the building, ventilation is provided by pivoting windows across the façade of the building. At the time of the fire, the building was occupied by a project team from London Underground Limited (London's underground railway operator).

1. USFA (2002), *Topical Fire Research Series Volume 2 Issue 18: High Rise Fires*

Steve Dudeney, who at the time of the fire was an ADO (Battalion Chief) with London Fire Brigade, picks up the story:

Three engines and an aerial ladder were despatched at 2044 hours to a fire alarm actuating at Telstar House, Eastbourne Terrace, Paddington, London, W2.

The first two engines and the aerial ladder arrived three minutes after the initial call. Crews arriving at the building could see no signs of fire and had no reason to suspect that this call was any different to the fourteen other automatic fire alarm calls that had received a fire service response to Telstar House in the previous twelve months.

The primary response incident commander approached the entrance to the building and was informed by an on-duty security guard that the alarm panel was indicating a fire on the seventh floor of the block. Asked if everyone was accounted for, he could not be sure but suspected that one person, allegedly working on the floor below the fire, had not been accounted for.

The station officer sent his deputy and a crew up to the fire floor. They exited the elevator on the floor below the fire and were met by a janitor who pointed upward and continued down the stairs. As they approached the entrance to the seventh floor office, the crew saw smoke and flames behind the door. The sub officer radioed the IC and informed him that there was a working fire in progress and that the riser (standpipe) should be charged with water. He requested two SCBA wearers to connect the 45 mm ($1\frac{3}{4}$ inch) hose and nozzle and mask up their SCBA. Meanwhile he and another firefighter entered the fire floor without SCBA and attempted to attack the fire in the first office workstation on the left, using a 19 mm fixed hose-reel fitted inside the building. This had no effect on the fire that was burning from floor to ceiling and had entered the false ceiling panels overhead. They retreated to the firefighting lobby and closed the door behind them. At this time the two SCBA firefighters who had connected the 45 mm ($1\frac{3}{4}$ inch) hose-line to the riser outlet on the floor below joined them.

It is known that a single unit office workstation will, when fully involved in fire (as this was), produce a peak heat release rate (HRR) of around 5 MW or more. However, the fire was further involving ceiling tiles and carpeting, which would increase the HRR to a greater extent.

At this time the third arriving engine pulled onto the block and firefighters observed some smoke coming from the seventh floor windows at the eastern end on the south side.

Only a matter of minutes after the first two firefighters had withdrawn, the SCBA crew entered the fire compartment and were immediately faced with a severe fire that was rapidly consuming the 1,500 sq m (16,000 sq ft) seventh-floor open plan office space. The heat was reported as 'unbearable' and they were immediately forced onto their stomachs where vision was nil and heat was beginning to penetrate their PPE.

They withdrew and requested that the aerial ladder be deployed to ventilate the fire floor.

At about 2053 hours the fire blew a window out on the seventh floor and flames began to run up the face of the building. The fire service had been in attendance for no more than six minutes. Several additional alarms were called in during the next few minutes as the fire rapidly escalated.

At this point most of the windows on the fire floor had failed and flames were licking the eighth floor when a second crew was committed into the fire floor.

However, the fire was now too advanced for the hose-line in use and the limited flow-rate of around 350 liters/min (90 gallons/min) was unable to deal with the fire's development.

An additional hand line was got to work by inserting a dividing breeching (siamese) into the sixth floor riser; this involved a temporary loss of supply to one of the lines. The first crew was then forced to withdraw due to low-pressure warnings on their SCBA, leaving a single crew of two on the fire floor as no crew rotation or relief system had been set up in time. The second crew was also forced to retreat without relief being available in time. This retreat was slow as a large amount of electrical cabling was now hanging from the ceiling and the crew was becoming entangled in it. They were also feeling the first signs of heat exhaustion.

At this point the first assistant divisional officer (battalion chief) arrived on scene and took over command of the incident from the station officer (captain). As fire was seen to be extending into the eighth floor, on-scene engines were increased to ten and additional aerial appliances were requested.

Up on the seventh floor the retreating crew had actuated their PASS alarms (ADSUs) and two emergency crews (Rapid Intervention Teams) were sent in to rescue them. One of the firefighters had become disoriented and wandered back into the fire where he soon collapsed. His partner quickly located him and was able to drag him back toward the exit where they were rescued by the RIT teams. It is worthy of note that at this time the firefighter who went to the aid of his stricken colleague was also suffering from severe heat exhaustion and had only just qualified out of recruit training school three weeks before.

The first firefighter was admitted to hospital (intensive care) with severe heat exhaustion and burns, and his partner was also admitted for heat exhaustion.

London Fire Brigade Deputy Assistant Commissioner (then SDO) Terry Adams picks up the story:
I was ordered on at twelve engines and on arrival was told that the incident commander had just increased the assignment to twenty engines and four aerials. As we were discussing the tactics being deployed, I was informed that a crew of four was reported missing on the seventh floor and that the internal attack had ceased as crews were being withdrawn.

This proved to be a critical decision as it was at this time that the fire development was at its most intense. All of the firefighters were found quite quickly but two had been overcome by heat stress and were removed to hospital (after just fifteen minutes' fire exposure). One was 'lucky to survive' we were told by hospital staff, and this was mainly due to his level of fitness.

Whilst we recovered these people from the seventh floor, we lost impetus with the firefighting operation and the fire spread rapidly to involve upper floors, as it was very windy. As the windows failed, the fire looped back into the floor above. The fire loading was significant as the floors were open plan, measuring about 50 m by 30 m. Most firefighting at this time was undertaken from the exterior, using hand-lines on the ground or aerial streams. Beyond the eighth floor the fire streams were not hitting the fire. We even had a portable ground monitor working from the sixth floor roof of an adjacent building. I had handed over incident command to an assistant chief and was I adamant that the only way we were going to put the fire out was from the inside, but he was considering the reports from sector commanders inside that the situation was particularly hot and conditions were untenable. I was not convinced. Street water supplies were adequate but the building itself only had one 100 mm (4 inch) dry rising main – strange for such a large building.

After about another forty-five minutes I persuaded him to let me resume the internal attack from the fire-protected stairwell. We established that we could have four aerials working to prevent lateral fire spread at the other end of the floor – from where we would commence the internal attack from the stairwell – although the aerial monitors were also struggling to hit the fire because of the projection required (they were too far from the building to be really effective). They could not get closer because of obstructions. The fire streams on the ground also had to be withdrawn because of falling glass.

We re-established a bridgehead on the sixth floor where crews could gather before being committed. The fire had started on the seventh floor and had spread to the tenth floor by this time. The plan was to deploy two hand-lines from the lobby doors into the fire, commencing with the tenth floor as this should stop the vertical fire spread. Crews were rotated after about ten minutes on a fire floor, regardless of how they felt, as conditions were very strenuous. We worked the two hand-lines in together with a sector commander at the doorway. This was having success and was stopping further upward spread after about thirty minutes. We re-assessed the attack and came back down to the eighth floor and commenced the same plan.

We would work from the doorway and then progress into the fire as the fire was controlled. By this time, certainly the seventh, eighth and ninth floor were in the diminishing phase of burn so one hand-line was sufficient as we were really mitigating further damage. This was achieved by committing one crew at a time into the fire floor with a hand-line and again rotating crews frequently.

Fire eventually consumed the whole of the seventh to tenth floors. Four aerial ladder water towers and a high flow ground monitor, used from the roof of an adjacent building, assisted in controlling the fire on the lower floors while a valiant and successful effort by SCBA crews eventually limited the spread of fire to 10% of the eleventh floor by 0200 hours the following morning. Over 150 firefighters on more than thirty-five units fought the battle through the night.

Learning outcomes from fire.
Deputy Assistant Commissioner Adams reports:

- We have looked at and will be placing on the run within three months (early 2008) on all pump ladders (quads) a rapid attack monitor (Akron Mercury) that can flow 900, 1,400 or 1,900 liters/min (240, 370 and 500 gallons/min) through 70 mm ($2\frac{3}{4}$ inch) attack hose. This will give a better application rate that we can use from the lobby doors with a 50 m (160 ft) throw that should have sufficient water to have real effect on an open-plan floor area with high fire loading. This will mean that crews will not need to advance into the fire floor initially, providing a much safer option. The initial crew almost had the fire, but were overwhelmed as our maximum flow from a hand-line is just 475 liters/min – ideal for most applications but not very large areas.
- Falling glass is a real problem.
- In a large un-compartmented building a single dry rising main standpipe is not enough.

- The only way to put out a high-rise fire is through internal attack!
- Crew rotation is particularly important and, often, ten minutes working time on the fire floor is all firefighters can safely manage. This has severe resource implications on the staffing required.
- Medical teams should be available in the bridgehead recovery area to monitor crews' physiological status and ensure that crews drink water both before being committed and immediately after being withdrawn.

London's high-rise SOP is currently in revision (2008) and is at a final stage of draft, but there are also national revisions being made to high-rise procedures following a series of events in other parts of the UK. Main changes to the London policy will be **two hand-lines to be deployed, one as a back up line,** and the attack will be initiated from **two floors below** the fire floor.

11.3 EDIFICIO WINDSOR FIRE (MADRID, SPAIN 2005)

On the evening of Saturday, 12 February 2005, a fire broke out in the thirty-two-story (three below ground) Windsor Tower in Madrid, Spain. The building is located in the heart of Madrid's commercial and banking center and is one of the tallest in the capital. Built in 1978, the building had a concrete central core. The structure above ground was characterized by two transition floors at the third and seventeenth floor levels, which housed plant and services. The transition floors, which carried much of the weight of upper floors, were formed with solid RC slabs and deep beams.

Spanish codes, in common with most continental European codes, place more emphasis on passive control measures than on active measures. The Madrid regional code does not require sprinkler protection for buildings with an evacuation height of less than 100 m, so active measures were limited to automatic detection and alarm, fire hose reels and a dry riser system. The building was being refurbished at the time of the fire.

Fire safety design in many countries relies heavily on sprinkler protection to prevent fire growth, and thereby limit possibilities for fire spread via the façade. The lack of sprinklers, along with the failure of compartmentation, appears to be an important factor in this case.

Company Officer José Antonio Gómez Milara was stationed in Madrid Central Station 1 on the night of the fire. He and his crew were the first firefighters to arrive at the fire floor. These views, along with the narrative for the fire, are his own, and do not represent an official view of the Madrid Fire Brigade.

The security-monitored control room in the basement of the Windsor building registered a fire signal from the twenty-first floor at 2305 on the night of the fire, but a call to the incident wasn't received at Station 1 until 2319 hours. On arrival at 2323 hours, there were no visible signs of fire to be seen from the exterior of the building. Security personnel claim that a period of ten to fifteen minutes further elapsed from the time the Fire Brigade arrived, to water being applied on the fire, i.e. thirty to thirty-five minutes after the initial fire alarm was registered in the control room.

Officer Milara and his crew immediately ascertained that the fire alarm was indicating fire on the twenty-first floor and ascended via elevator to the eighteenth floor in order to make their final approach by stairs. However, as they exited the

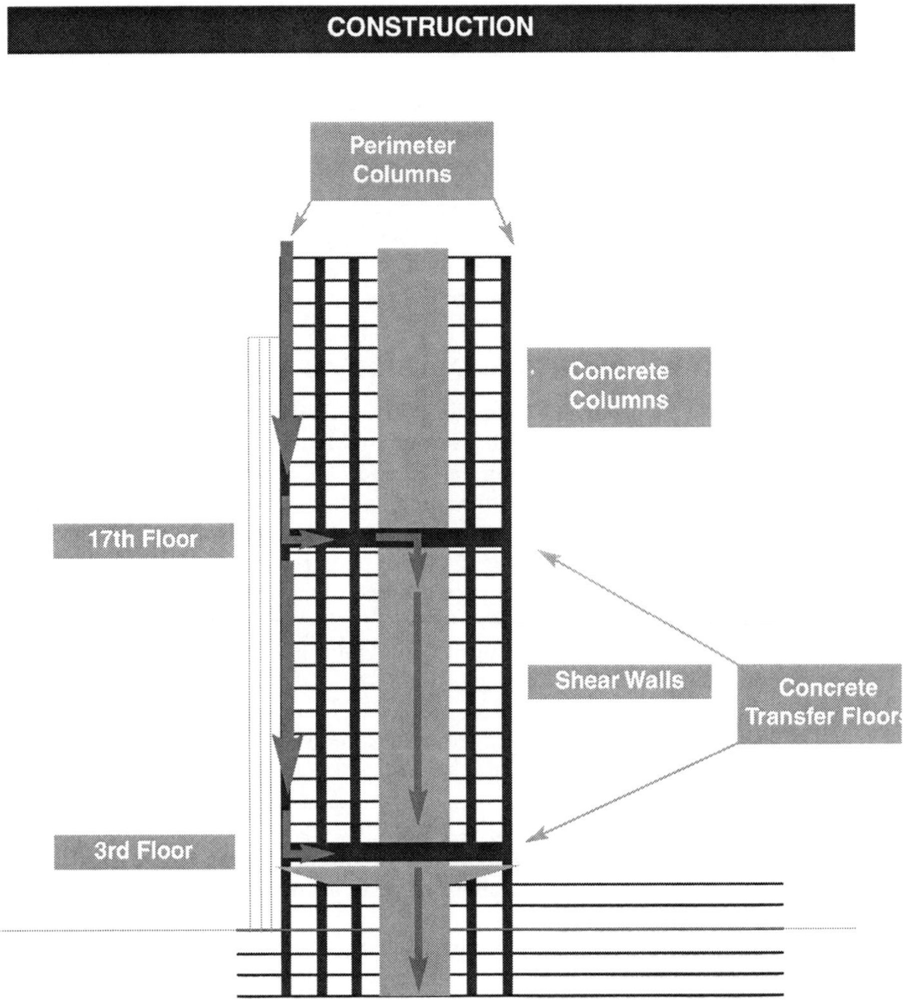

CONSTRUCTION

Perimeter Columns

Concrete Columns

17th Floor

Shear Walls

Concrete Transfer Floors

3rd Floor

Fig. 11.1 – Edificio Windsor building showing concrete transfer floors at third and seventeenth levels.

elevator at level eighteen they were unable to locate the stairway and were forced to return to the elevator to make their final approach to the reported fire floor.

On their arrival at level twenty-one, as they exited the elevator into the fire floor, they were confronted with a vast amount of dark smoke that hung at the ceiling, around 1.5 m from the floor. To the left of them was a door leading to the accommodation, from which heavy dark smoke was emitting. Officer Milara estimates that they arrived at the fire floor by elevator at around 2326 hours.

The firefighters were informed that a building 45 mm hose-line had been deployed into the fire-floor earlier by security personnel, and opted to locate this line to attempt an attack on the fire. Officer Milara reports that inside the fire floor, visibility was almost zero, with the smoke layer almost down to the floor. They could not see or locate the fire, so Milara asked his colleague to stop and listen. Beyond the sound of breathing from their SCBA they could hear the crackling of fire to their right, so they turned and advanced in this direction.

Outside in the street, the fire had erupted from the windows of the fire floor and spread up to involve another two floors above. Calls for further assistance were immediately despatched.

Fire came heading directly at Milara's crew as the fire gases above their heads ignited. He took the nozzle himself and applied a series of nozzle pulsations into the flaming gases. This had no effect on the fire so he resorted to a series of long bursts of water-fog, trying to reach the entire volume of flaming combustion gases that filled the overhead. There were desperate attempts to halt the fire at this point in an effort to stop flaming gases from getting above and behind the firefighters, but the water was evaporating within 3 ft (1 meter) of leaving the nozzle, and the radiative heat levels were exhausting. With fire spreading above them in the false ceiling, further attempts were made to halt the fire's progress by changing to straight stream in an effort to displace the ceiling tiles and get at the fire, but it was noted that the pressure from the building's hose-line was very poor and the stream reach inadequate.

At this point Milara urgently requested some relief for his crew and another team of firefighters (Station 6) advanced in to take the nozzle from them. At this point they had been on the fire floor for around fifteen minutes and had consumed over half of their working consumption of available air.

In the stairway outside of the fire floor Milara explained to the fire floor commander that they urgently needed more hose-lines with greater flows and pressure if they were going to be able to take this fire. At this point there were communication problems and requests for assistance from the fire floor went unheard. Officer Milara and the fire floor commander both then entered the fire floor with fresh SCBA cylinders in order for him to assess the fire's growth and to evaluate the interior crew's safety. The fire was becoming much hotter and spreading rapidly. Milara had never seen anything like this and described it as 'looking into hell.' At this point a minor collapse of interior partitions caused some serious problems.

The firefighters on the nozzle, Milara, and his commander all became separated by the collapse as heavy smoke increased, causing a dramatic reduction in visibility. Then Milara could hear a firefighter screaming, 'Help I am dying.' He quickly located this firefighter who had lost his helmet and SCBA mask during the interior structural collapse. As the firefighter reached Milara he grabbed at him, taking Milara's helmet and mask off as well.

Milara writes,

In that instant, I felt as if something had hit me in the face. The fire must have been close and the temperature up around 300–400 °C (570–750 °F). As I breathed in the hot gases I felt as if I had suddenly been KO'd in a boxing ring!

The other firefighter was now unconscious on the floor and Milara was himself calling for assistance. One firefighter was able to assist Milara's collapsed colleague out towards the stairway, but it was left to Milara to find his own way out. Crawling along on his hands and knees, hardly able to breath, he suddenly realized he had become disoriented and lost. He managed to find his way inside an office but he felt weak and was fast giving up the fight to escape. At this point all he could think of were the images of his smiling children's faces and that of his wife. He writes,

Alone, I thought everything is finished, this is my end. I still don't know how I managed to get out of that office in this blazing high-rise and make it along the corridor without my mask. Suddenly my colleagues located me near the exit to the stairway and pulled me to safety.

Milara himself, suffering from minor burns and smoke inhalation, was then transported down in the elevator and placed on oxygen whilst several other firefighters were rescued and taken to the local hospital. He could not bear to look at his colleague next to him, as he sincerely believed the worst had happened. However, Milara later returned into the building with new PPE and fresh SCBA and began the long ascent via stairs back up to the fire floors with his own crew, as the elevators were heavily smoke logged. As they reached the fourteenth floor a tremendous 'roaring' sound was heard and then nothing. Listening to the fire-ground radio, again Milara feared the worst. However, nothing was heard about evacuating and the firefighters continued their ascent to level twenty-one, where twenty-four other firefighters from four fire stations were now mounting the attack.

It was noted, at this point, that cracks were appearing in the walls of the stairway and finally the call for building evacuation came to all firefighters. By about 0115 hours the fire had spread to most of the floors above the twenty-first floor, resulting in a ten-story blaze. Soon afterwards the first chunks of façade started falling off, taking the perimeter bay of the RC slab with it in places. The spread of fire downwards was gradual at first, probably due to burning embers dropping through services penetrations, through slab edge openings, and through other openings in the concrete slabs caused by core wall expansion.

The fire led to the collapse of virtually all the slab edge bay above the seventeenth floor as well as one internal bay on the north side. The transition floor resisted the impact of the partial collapses. Below this level there was substantial structural damage and deformation, but no significant collapse.

It is good to know that all of Madrid's heroic firefighters who fought the Windsor fire on that night lived to tell the tale! The building suffered severe collapse where the top twelve stories of the twenty-nine-story building fell. There are several dramatic videos available online that record these events.

There were many lessons to be learned:

- Firefighters reportedly unfamiliar with the building layout;
- No building plans reportedly available to assist firefighters;
- Local security reportedly unfamiliar with the building's firefighting systems;

- Delayed response by security to the fire department who attempted to extinguish the fire themselves before calling for help;
- Very heavy fire load on floors of 1,000 sq m (10,800 sq ft) where offices were separated by wooden partitions with layered PVC linings;
- Inadequate pressures and flow-rate on the initial attack hose-line;
- Inappropriate nozzle techniques used against heavy fire in the overhead (see also the Telstar House fire in London);
- Problems with subsequent dry stand-pipe locations, pressures and flow-rate (there were four inlets for four standpipes but these were not marked as to which outlets or stairways were being fed); and there was an open outlet on floor twenty-four above the fire that lost pressure to the fire floors below;
- Buildings in Madrid below 100 m (328 ft) do not need sprinklers installed – the Windsor building was 97.5 m in height;
- There were voids and unstopped gaps where fire was able to spread to upper floors, particularly in the curtain wall.

Officer Milara offers some personal learning points from his own experience that he feels may assist firefighters in similar situations:

- He felt that by remaining more calm he would have been able to replace his mask and helmet but, in the moment, all he wanted to do was get out. (*Author's note: This is totally understandable!*);
- He believes he should have followed procedure and activated his PASS alarm;
- He considers he should not have left his partner during the second entry into the fire floor;
- He should have taken more note of his exit route should he have needed to exit in zero visibility and in a hurry;
- The need for a high flow-rate with open floor spaces is critical.

He further comments that it is his professional desire that the experiences of what happened at this fire should be shared, so that other firefighters may be provided with some 'vision' on how to react under similar circumstances. Mr Milara also wishes to record thanks and appreciation to the Fire Chief of Madrid who had the courage to evacuate his firefighters from such a prominent structure in the city despite the potential for 'social repercussion.' This was undoubtedly an excellent decision made by a professional fire chief who was fortunate in being able to recognize signs of collapse in time to enable a complete evacuation of firefighters from the otherwise unoccupied structure to take place.

11.4 TWO MAJOR OFFICE FIRES – SIMILAR EXPERIENCES!

These two serious office high-rise fires that occurred in London (Telstar) 2003 and Madrid (Windsor) 2005 repeated fire-ground experiences that have already been learned previously by firefighters in New York, Los Angeles and Chicago.

- Call to automatic fire alarm with nothing showing from the exterior;
- Time delay in getting fire service attack hose-line operating;
- Firefighters not familiar with the building layout and installations;
- Rising main standpipes unable to provide sufficient water or pressure;
- Heat exhaustion within fifteen minutes on the fire floor;

- High resource and staffing issues;
- No sprinkler installations;
- Available flow-rate on interior hose-lines unable to match involved fire load.

It is well established that in order to prevent firefighters suffering from heat exhaustion in high-rise office fires, they should not be deployed into fire floors for periods greater than fifteen minutes at a time. This requires a relief system that is termed 'three team rotation,': where Crew One is actively operating for fifteen minutes on the fire floor, Crew Two is waiting to relieve them from the bridgehead or stairway, and Crew Three is two or three floors below at forward command, awaiting to relieve Crew Two. In order for this system to function effectively, Crew Two must be advancing in to relieve the fire-floor crew (Crew One) so that there is no down time at the nozzle. It can be seen there is a constant rotation cycle from forward command, to stairway, to fire floor to rehab. In order to maintain the relief cycle, at least twelve to sixteen firefighters are needed for each hose-line in use. This is in addition to multiples of other roles that need filling at such an incident. It can be seen that without such resources and staffing available on-scene from a very early stage, firefighters will become disoriented and suffer heat exhaustion, or there will be extended periods where water is not being applied on the fire.

In the author's 1992 book, *Fog Attack*, his twenty-eight-page chapter on high-rise fires described several similar incidents in the USA where all the above factors served to hinder the fire service response. He further explained how 100 mm (4 inch) rising main standpipes were incapable of supplying sufficient amounts of water or pressure to deal with developing fires involving office workstations in open-plan floor space. In 1999 he further advised the 'operations' group and 'research and development' group in London Fire Brigade that reducing flow-rates on primary attack hose-lines by replacing 12.5 mm, 19 mm and 25 mm ($\frac{1}{2}$ inch, $\frac{3}{4}$ inch, and 1 inch) smooth-bore nozzles with 475 liters/min (125 gallons/min) combination fog/straight nozzles, was a strategic error. In effect, this actually halved the potential flow-rate from a high-rise primary attack hand-line. The 475 liters/min (125 gallons/min) combination nozzles (or similar) are under-pumped by 89% of the UK's fifty-eight fire brigades (see Chapter Nine) and in high-rise buildings with 100 mm (4 inch) rising main standpipes it is not possible to supply 475 liters/min above the seventh floor without exceeding safe limits of single-jacket European hose pressures or UK BS rising main standards. This is due to the fact that these nozzles require 7 bars (100 psi) at the tip to fully flow and the maximum working limits placed on hose and rising mains prevents such flows being achieved.

The strategic concept of London firefighters using high-flow rapid attack monitors in high-rise office floors is exciting, but slightly unrealistic unless smooth-bore nozzles are utilized. Even then, the equipment will not be available on the primary attack line as the 475 liters/min nozzles are carried in the high-rise bag. What is needed in the rapidly developing open-plan office floor fire is a high-flow hand-line supplying at least 700 liters/min (180 gallons/min) at low nozzle pressures around 3 to 3.5 bars (45–50 psi) (See Figs 11.1 and 11.2).

11.5 HIGH-RISE COMMERCIAL FIRES – PAST LESSONS LEARNED

A 5,000 sq m fire located high up within the confines of a downtown office high-rise structure is a lot different to the same fire located on the second floor! The logistical demands placed on firefighters have demonstrated that incident command needs to function well in advance of actual needs, for as a plan is initiated there is a lengthy time delay before directives become actions, and before strategy becomes reality. At two fires in 1988 (Interstate Bank) and 1991 (Churchill Plaza), US and UK firefighters were faced with serious working office fires on upper levels that demanded a fresh 30-minute SCBA cylinder every 33 seconds for the entire duration of the Interstate fire and similar requirements for a fresh 45-minute cylinder every 80 seconds in the Plaza fire! Similarly, in both fires, hundreds of firefighters were required to undertake a wide range of duties, estimated at both incidents as one firefighter for every 25 sq m of fire involvement.

Common problems:

- Initial response approach by firefighters is often complacent;
- General non-familiarity with buildings and firefighting installations;
- Often initial confusion as to the exact floor of fire's location;
- Building plan and installations **aide-mémoire** non-existent or not sought by fire department on arrival;
- Incident command only as good as the SOP and its implementation;
- No sprinkler installations!;
- Communication problems (both human and technological issues);
- Severe time-lag between strategy becoming actions;
- There may be long delays in getting a primary **fire department** hose-line into operation on the fire floor;
- Critical areas of the building are not being searched quickly enough;
- This includes fire floor, floor above the fire and **all stairways;**
- Search and rescue operations are often ad hoc;
- Evacuation of occupants – **who is responsible** for supervision?;
- Crew rotations in the first 30 minutes are very disorganized;
- Maximum 15 minutes of working time per firefighting crew;
- There are lengthy periods when water is not being applied on the fire;
- Getting adequate staffing on-scene and in place is a major task;
- Rising main standpipes often unable to supply sufficient water;
- Primary attack hose-lines generally inadequate in flow-rate;
- Fire load on open-plan floors leads to rapid fire development;
- Often, such fires cannot be extinguished without support from exterior streams or by allowing the fire to progress into decay stages.

On 17 October 2003, a fire at the Cook County Administration Building **(CCAB) in Chicago** killed six people and seriously injured several others. There were many lessons to learn from this experience. It was a fire that tested the Chicago Fire Department's incident management system to its absolute limits and demonstrated flaws in a high-rise procedure that had evolved over several decades of high-rise firefighting experience.

At around 1700 hours a fire broke out in a storage room adjacent to a 243 sq m (2,620 sq ft) open-plan office suite on the twelfth level of the thirty-seven-story CCAB building. As security officers responded to the twelfth floor they found a fast-developing fire spreading rapidly out of control. A decision was immediately made by the on-scene security staff to initiate a complete evacuation of the structure and as the building was already in the process of naturally emptying, at the end of a working day, it seemed logical to continue this process. A public address system was used by security to direct people towards the two stairways situated in the NW and SE sections of the structure. Whilst both stairways had standpipe rising mains, the SE stairway was also fitted with a smoke tower. The purpose of a smoke (fire) tower is to remove any smoke that enters the lobby between the office space and the stairway, with the objective of keeping an evacuation stairway clear of heavy smoke contamination.

The initial response from the Chicago Fire Department (CFD) brought several engines and ladder companies to the scene at 1707 hours, where it was noted that heavy dark smoke was issuing from an upper level window. A group of firefighters ascended to the twelfth floor using a lift and then the SE stairs. There was criticism that the fire department failed to take control of the evacuation process and direct occupants away from the SE stairs, which was now the 'attack' stairway.

It was noted by some firefighters that the doors on the stairway were self-locking and therefore were inaccessible from the stairway side. In noting this they wedged some of the doors open below the twelfth level to enable them to access the stand-pipe outlets. Just prior to forcing entry into the fire floor at approximately 1716 hours (when they were registered as flowing water) they noted heavy dark smoke pushing out around the sides of the closed stairway door. On entry they were unable to advance the attack hose-line more than a few feet into the floor due to the severity of conditions.

There were unconfirmed reports by some occupants that they were asked to ascend back up the SE stairway as they reached the twelfth level, as firefighters were masking up in preparation for gaining entry into the floor. The CFD view is that these statements are unconfirmed and have no foundation.

Two firefighters reportedly suffered a near miss (caught by heavy smoke and heat) as they were attempting to force entry into a floor directly above the twelfth level to search for occupants and fire spread.

Then, from 1718 hours there were several 911 calls received (one lasting over eight minutes) from occupants stating they were locked and trapped in stairways above the fire and that the heavy smoke conditions were making it difficult for them to breath. At 1719 hours the 911 alarm office passed information to the central communications van on the fire-ground that there were occupants reporting they were locked in the stairways. At 1725 hours the occupant in the SE stairway on the phone to the 911 alarm office (for 8 minutes 14 seconds) stopped responding to the dispatcher.

The fire-ground communications van made several efforts to inform fire com-manders that there were occupants reported as 'trapped' in several areas of the building, but more specifically at the twenty-seventh floor and in the NW stairs. They heard someone say 'message received' across very busy channels but no call signs were used, as they should have been, according to CFD radio procedure. Therefore, there was no certainty that this transfer of critical information (several messages) was ever received.

At 1751 hours crews were withdrawn from the twelfth floor (attack lines were now advancing from both stairways on a 30 MW fire involving 14% of the twelfth floor) in order to implement an exterior attack, to get some knock-down of the fire and to prevent fire lapping into upper floors. The exterior hose streams were just able to reach the fire floor.

At 1806 hours the interior attack was reinstated from both stairways and the exterior streams were shut down. By 1840 hours the fire on the twelfth floor was under control and suppressed. At 1850 hours (over 90 minutes after the twelfth floor door was breached by the fire department), several bodies were found in the SE stairs, between the sixteenth and twenty-second floors. Other occupants had managed to ascend as high as the twenty-seventh floor, where a door from the SE stairs was unlocked, to enter the relative safety of the office space.

There was much criticism levelled at the CFD and they responded by investigating the circumstances that surrounded possible failings in their operational approaches, Incident Command System, and communications between the alarm centre, the fire-ground control van and on-scene fire commanders. The CFD high-rise procedures, incident management system, and several other relevant procedures were updated as a direct result of this fire.

Some of the main changes in procedure brought additional battalion chiefs on the initial Box Alarm (working fire) into specific **'forward fire command'** and **'search and rescue'** assignments. The concept of **Rapid Ascent Teams (RATs)** was also introduced to complete 'top to bottom' stairway searches. A further 'golden rule' found its way into the new CFD high-rise procedure, that **'before beginning fire attack operations, check the stairwell above the fire floor for occupants. It is important to hold the fire attack until all occupants are clear of the stairway for a minimum of five (5) floors above the fire floor.'** This is a directive that has appeared in the FDNY high-rise procedure for more than a decade. It should also be representative of a very basic rule of firefighting in all buildings not to open a door onto a stairway, immediately above which there may be firefighters or occupants in the stairway.

The model approach at this fire should have accounted for the following points:

- A full evacuation of the building was already underway on fire department arrival – the building was also naturally emptying at the end of a working day as the fire occurred;
- The strategy, under such circumstances, should be to protect and maintain the integrity of occupant escape routes;
- Firefighters were aware, prior to forcing entry into the twelfth floor, that the stairway doors were self-locking;
- The door from the SE stairway to the twelfth floor should not have been breached until it was certain that the stairway was clear of occupants and fire-fighters, for at least five floors above the point of entry, and that the evacuation was under the direct control of the fire department with occupants being directed away from the attack stairway;
- Search and rescue responsibilities in a high-rise should be assigned as a specific command task and not as part of the IC's responsibilities;
- Effective communications are essential, and the **transfer of critical information**, from the alarm office to the fire-ground and onto relevant fire commanders, must follow stringent protocols.

11.6 IS THERE A TRAINING NEED?

Here are some statements made following several recent tragedies:

*Since the fatal 2001 Four Leaf Towers residential tower block fire in Houston, Texas, the fire department has placed a **stronger emphasis on training** and staffing their fire force. The fire resulted in the deaths of a firefighter and a building occupant.*

*Since the fatal 2002 Dolphin Cove residential condominium high-rise fire in Clearwater, Florida, the **fire department has placed a stronger emphasis on training**. The fire resulted in several deaths and injured firefighters.*

*Since the fatal 2003 CCAB high-rise office fire in downtown Chicago, the **fire department has placed a stronger emphasis on training**. Greater efforts to improve communications, transfer of information, incident command and firefighting tactics are now being seen. The fire resulted in the deaths of six occupants.*

*Since the fatal 2005 Harrow Court residential tower block fire in Stevenage, UK, the **fire department has placed a stronger emphasis on training** in incident command, transfer of information, and high-rise firefighting tactics. The fire resulted in several deaths including an occupant and two firefighters.*

11.7 HIGH-RISE RESIDENTIAL FIRES

The hazards of large open-plan floor fires are not generally associated with residential tower blocks. These buildings are usually more effectively compartmented and separated by fire-resisting boundaries. Any fire development is therefore much slower and more easily confined. Nevertheless, these fires bring their own specific hazards and problems and the tactical approach should **never be underestimated**. Such fires in the USA, UK and throughout Europe have repeatedly killed and injured both occupants and firefighters and this entire book could easily be filled with examples and case histories. In the UK, firefighters in London, Glasgow, Manchester, Edinburgh, West Midlands, Essex, Kent and Hertfordshire have all experienced tragic circumstances involving serious fires in high-rise residential tower blocks.

At approximately 0500 hours on the morning of 28 June 2002, a fire started in the kitchen of a condominium located on the southwest corner of the fifth floor of an eleven-story residential high-rise, built of fire resistive construction in **Clearwater, Florida**. The delayed alarm resulted in the death of two occupants of the building and injury to five Clearwater firefighters, which included serious burns to one firefighter that required over three months of recovery. Challenges encountered by arriving firefighters included:

- Delayed call to the fire department as occupants attempted to extinguish the fire;
- A lack of knowledge of a working fire by the first engine company, due to communication difficulties;
- A shutoff standpipe riser;
- An out-of-service hydrant on the south side of the building;
- Tactical deviation from the Clearwater SOP for alarms in high-rise structures;
- Long delay in getting water on the fire;
- Firefighters evacuating occupants from the fire floor caught by rapid fire progress;
- Insufficient flow-rate in the primary attack hose-lines.

After addressing these initial difficulties, the operation recovered and the firefighters extinguished the fire with a combined exterior and interior attack. Three alarms were called because of the injured firefighters, the need to search and evacuate a large building with a wide occupant age range and evacuation capabilities, and the early problems encountered with insufficient fire flow to extinguish the fire.

On 18 December 1998, several fire companies and firefighters responded at 0454 hours to a reported fire on the tenth floor of a ten-story high-rise apartment building for the elderly in **Vandalia Avenue, New York City**. The fire had been burning for twenty to thirty minutes before it was called in because the resident attempted to put the fire out with small pans of water. Whilst engine company members were hooking up the attack hose-line, three ladder company members moved ahead of the line into a smoke-logged hallway, outside the apartment on the tenth floor, responding to reports of a trapped occupant. They reported the door to the apartment as being 'partially open,' but on opening the door fully they were caught in the hallway by wind-assisted 'rapid fire progress.' They called a 'mayday' but were found unresponsive some minutes later.

- Delayed call to the fire department as occupants attempted to extinguish the fire;
- Delays in getting water onto the fire due to fire development and standpipe issues;
- Occupant reported trapped was removed from an adjacent apartment by firefighters;
- Victims were attempting to 'locate' the fire and search the involved apartment for occupants;
- Exterior wind of 26 mph gusting to greater strength;
- Two hose-lines (one $2\frac{1}{2}$ inch and one $1\frac{3}{4}$ inch) needed side by side to gain headway along a 40 ft hallway towards the fire involved apartment.

On 13 October 2001, a forty-year-old captain died and another captain was injured while fighting a fifth-floor high-rise apartment fire in **Houston, Texas**. Originally called to a fire alarm actuating, firefighters were informed on arrival that occupants were believed trapped by fire in a fifth-floor apartment.

- Initial confusion as to floor of origin;
- Support crews reporting to wrong floors for operations;
- Attack team applying water to fire inside apartment;
- Crew integrity/continuity was not maintained on more than one occasion;
- High winds affecting building's air dynamics;
- Approach hallway became heavily smoke-logged;
- Hose-line was coiled and looped in hallway, not providing an easy method of locating the way back to the stairs;
- Victim and colleague disoriented on trying to exit fire floor;
- RIT not immediately available to assist trapped firefighters;
- Two firefighters sent to rescue the remaining victim became lost and trapped, but eventually escaped unaided.

On 11 April 1994, at 0205 hours, a call was placed to the **Memphis (Tennessee) Fire Department** from the security service for a high-rise apartment building, reporting a possible fire on the ninth floor. Five firefighters wearing SCBA entered the elevator and headed for the ninth floor knowing that smoke was issuing from a ninth-floor window. When the elevator doors opened on the ninth floor, thick black smoke filled the elevator and firefighters struggled to mask up in a hurry.

As four firefighters exited the elevator the fifth firefighter remained in the car with the hose-pack, as he also struggled to mask up. However, before he could exit with the hose-pack, the elevator doors automatically closed and returned to the ground level, leaving four firefighters on the fire floor.

At this point one of the firefighter's SCBA malfunctioned and he radioed that he was in trouble. He, along with another firefighter, attempted to locate the stairway. Both these firefighters had to be rescued along with a resident from an apartment window on the ninth floor. It was not known at this time that a female resident was also in the same apartment, whose deceased body was found later. The other two firefighters also became disoriented in attempting to locate the stairway and were later found deceased on the ninth floor.

- Failure to take the elevator, according to SOPs, to a floor level some way below the reported fire floor.

On 12 October 1998, while attempting a rescue, two **Memphis (Tennessee)** firefighters became disoriented in a smoke-charged hallway on the twenty-first floor of a high-rise apartment building and ran out of air. The fire started shortly after 0900 hours, and the resident unsuccessfully attempted to extinguish the fire with glasses of water for a period of time. She then called the front desk, reported the fire and exited the apartment, leaving the door open, and took the elevator to the lobby. By the time the fire department arrived, the entire hallway on the twenty-first floor was fully charged with thick black smoke, and the fire had escalated, breaking out the apartment windows and allowing the wind to blow the apartment door shut.

The two firefighters were attempting to rescue a trapped occupant, despite being at the alarm stage of the low-air warning on their SCBA. As they ran out of air, one firefighter managed to locate an exit stairway and escape whilst the other collapsed into a trash room closet. He was deceased when found some thirteen minutes later.

In the early hours of 2 February 2005, a fire occurred in an apartment situated on the fourteenth floor of a seventeen-story residential tower block in **Stevenage, Hertfordshire UK**. Two firefighters and one occupant were tragically killed during this incident following an event of abnormal 'rapid fire development' (ARFD)[2] (rapid fire progress).

On arrival firefighters observed heavy smoke issuing from a window high up in the structure. A team of three firefighters, including a company officer without SCBA, deployed directly to the reported fire floor via elevator (there were nine firefighters, including officers, on-scene at this stage). As they exited the elevator into the twelfth floor there were no obvious signs of fire that appeared well contained inside an apartment unit. Whilst the officer attempted to hook the attack hose-line to the standpipe, the two other firefighters apparently heard a call for help from within the apartment and immediately forced entry to deploy themselves in a 'snatch rescue' attempt.

The fire at this stage was confined to one room and they were able to locate and rescue an occupant from a point beyond the fire. He informed them that his girlfriend was still trapped inside the bedroom where heavy fire was developing and they re-entered in an effort to also bring her out to safety. During this second entry, the abnormal rapid fire progress (ARFD) occurred, trapping both firefighters inside the apartment, along with the remaining occupant. CCTV and witness statements suggested the fire developed rapidly to involve the entire 65 sq m (700 sq ft)

2. ARFD is a term used in the UK to describe various events associated with rapid fire progress

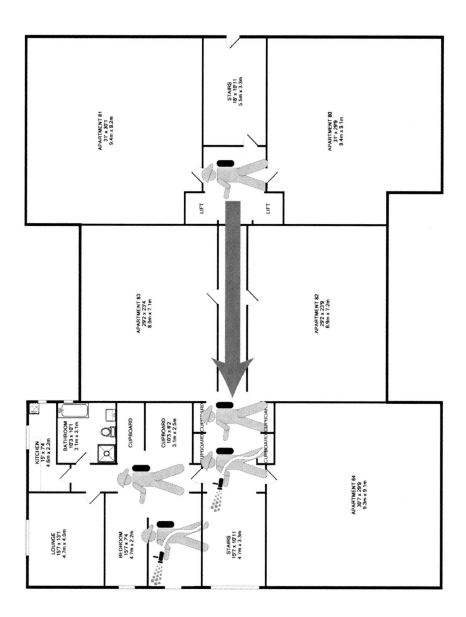

Fig. 11.2 – The Harrow Court Fire showing two firefighters searching in the apartment and a further two firefighters out in the lobby/hallway trying to advance a 230 liters/min hose-line in to rescue them seconds after the rapid fire phenomena occurred. The intensity of the fire forced the two firefighters on the hose-line to back along the corridor and attempt an entry from the other stairway. One of the firefighters who died was located a few minutes later outside in the lobby area where the two firefighters are seen advancing the line.

five-roomed apartment in less than sixty seconds. A Rapid Intervention Team of two firefighters was immediately committed in an effort to rescue the trapped firefighters, but the exterior wind was forcing a very intense fire out into the lobby. They attempted to advance into the apartment with a charged 45 mm ($1\frac{3}{4}$ inch) hose-line but the flow-rate, estimated at just 230 liters/min (60 gallons/min), was inadequate and prevented the firefighters from entering the apartment.

One of the trapped firefighters was killed almost instantly by the ARFD but there was argument in court that the second firefighter may have found some shelter in one of the last rooms of the apartment to develop into flashover. He was eventually located outside in the lobby where the RIT had attempted to gain entry. He was entangled in fire alarm cables that had melted and fallen from the ceiling and had apparently struggled to release himself. His SCBA and mask were still in place but his cylinder was empty. The scenario proposed in court by the author was that he was able to escape the ARFD in the apartment **after** the RIT had retreated in an effort to make an advance from another stairway. During this period it was thought that he might have still been able to escape by crawling under the fire plume in the overhead of the hallway, which was leaving the bedroom and heading along the hallway before exiting the kitchen window.

- Failure to communicate information received on arrival between officers on-scene;
- Not all firefighters ascending in the elevators were required to wear SCBA;
- Self-deployment by attack team via elevator, leaving search and support team, along with **vital equipment**, at ground level;
- Unfamiliarity with the building, and knowledge that the rising main outlets were chain-locked, severely hampered the charging of the primary attack hose-line;
- Failure to take the elevator, according to SOPs, to a floor level some way below the reported fire floor;
- Inadequate flow-rate available on primary attack (rescue) hose-line;
- Only one firefighter on the primary response had ever experienced a fire in a high-rise building before;
- None of the firefighters or company officers on the primary response had ever received practical training in high-rise fire procedures;
- None of the firefighters or company officers had ever taken part in an exercise in a high-rise building;

On 17 November 2003 the **Strathclyde Fire Brigade** in Scotland attended a high-rise fire that almost resulted in multiple firefighter fatalities. The fire at Petershill Court involved an apartment on the twenty-first floor of a twenty-four-story building.

On arrival the incident commander took an elevator with three other firefighters (two wearing SCBA) directly to the reported fire floor. CCTV shows the two firefighters masking up as the elevator ascends to the twenty-first floor. An additional two SCBA wearers take an adjacent lift also directly to the fire floor. Again, these firefighters masked on the way up.

As the first elevator arrived at the fire floor, the incident commander along with the other firefighter not wearing SCBA, was immediately overcome by hot dark smoke that entered. Despite some desperate attempts to close the elevator doors, they remained open. The second elevator then arrived at the twenty-first floor and immediately filled with smoke, but the firefighters were able to close the doors and descend to the twentieth level. What followed was a desperate rescue attempt by firefighters wearing SCBA to rapidly locate the stairway and evacuate their two colleagues who were not wearing SCBA. The IC himself required resuscitation and CPR but both firefighters survived.

Following on from this, an elevator was used by a firefighter to transport paramedics to the upper floors to assist the casualties. However, the elevator was again taken directly to the fire floor on level twenty-one. Again smoke entered but this time all personnel were able to make their escape via the stairway.

- Failure to take the elevator, according to SOPs, to a floor level some way below the lowest reported fire floor;
- Not all firefighters ascending in the elevators were wearing SCBA. (There was no directive for all to do so in the SOP);
- The first actions by firefighters were to rescue their own, before the actual fire could be dealt with;
- Inadequate pressure and flow-rate was reported by firefighters on the fire floor where a 12.5 mm ($\frac{1}{2}$ inch) smooth-bore nozzle was fitted on a 45 mm ($1\frac{3}{4}$ inch) attack hose-line;
- Previous experience in this area of serious fires where heavy smoke had moved down two floors below the fire floor.

On 9 August 1998, a fire occurred in an apartment[3] located on the fourth floor of the Westview Towers Building in North Bergen, New Jersey. Four residents of the building died and thirty-two people, including seven emergency responders, were injured seriously enough to require transportation to a hospital. Twenty-two fire-fighters and an undetermined number of residents were also treated at the scene for minor injuries due to heat, smoke inhalation, minor burns, cuts, and bruises.

Two of the fatalities occurred in the apartment of origin, number 4E, when the victims were trapped by heat and smoke conditions that forced them to retreat to the balcony of their apartment. The balcony was inaccessible by an aerial device and firefighters attempted to reach the trapped victims with ground ladders. One victim fell to her death during this attempt because she lunged at a ladder, which had not been secured to the balcony railing. The other victim succumbed to the intense heat and smoke conditions on the balcony before firefighters could reach her.

Two additional victims were discovered in a stairwell between the sixth and seventh floors. They were residents of the tenth floor and were overcome by smoke as they attempted to escape down the stairwell. Firefighters removed them to an apartment on the sixth floor, but were unsuccessful in their attempt to revive them.

On 12 October 1998 in Saint Louis, Missouri a fire started in apartment 2103 of the Council Tower Apartments Building and communicated out the windows to the apartment immediately above on the twenty-second floor. No one died in the eight-alarm blaze, but thirteen residents and three firefighters received sufficient injuries to require transportation to a hospital. Most of the injuries proved to be minor. However, a Fire Department captain suffered severe burns to his respiratory tract and is not expected to be able to return to active duty as a firefighter due to the extent of his injuries.

3. USFA Report, (2001), *TR/119 A Comparison of Two Fires*

The extinguishment effort in the twenty-seven-story building required the response of over 150 firefighters, and all but three of the department's fire companies. Mutual-aid companies were called in to fill vacant St. Louis fire stations and off-duty personnel also responded. The fire was similar to the fire in New Jersey because of the presence of three oxygen cylinders in the apartment of origin. At least two of the cylinders ruptured, which intensified fire conditions and allowed heat and smoke to spread to an adjacent apartment on the fire floor. The Council Tower Apartments Building was also built of fire resistive construction and was not fully sprinklered.

In December 2008 twenty-one people were killed in a fire that tore through a twenty-eight-story apartment building in eastern Wenzhou, **China**. The Xinhua news agency said more than 200 firefighters battled the fire for three hours. About 200 residents were rescued or evacuated. Other high-rise occupancy types may present an even wider range of problems still.

11.8 BDAG RESEARCH UK – SAFE WORKING PRACTICE IN HIGH-RISE FIRES

Following the tragic events in New York City in 2001 when the WTC Twin Towers were brought to the ground, many other cities and countries around the world responded with research and reviews of their own positions in high-rise fire and terrorist operations. In the UK, part of this response was to initiate a research project to review the interaction of building design with the operational response by fire and rescue agencies to fires and other emergencies.

The Building Disaster and Assessment Group (BDAG) were/are undertaking this research and have produced several excellent reports. However, their work is flawed in one aspect and that is in their failure to establish safe working practices in relation to minimum flow-rates needed to suppress fires.

The author's 1992 book *Fog Attack* first identified the limitations of UK rising main standpipes, in line with hose and pump combinations, as being incapable of transporting adequate amounts of water to the upper floors of high-rise buildings involved in fire. More specifically, the installation of 100 mm (4 inch) risers with a maximum operating pressure of 10 bars (150 psi) into non-sprinklered and non-compartmented office floors is totally inadequate and potentially dangerous for fire-fighters. In residential tower blocks there should also be a minimum 150 mm (6 inch) rising main for buildings over 30 m (100 ft) in height or the 150 mm (4 inch) risers should be strengthened to allow much higher pumping pressures and flow-rate. Where rising main standpipes are in excess of 60 m (200 ft) in height, additional features are required.

The above examples and case histories have clearly demonstrated a need for greater flows to deal with rapidly developing fires in tall buildings and yet UK (and many European) firefighters are still forced to squeeze small amounts of water through 'pinholes' to fight large fires. This is a lesson learned long ago in the USA.

If the London Fire Brigade have a fire that takes out five floors of a London high-rise (Telstar House) and state that the 100 mm (4 inch) rising main was unable to practically provide adequate amounts of water to effectively supply 475 liters/min (125 gallons/min) hand-lines then how will such installations support their desire for flows of 1,000 (265 gallons/min) to 1,900 (500 gallons/min) liters/min at the fire floor to deal with such fires in the future?

Pressure	Bar	Bar	Bar	Bar	Bar
Pump pressure	5	7	10	12.5	15
Base of DRM (inlet) pressure	4.5	6	10	12.5	15
Pump shaft speed (revolutions)	2,100	2,500	2,900	3,100	3,400
Top outlet pressures					
Static	0.5	2.5	3.9	6.1	8.25
Running One Nozzle (Akron 1720)					
Set @ 115 liters/min	X	X	3.5	6	8
230 liters/min	X	X	X	6	8
360 liters/min	X	X	X	6	7.5
475 liters/min	X	X	X	5	7
Running Two Nozzles (Akron 1720)					
Set @ 115 liters/min	X	X	3.5	6	8
230 liters/min	X	X	X	6	8
360 liters/min	X	X	X	6	7.5
475 liters/min	X	X	X	5	7
Top outlet flows					
One Nozzle (Akron 1720)					
Set @ 115 liters/min	X	X	115	115	115
230 liters/min	X	X	X	230	230
360 liters/min	X	X	X	360	360
475 liters/min	X	X	X	475	475
Two Nozzles (Akron 1720)					
Set @ 115 liters/min	X	X	200	200	300
230 liters/min	X	X	X	300	450
360 liters/min	X	X	X	550	700
475 liters/min	X	X	X	800	900

Fig. 11.3 – Typical pressures and flows achieved from a UK 150 mm (4 inch) rising main standpipe. Note that X denotes firefighting streams that were considered inappropriate for recording through inadequate reach or flow-rate. It can be seen that to achieve full working performance of a 7 bar (100 psi) combination fog nozzle, the designed safe riser main working pressure (10 bars) must be exceeded by 50%.

To achieve sufficient pressures (2–3 bars) to flow smooth-bore nozzles on upper floors of high-rise structures, firefighters in New York will use the following pump pressures when supplying/augmenting standpipes:

Floors	Pump Pressure	Pump Pressure
1–10	10 bars	150 psi
11–20	13.5 bars	200 psi
21–30	17 bars	250 psi
31–40	20 bars	300 psi
41–50	24 bars	350 psi
51–60	27 bars	400 psi

*Fig. 11.4 – FDNY pumping pressures through 150 mm (6 inch) standpipes to achieve 2–3 bars at upper floor levels to fully flow smooth-bore nozzles on attack hose-lines. If 7 bar **nozzle** pressures are needed to flow combination fog nozzles, the pump pressures will need to be much higher still.*

High-rise flow-rates

The use of 7 bar (100 psi) combination fog nozzles above the seventh floor is limited by the ability of the rising main standpipe and supply hose-lines to handle high pressures. In the UK (for example) these pressures are restricted to just 10 bars (150 psi) under BS design codes. Therefore realistically, 475 liters/min (125 gallons/min) rated nozzles will not flow to their full capacity on the fire floor where actual flow-rates will be nearer 250–300 liters/min (60–80 gallons/min).

In this case, using a smooth-bore nozzle on the attack line can actually **treble flow-rates**.

In the theme of gas-phase cooling and fuel-phase firefighting, as important as droplet size and fog patterns are, **nothing is more important than flow-rate!**

Here we have two inner-city fire brigades in Europe who suffered flow-rate deficiencies on the fire floors of serious high-rise office fires. In both cases firefighters were overcome by the fire and nearly lost their lives and dramatic destruction to the buildings resulted. In both cases the first in firefighters on the nozzle felt that if they had received an effective flow-rate at the nozzle, the fire might have been suppressed much earlier and not spread beyond a single floor.

Even so, we must establish realistic limitations where modern open plan office floors are involved in fire. In Chapter Nine we see that the European tactical (metric) and US NFA fire flow formulas are two methods of calculating needed fire-flow rates for suppressing structure fires. Both formulae were derived from independent studies of actual flow-rates needed to suppress fires and were both confirmed by operational fire officers for their tactical accuracy. These formulae both arrive at the same conclusions that in true practical terms, one hose-line flowing 700 liters/min (185 gallons/min) will be able to deal with 120 sq m (1,300 sq ft) of open-plan office space involved in fire. This hose-line should further be supported by a secondary back-up hose-line, of equal flow, to protect the exit route for the primary attack team working just ahead of them. If the flow-rate at the nozzle is restricted by two thirds due to rising main standpipe inefficiencies or hose/nozzle combinations, then the single hose-line in use is capable of dealing with just 40 sq m (430 sq ft) of fire.

It is for this reason that those firefighters who have a vast knowledge of fighting serious fires in open-plan office high-rise space prefer to have a smooth-bore, or other high-flow nozzle (at low nozzle pressure) in their hands, in order to achieve maximum cooling effect of the massive fire load involvement that threatens to devour the entire floor with great speed.

Deputy Chief Vincent Dunn, a veteran of the New York City Fire Department's (FDNY) Manhattan District, suggested a single 2.5 inch (63 mm) hose-line, flowing 300 gallons/min (1,134 liters/min) through a 1.25 inch (32 mm) nozzle, could handle up to 2,500 sq ft (232 sq m) of open-plan office space fire involvement.

- **Chief Dunn suggests 300 gallons/min (1,134 liters/min) will deal with up to 2,500 sq ft (232 sq m) of fire.**

Another interesting suggestion, based on research in the USA by Chief Bill Peterson of the Plano Fire Department, stated that when a compartment fire reaches 925 sq ft (86 sq m) in size, the interior fire attack stood a 50% chance of failing. Statistics demonstrate that only a very small number of fires progress to 1,000 sq ft (100 sq m) or beyond.

- **Chief Peterson suggests 50% failure rate to control fire from the interior after fire size exceeds 925 sq ft (86 sq m).**

According to the **NFA flow formula**, a fire involving a floor area of 1,000 sq ft would require two hose-lines (primary and back-up), each flowing at least 165 gallons/min (1,000/3 = 333 gallons/min shared between two hose-lines).

The same example using the author's **metric formula (tactical flow-rate)** would approximate to a fire involving 100 sq m of floor area, which would require an attack flow of 100 × 6 = 600 liters/min (160 gallons/min) (a back-up hose-line of equal or higher flow is additionally recommended) – Total flow requirement = 1,200 liters/min (320 gallons/min).

- **NFA fire-flow formula – Area sq ft/3 = gallons/min**
- **Grimwood's metric formula – Area sq m × 6 = liters/min**

The NFA method of calculating needed flow-rate is based upon an interior aggressive fire attack and the formula may become increasingly inaccurate where fire involvement percentages above 50% of large floor spaces might not offer any opportunity for such an approach. The accuracy of the NFA formula may therefore be questionable in compartments larger than 6,000 sq ft (560 sq m), demonstrating in excess of 50% fire involvement. The NFA approach to fire-ground flow-rate calculation is designed upon direct attack (fuel surface) applications in commercial structures, where the upper flow-rate does not exceed 1,000 gallons/min (3,780 liters/min) and the property is not over-sized. It is acknowledged by those who produced the formula, in its fire-ground format, that the NFA calculation provides more water for suppression than would be necessary if the building were to remain unvented and tightly closed.

The NFA formula recognizes that an aggressive interior attack has a probable upper limit in flow-rate of 1,000 gallons/min, or 50% involvement, before structural integrity is dangerously compromised (rule of thumb)

Optimizing the interior attack hose-line and back up hose-line means getting as much flow-rate and stream velocity (reach) to the nozzle as a crew of firefighters can handle, without making the line so inflexible through over-pressure that it can't flex around corners, or overly flexible so that it will kink and reduce flow through under-pressure. It must be lightweight and easy to maneuvre for the number of firefighters staffing the line, and nozzle reaction should be such that the line can easily be advanced with water flowing. Where a line is staffed with four or more firefighters, then higher flow-rates are generally achievable from a constantly flowing hose-line. Where three or fewer firefighters are staffing the interior hose-line then there may need to be a trade-off with flow-rate and/or stream performance if the nozzle is to be constantly flowed during the advance. It may be possible to utilize the flow-control handle to maintain a high flow-rate and close the line down during each occasion the line is advanced, although this may not be ideal in some situations.

There have been many studies in relation to the ideal attack flow-rate, the optimum diameter attack hose, and the most effective nozzle for firefighting. In this respect the author has no intentions of presenting a biased view! All nozzles have a place on the fire-ground and it is simply finding that place and appreciating what works best for you.

It has become clear through much research that 2 inch (51 mm) hose-lines offer the most water with least friction losses for an attack line that is manageable and easy to maneuvre, whilst taking all the above factors into account. There has been extensive empirical and physiological research into this in the UK and I would suggest that 2 inch (51 mm) hose-lines may suit hose-line staffing up to three persons and may possibly serve as the optimum choice where staffing is four or higher, although the 2.5 inch (63 mm) line has long been the weapon of choice in these situations.

Nozzle Tip	Nozzle Tip	Nozzle Pressure	Flow-rate
12.5 mm	$\frac{1}{2}$ inch	7 bar (100 psi)	283 liters/min + (75 gallons/min)
16 mm	$\frac{5}{8}$ inch	7 bar (100 psi)	438 liters/min + (115 gallons/min)
19 mm	$\frac{3}{4}$ inch	6 bar (90 psi)	585 liters/min (155 gallons/min)
22 mm	$\frac{7}{8}$ inch	4 bar (60 psi)	665 liters/min (175 gallons/min)
24 mm	$\frac{15}{16}$ inch	3.5 bar (50 psi)	700 liters/min (183 gallons/min)
25 mm	1 inch	3 bar (45 psi)	756 liters/min (200 gallons/min)
28 mm	$1\frac{1}{8}$ inch	2.5 bar (35 psi)	850 liters/min (225 gallons/min)
32 mm	$1\frac{1}{4}$ inch	2 bar (30 psi)	960 liters/min (255 gallons/min)

Nozzle Tip	Nozzle Tip	Nozzle Pressure	Flow-rate
12.5 mm	$\frac{1}{2}$ inch	3 bar (45 psi)	189 liters/min (50 gallons/min)
16 mm	$\frac{5}{8}$ inch	3 bar (45 psi)	295 liters/min (78 gallons/min)
19 mm	$\frac{3}{4}$ inch	3 bar (45 psi)	423 liters/min (112 gallons/min)
22 mm	$\frac{7}{8}$ inch	3 bar (45 psi)	578 liters/min (153 gallons/min)
24 mm	$\frac{15}{16}$ inch	3 bar (45 psi)	662 liters/min (175 gallons/min)
25 mm	1 inch	3 bar (45 psi)	756 liters/min (200 gallons/min)
28 mm	$1\frac{1}{8}$ inch	3 bar (45 psi)	N/A
32 mm	$1\frac{1}{4}$ inch	3 bar (45 psi)	N/A

Nozzle Tip	Nozzle Tip	Nozzle Pressure	Flow-rate
12.5 mm	$\frac{1}{2}$ inch	2 bar (30 psi)	155 liters/min (41 gallons/min)
16 mm	$\frac{5}{8}$ inch	2 bar (30 psi)	242 liters/min (64 gallons/min)
19 mm	$\frac{3}{4}$ inch	2 bar (30 psi)	350 liters/min (95 gallons/min)
22 mm	$\frac{7}{8}$ inch	2 bar (30 psi)	472 liters/min (125 gallons/min)
24 mm	$\frac{15}{16}$ inch	2 bar (30 psi)	540 liters/min (143 gallons/min)
25 mm	1 inch	2 bar (30 psi)	616 liters/min (163 gallons/min)
28 mm	$1\frac{1}{8}$ inch	2 bar (30 psi)	778 liters/min (206 gallons/min)
32 mm	$1\frac{1}{4}$ inch	2 bar (30 psi)	960 liters/min (254 gallons/min)

*Fig. 11.5 – Smooth-bore nozzle flow-rates with optimum (top) and typical pressures that may be encountered from rising main standpipes. In comparing these flow-rates with those in Fig. 11.3 above, it can be seen how flow-rate can be trebled in some situations. In the theme of gas-phase cooling and fuel-phase firefighting, as important as droplet size and fog patterns are, **nothing is more important than flow-rate!** The highlighted areas of 19 mm, 22 mm and 24 mm nozzles demonstrate ideal options for interior fire suppression.*

11.9 MODEL HIGH-RISE PROCEDURE FOR LIMITED-STAFFING RESPONSE

It is clear that there are several tactical errors repeated over and over again during routine fire service responses to both office and residential high-rise fires. These errors have sometimes resulted in fatalities and LODDs. What follows is a basic high-rise response plan for low staffing areas. This model procedure offers a guideline for key roles and critical tasks that need addressing on the initial response to a high-rise fire.

High-rise firefighting Standard Operating Procedure (SOP) *Basic SOP for limited-staffed responses*	Draft Model Version 2/2007 GRIMWOOD. P Fire2000.com

Contents
1. Pre-planning
2. Information retrieval
3. Risk assessment
4. Critical control measures
5. On Arrival – Key tasks
6. Primary incident command assignments
7. Deployment of the reconnaissance team
8. Equipment of the reconnaissance team
9. Establishing a bridgehead
10. Purpose of the bridgehead
11. Search, rescue and evacuation
12. Rapid Deployment Measures
13. Critical task assignments on the secondary response
14. Incident Command System – Secondary assignments
15. Crew rotation system at serious working fires
16. Additional equipment required at the bridgehead and staging area
17. Training for high-rise response
18. Air dynamics and wind effects at high-rise fires
19. Fire suppression flow-rates in high-rise buildings
20. Communications at high-rise fires

A   On arrival

- Ten firefighters (minimum) are needed on-scene to implement the **primary response** plan fully:
  1. **Forward fire commander**
  2. Pump operator
  3. Fire attack
  4. Fire attack
  5. BA entry control officer
  6. **Lobby commander**
  7. Primary search
  8. Primary search
  9. Water support
  10. Lift controller
- A reconnaissance team should take control of a fire lift and be deployed by the incident commander, following briefing, to a location at least four floors below the reported fire floor.

- The objectives of this reconnaissance team should be to locate the fire, advance the risk assessment and establish a bridgehead in a smoke-free and protected area, at least one floor below the fire.
- As control measures they should establish a fire attack as a primary action; launch firefighting actions from the bridgehead, or from a protected firefighting lobby fitted with a rising main outlet; advance towards the fire compartment behind the protection of a charged hose-line; and utilize effective door entry and compartment firefighting techniques.
- Where the first arriving pump is several minutes ahead of the second arriving pump the reconnaissance team should comprise the incident commander and two firefighters, all equipped with BA and a high-rise equipment pack.
- Rapid Deployment Procedure is a last resort and should be undertaken with a fire attack hose-line wherever possible.

B. Incident command on arrival

- The two primary roles to be undertaken in the incident command function are those of:
  1. Lobby commander
  2. Forward fire commander
- The lobby commander will be the senior officer in a two-pump attendance that arrives together, or closely within a minute.
- Where the second pump is delayed by several minutes, the first officer on-scene will undertake the forward fire commander's role and form part of the initial reconnaissance team.
- The second arriving officer will become the lobby commander and will deploy the remaining crew members to make up the six-person reconnaissance team and establish control of the fire lifts.
- Where the incident commander has deployed as the forward fire commander, then any subsequent handover in incident command must be undertaken at the bridgehead, where the handing over officer may remain as forward fire commander and the new incident commander will return to the ground floor to take over the lobby command post.

1. Pre-planning

- Fires on the upper floors of high-rise buildings present individual challenges to firefighters not encountered in low-rise structures. Office towers or residential tower blocks each present their own specific challenges and efforts must be made to familiarize with local structures. The layout of buildings, means of access and egress, building security issues, inbuilt fire protection features, local water supplies and occupancy loads are all relevant factors.

2. Information retrieval

- It is essential that a process of information retrieval is begun immediately on, or even prior to, fire service arrival.
- Locate any fixed sources of information such as building plan boxes, hydrant plates etc.
- Building occupants and officials must be questioned to gain important information and confirmation of fire location, occupant status, building systems etc.
- It is important that highly relevant information is immediately shared and communicated to all fire-ground commanders and key operational staff on-scene.

3. Risk assessment

- The risk assessment begins on approach, taking in all known information about the structure, observing signs of smoke or fire penetrating the outer skin of the building, taking into account the volume of any such indicators and the velocity of smoke.
- Ideally, it is important to gain an angled view of a high-rise structure to begin the risk assessment process. It is unlikely that a complete 360 degree walk-around will be immediately available but a corner position will give a good view of at least two sides of the structure.
- Where flaming combustion or smoke is observed issuing under pressure, an immediate assistance message should be sent as the resources needed to deal with a working fire in a tall structure are greatly magnified in comparison to a similar fire in a low-rise structure.
- The forward fire commander should take every opportunity to expand the risk assessment at the fire's location by checking for obvious warning indicators such as fire and smoke behavior, feeling for heat at various heights on doors serving the suspected fire compartment, checking letter post boxes etc.

4. Critical control measures

- Approach the fire with adequate staffing and equipment, strictly according to procedure.
- Ensure the procedure is strictly adhered to; any deviation from procedure must be accountable with sound reasoning at a later stage.
- Keep crews together and under supervision.
- Establish a confirmed bridgehead at a safe location, at least one floor below the fire.
- Launch firefighting operations from a protected lobby or area with rising main outlet, attempting to maintain the integrity of the stair-shaft at all times.
- Advance towards the fire behind the protection of a charged hose-line.

- Utilize effective door entry and compartment firefighting techniques in line with the type of nozzle in use.
- Rapid Deployment Procedure should be avoided unless staffing is limited on the first attendance and there are exceptional circumstances.
- It should be acknowledged that a prompt fire suppression action may serve to save lives.
- Stair-shaft integrity should be maintained as far as possible. Whilst it is recognised that building design may, in some situations, place the rising main outlets in a stair-shaft, every effort should be made to keep doors to the stairs closed as much as possible.
- Prior to opening a door into the stair-shaft from the fire floor, a check should be made for occupants in the stair-shaft for at least five floors above the fire floor.
- Any stair-shaft contaminated by smoke should be prioritized for secondary searches on arrival of the secondary (assistance) response.

5. On arrival – Key tasks

- Dynamic risk assessment of the exterior of the structure – it is important to gain an angled view from at least one corner of the building to view at least two sides of the structure.
- Gather information from evacuating occupants or building officials as to the fire's exact location – communicate this immediately to the assigned forward fire commander, pump operator and brigade control, if different from the initial response call-sheet.
- Gain control of the fire lift(s).
- Establish the status of building evacuation and immediately monitor or take control of this function.
- Deploy a fire reconnaissance team to four floors below the reported fire floor, via the fire lift, to confirm the fire's location and then establish a bridgehead in a smoke free and protected area, at least one floor below the fire floor.
- It should be the general aim to ensure a water supply is effected to the rising main within **three minutes** of arrival and further augmented into the fire pump from a hydrant within **five minutes** of arrival; or within **seven minutes** of arrival if a second pump is used in line.
- A particularly high standard of pump operating is required at high-rise fires. With this in mind it is essential that the pump operator is not expected to fulfil other roles or tasks and remains at the pumping panel to observe and adjust flow demands as necessary.
- A 'water support' firefighter (possibly the second pump operator) will be assigned to augment the water supply and support the pump operator, and primarily to act as command support in sending informative or assistance messages to control.
- Establish a lobby command post in the ground floor entrance lobby, or suitable area nearby if this is not considered safe or viable.

- Send an informative and/or assistance message to control within three minutes of arrival.
- **Staffing assignments on the initial response are as follows:**
  11. Forward fire commander
  12. Pump operator
  13. Fire attack
  14. Fire attack
  15. BA entry control officer
  16. Lobby commander
  17. Primary search
  18. Primary search
  19. Water support
  20. Lift controller
- Therefore, to implement the primary response plan fully, a minimum of ten firefighters is needed on-scene.
- The status and integrity of the rising main outlets should be checked at the earliest opportunity, as resources allow, to ensure pressure and flow is not lost from the main due to previous vandalism.

6. Primary incident command assignments

- Lobby commander (lobby)
- Forward fire commander (FFC) (reconnaissance and bridgehead)

Dependent on the arrival times of first arriving appliances, the incident commander may take either role.

7. Deployment of the reconnaissance team

- The reconnaissance team should consist of a minimum of six firefighters wherever possible
  1. Forward commander
  2. BA wearer
  3. BA wearer
  4. BA wearer
  5. BA wearer
  6. BA entry control officer
- A seventh crew member should immediately be assigned the role of **lift controller.** They should immediately take control of the fire lift(s) – the lift controller will remain in charge of the lift until relieved of this role and return to the ground floor lobby after each transport. This firefighter will have fire-ground radio communication at all times.
- They should take the high-rise pre-packs of equipment and report immediately to the ground floor lift lobby.
- When ordered to by the incident commander, the reconnaissance team should ascend via the fire lift, remaining together as a six-person crew where possible, to a position at least four floors below the reported fire floor – all firefighters ascending to the upper floors will don but not start SCBA breathing apparatus.

- Where the fire lift will not accommodate seven firefighters together with equipment pre-packs, or where a single fire appliance arrives on-scene some minutes ahead of the next responding appliances, a three-person reconnaissance team will ascend and exit the lift at least four floors below the fire floor.
- Three-person reconnaissance team – **In the case of a single appliance arriving alone,** several minutes ahead of further appliances, the IC will become the forward fire commander **and** lift controller and will exit the lift with at least two firefighters at a point four floors or more below the reported fire floor. The IC will then immediately return the lift to the ground floor on exiting with the crew. Where the remaining reconnaissance team members are waiting to ascend, they should do so immediately and communicate to the initial crew that they are closely following.
- Each crew of three reconnaissance firefighters will be equipped with BA and carry a high-rise pre-pack containing at least: one nozzle, 30 meters of hose, a breaking-in tool, lift/riser keys, Stage One BA board with rapid deployment facility, and bolt croppers. They will also be equipped with radio communications.
- On arrival of additional fire pumps a further three reconnaissance firefighters will immediately make their way to a point at least four floors below the reported fire floor and if staffing will allow, a lift controller should also be assigned. The assignment of lift controller should be a permanent position as soon as ten firefighters are on-scene.
- The crew commander of the second arriving appliance will automatically become the lobby commander whilst the forward fire commander will remain with the reconnaissance crew(s) on the upper floors.
- Where the most senior officer has initially ascended as part of the reconnaissance team he/she will retain the role of IC until relieved of such command by a more senior officer arriving on the fire-ground. Handover of command and control should be undertaken by 'face to face' contact. The new IC should then return to the lobby and relieve the lobby commander.
- Crews on the fire floor should **always** remain under the direct supervision and control of a forward command officer sited at the bridgehead, or a fire attack commander working in SCBA beyond the bridgehead, once assigned.
- The first arriving members of the reconnaissance team should then ascend via a protected stairway to the reported fire floor whilst under the forward fire commander's control, briefly checking each floor for smoke as they go.

8. Equipment of the reconnaissance team

- All members of the reconnaissance team, including the team commander and the lift controller, should be rigged in BA, donned but not started.
- **Hose and equipment pre-packs** (may be split across two packs) should carry the following items:

- 60 m of 51 mm hose (or a mix of 70 mm and 45 mm) in 15-meter lengths;
- A suitable fog/jet combination nozzle with flow control handle and pistol grip, providing a minimum flow-rate of 500 liters/min at a branch inlet pressure of 4 bars;
- A suitable solid stream smooth-bore nozzle with flow control handle and pistol grip, providing a minimum flow of 470 liters/min at a branch pressure of 2 bars;
- BA entry control board with rapid deployment facility;
- Breaking-in tool;
- Lift/riser keys;
- Bolt cutters;
- Thermal imaging camera.

9. Establishing a bridgehead

- As soon as the fire's exact location has been confirmed and communicated to the lobby commander the reconnaissance team should return to the relative safety of a **protected area** to set up a bridgehead, taking any opportunity to evacuate occupants in the immediate area as they go.
- At no time should the reconnaissance team split apart from each other. There should always be a minimum of two or five firefighters under the direct supervision of, and in visible contact with, the forward commander, until they begin their advance towards the fire compartment behind the protection of a charged hose-line.
- At this point the forward fire commander should meet up with the remaining members of the reconnaissance team, selecting a suitable location for setting up the bridgehead. This should be in a protected smoke free area **at least one floor below the fire floor**. This location should immediately be communicated to the lobby commander.
- On the bridgehead being established, the reconnaissance team is immediately disbanded and re-organized into (a) fire attack, (b) primary search, and (c) BA entry control assignments, under the direct supervision of the forward fire commander.
- Firefighting operations should then be launched from the bridgehead where BA control will be sited.
- The **fire attack team**, consisting of two firefighters donned in BA under air, will advance from the bridgehead, or a protected firefighting lobby fitted with rising main outlet, behind the protection of a charged hose-line. The integrity of the stair-shaft must be maintained smoke free as far as possible.
- A further two firefighters donned in BA under air will accompany them to assist in advancing the hose-line, forcing entry to the fire compartment and then initiating a **primary search** of the immediate fire area.
- The forward commander will take a position adjacent to the BA control officer and relay important communications to the lobby commander.

10. Purpose of the bridgehead

- The bridgehead provides a safe working 'platform' from which to launch firefighting and rescue operations to the fire floor, using a controlled approach.
- The bridgehead serves as a sector command post in the Incident Command System where the forward fire commander is located.
- The bridgehead serves as a safe location near the fire floor from where BA control may be implemented. The BA entry control officer is able to carry out entry control procedures from this position in a safe and effective manner.

11. Search, rescue and evacuation

- Primary search is undertaken by two members of the initial reconnaissance team under the direct instructions of the forward fire commander. They will also assist the fire attack team with advancing the primary attack hose-line and forcing entry to the fire compartment.
- The primary search zone is defined as those areas on the fire floor closest to the fire, or most affected by smoke and heat.
- The primary search team will work in close coordination with the fire attack team and not work ahead of the hose-line without their agreement.
- Efforts must be made to either control, suppress or isolate the fire, prior to the primary search team advancing into areas ahead of, or beyond the safety of the hose-line.
- Areas such as corridors and lobbies leading to the fire compartment may also become heavily contaminated by heat and smoke and should form part of the primary search zone at the earliest opportunity.
- Search and evacuation of areas adjacent to the fire floor should be considered as 'secondary' search zones. These areas should include (a) stair-shafts, (b) floors above the fire, (c) roof area etc. These areas are primarily and generally the responsibility of secondary response units.
- In some buildings an evacuation system may be in place that utilizes a building or zoned alarm or Public Address system. This system should be immediately monitored at source by the lobby commander on arrival for operation and effectiveness, in accordance with tactical approaches being made. For example, one stair-shaft may be in use as a fire attack stairway, leaving the other as an evacuation stairway. Such information is relevant to any message being sent around the building. The responsibility of building evacuation is handed over to a senior officer on the secondary response.

- It is essential that immediate and urgent attention is given to all secondary search areas as soon as resources allow and it is the responsibility of the lobby commander to ensure this role is assigned and supervised at the earliest opportunity.
- Some buildings may have security features that prevent access to floors from the stair-shaft. These self locking doors may serve to hinder the firefighting efforts and may even trap building occupants in stair-shafts that later become heavily contaminated by smoke and heat. This is the type of hazard that should be pointed out during building familiarization visits.

12. Rapid Deployment Measures

- *TB 1/97* deals with rapid deployments under BA control procedure where persons are known to be in urgent need of assistance inside a building.
- Such a deployment is only undertaken in exceptional circumstances where resources and staffing on the initial attendance are limited and normal approaches under Stage One BA procedure cannot be effectively implemented.
- Any such deployment should be closely monitored and supervised by an outside commander who is in radio contact with the crew prior to, and during, the entry.
- Rapid deployments are generally the result of firefighters being placed into positions where they may face the 'moral dilemma' of seeing, hearing or knowing there are confirmed occupants trapped within.
- The primary consideration should be to assess the occupants' viability as a live victim, prior to any such deployment.
- To prevent firefighters being placed into such a position in the first place, every opportunity should be taken to (a) keep crews together whilst under direct supervision, (b) launch firefighting operations from a protected area, (c) advance towards the fire behind the protection of a charged hose-line, (d) encourage firefighters that a fire suppression action or a simple fire isolation action might serve to save lives on their own.
- Where a rapid deployment becomes necessary, correct procedure should be followed, using the rapid deployment BA control board facility.
- Consideration should be given to the three-person rapid deployment procedure (See Chapter Eight), utilizing a door control assignment in an effort to control and monitor the fire development.

13. Critical task assignments on the secondary response

- This procedure requires at least ten firefighters to be on-scene in order to be able to implement the primary functions of reconnaissance, lift control, bridgehead formation, fire attack, primary search, water supply and basic incident command.
- The secondary response is defined as additional resources and staffing called in to assist the primary response.

- Critical roles and tasks of the secondary response include the following:
  1. Secondary support (back-up) hose-line on the fire floor;
  2. Additional attack hose-line(s) from alternative strategic positions;
  3. BA emergency team (RIT);
  4. Secondary search and evacuation of stair-shafts, unaccounted lift cars and all floors and areas above the fire, including the roof;
  5. BA relief support (three-crew rotation system);
  6. Logistics – transporting BA cylinders and equipment to the upper floors and setting up a staging post;
  7. Tactical ventilation support;
  8. Evacuation of building occupants according to strategic needs and self evacuation that may already be in progress;
  9. Medical triage post.

14. Incident Command System – Secondary assignments

- Fire attack commander (FAC);
- Further fire attack sector commanders as needed;
- Search and rescue commander (SARC);
- Safety or sector officer (exterior of building);
- Evacuation commander;
- Staging commander;
- BA control commander;
- Salvage commander;
- Command support.

15. Crew rotation system at serious working fires

- The three-crew rotation system recognizes the need for prompt relief of crews working on the fire floor and elsewhere under arduous working conditions.
- In high-rise fires, firefighter heart rates can soar in excess of 200 bpm and BA cylinder contents may be depleted rapidly.
- It has been estimated at previous working high-rise fires that a fresh 30-minute cylinder is needed every 33 seconds and a fresh 45-minute cylinder is needed every 80 seconds.
- It was also estimated at fires in the US and the UK that a firefighter was needed for every 25 sq m of fire involvement to ensure strategic objectives were achieved effectively and safely.
- The three-crew rotation system of relieving firefighters on the nozzle places one team on the nozzle, one team at the bridgehead, and one crew in rehabilitation at staging.
- The three-crew rotation system is needed for each crew working in BA. Therefore forty additional firefighters are needed to provide direct cover for twenty firefighters working in SCBA.

16. Additional equipment required at the bridgehead and staging area

- As already noted, at a serious working high-rise fire, there is a high demand on BA cylinders due to the shorter working durations caused through arduous working conditions.
- If a fire progresses through several floors of an office tower then anything from 100–300 cylinders may be required.
- 51 mm hose stocks ready and waiting at staging.
- Interior lighting working from portable generators or building electrical mains supply.
- Nozzles and hand-controlled branches of both fog and smooth-bore design with the ability to provide high-flows, effective throw, and fully-filled fog patterns at low pressures.
- Breaking-in and forcible entry tools.
- Thermal image cameras.
- BA control support.
- Oxygen resuscitators and triage support.

17. Training for high-rise response

- Training firefighters in high-rise firefighting should be a practical 'hands-on' process.
- The Standard Operating Procedure must be regularly covered in a class-room setting and firefighters should be fully familiar with their roles in the tactical plan.
- Familiarization visits and frequent exercises in local buildings are essential if firefighters are to develop and maintain an efficient and confident approach to high-rise firefighting and rescue operations.

18. Air dynamics and wind effects at high-rise fires

- The influences of natural air dynamics, stack effects and exterior wind forces have often created havoc in previous fires in tall buildings.
- It is certain that firefighters will most likely be unfamiliar with such effects and therefore find it almost impossible to anticipate how fires are likely to behave or how smoke is likely to spread in tall structures.
- The effects of exterior winds are greatly magnified at height and what is a seemingly light wind at ground floor level may create high velocities on upper floors. Many firefighters have been tragically killed or badly burned by such effects.
- The effect of opening internal doors and creating openings in windows, or at roof level in stair-shafts, can completely change the air dynamics associated with fire and smoke movements. Therefore, from a tactical standpoint, it is essential that firefighters receive at least some training in the basics of air dynamics in high-rise structures.

19. Fire suppression flow-rates in high-rise buildings

- It has been demonstrated in glass-fronted high-rise structures with fire involvement in excess of 200 sq m, that some auto-exposure with fire spreading upwards on the building façade is almost inevitable.
- In many situations, the only way this exterior spread of fire, from floor to floor, can be curtailed is through the application of exterior fire streams.
- The problems associated with loss of head pressure in rising mains and frictional loss in delivery hose-lines means that pressures available at the nozzle are generally much lower than in low-rise structural firefighting.
- Rising mains are generally designed to provide attack hose-lines with a flow-rate of 500 liters/min but due to the problems mentioned above, the actual flow-rate may only be around 100–250 liters/min. This means attack hose-lines are generally only around 20–50% as effective as routinely expected.
- British Standard rising mains are generally designed to operate at a maximum pressure of 10 bars and BS fire hoses to a maximum of 15 bars.
- Flow-rates based on the average UK residential fire loading have been estimated at around 450–500 liters/min (Local Government Association for example). This means a flat (65 sq m) or house (76 sq m) would require this flow-rate to suppress at full involvement in fire.
- A well-involved open-plan office floor may need higher flow-rates as the compartments are larger and the fire loading is greater. However, 500 liters/min remains the ideal attack hose-line flow-rate as it offers a perfectly manageable nozzle reaction for interior firefighting.
- Where exterior winds are spreading the fire, heat release rate and associated compartment temperatures are likely to be higher and therefore greater flow-rate will be needed to suppress fire during the growth stages of development.
- Working at 10 bars supply pressures, a twin fed 100 mm dry rising main will provide outlet pressures of 7.3 bars at the tenth floor and just 5 bars at the nineteenth floor.
- If 15 m hose-lines are used there will be 0.8 bar friction loss per length at 500 liters/min flow.
- Therefore, a 45 m run of hose (3 × 15 m) will lose 2.4 bar in friction loss from the rising mains outlet pressure.
- Automatic nozzles should be avoided as they lose anything between 1–4 bars more at 500 liters/min as the flow passes through the nozzle!
- Wet rising mains are designed to provide similarly low outlet pressures of 4–5 bar in the UK.
- Therefore, nozzles providing 450–500 liters/min with effective throws at low nozzle pressures are required.
- Compartment firefighting and door entry techniques should be adjusted depending on the type of nozzle in use.

20. Communications at high-rise fires

- Radio communications may be severely affected in high-rise situations due to (a) the height and design of the building, causing 'dead' spots; and (b) the amount of radio traffic.
- Pre-planning and research testing during live exercises may demonstrate some of the problem areas.
- Strict radio protocol should be used at all times and messages should be acknowledged and important content repeated as confirmation of receipt.
- In some situations it may be necessary to utilize firefighter 'runners' to despatch urgent messages that do not appear to be getting through or are not receiving confirmation.
- Additional use should be made of all communication systems such as internal phones or mobile phones where normal channels are unable to transmit.

Chapter 12

CFBT Training Modules

12.1 INTRODUCTION

UK Edexcel CFBT instructor (3 modules – 90 hours)

- Edexcel Level 3 BTEC **Award** in Compartment Fire Behavior Training
- Edexcel Level 3 BTEC **Certificate** in Compartment Fire Behavior Training

The Edexcel Level 3 BTEC Award and Certificate in Compartment Fire Behavior Training are designed to provide:

- Education and training for those in the fire and rescue services with a responsibility to provide compartment fire training to firefighters;
- Opportunities for fire instructors in the fire and rescue services to achieve a nationally-recognized Level 3 vocationally-specific qualification;
- The knowledge and understanding those in the fire and rescue services need to train firefighters in compartment fire behavior.

Structure of the Level 3 BTEC Certificate in Compartment Fire Behavior Training

---

Core units

Unit 1	Fundamentals of Compartment Fire Behavior
Unit 2	Application of Compartment Fire Behavior Training
Unit 3	Application of Positive Pressure Ventilation Training

Assessment and grading

The assessment for the Edexcel Level 3 BTEC Award and Certificate in Compartment Fire Behavior Training is criterion referenced, based on the achievement of specified criteria. Each unit contains contextualized pass criteria for unit assessment.

In the Edexcel Level 3 BTEC Award and Certificate in Compartment Fire Behavior Training, all units are internally assessed. Center assessment will be externally verified through the National Standards Sampling process.

The overall grading for the Edexcel Level 3 BTEC Award and Certificate in Compartment Fire Behavior Training is a pass, based upon the successful completion of all units.

- Learners must pass the **two core units** to achieve the Level 3 BTEC **Award** in Compartment Fire Behavior Training.
- Learners must pass the **three core units** to achieve the Level 3 BTEC **Certificate** in Compartment Fire Behavior Training.

The purpose of assessment is to ensure that effective learning of the content of each unit has taken place. Centers are encouraged to use a variety of assessment methods, including assignments, case studies and work-based assessments, along with projects, performance observation and time-constrained assessments. Practical application of the assessment criteria in a realistic scenario should be emphasized and maximum use made of practical work experience.

Assignments constructed for assessment by centers should be valid, reliable and fit for purpose, building on the application of the assessment criteria. Care must be taken to ensure that assignments used for assessment of a unit cover all the criteria for that unit as set out in the *Assessment Criteria* section. It is advised that the criteria which an assignment is designed to cover should be clearly indicated in the assignment to (a) provide a focus for learners (for transparency and to help ensure that feedback is specific to the criteria); and (b) assist with internal standardization processes. Tasks and activities should enable learners to produce evidence that directly relates to the specified criteria.

The creation of assignments that are fit for purpose is vital to achievement by learners and their importance cannot be overemphasized.

In the Level 3 BTEC Award and Certificate in Compartment Fire Behavior Training, each unit consists of thirty guided learning hours. The definition of guided learning hours is 'a notional measure of the substance of a qualification.' It includes an estimate of time that might be allocated to direct teaching, instruction and assessment, together with other structured learning time such as directed assignments or supported individual study. It excludes learner-initiated private study. Centers are advised to consider this definition when planning the program of study associated with this specification.

It is strongly recommended that *Unit 3: Application of Positive Pressure Ventilation Training,* is delivered after *Unit 1: Fundamentals of Compartment Fire Behavior* and *Unit 2: Application of Compartment Fire Behavior Training* are completed. In order to be able to achieve this unit safely, learners must have a firm understanding of compartment fire behavior and be able to carry out tactical ventilation techniques and cooling and extinguishing techniques to a standard that meets the requirements of *Unit 2: Application of Compartment Fire Behavior Training.* They must also be familiar with risk assessments and operational procedures in CFBT facilities.

12.2 UNIT ONE – FUNDAMENTALS OF COMPARTMENT FIRE BEHAVIOR

Learning outcomes

On completion of this unit a learner should:

1. Understand the principles of combustion and compartment fire behavior;
2. Understand how fire develops and spreads within a compartment and how it can be extinguished;
3. Understand the methods used by firefighters to deal with and prevent fire development within a compartment;
4. Understand the safety procedures relating to fire development within a compartment and how to implement them.

- **Understand the principles of combustion and compartment fire behavior**

  Combustion: triangle of fire (interaction of heat, fuel and oxygen); propagation (conduction, convection, radiation); process (pyrolysis); chemistry; types of combustion (complete, incomplete); products (carbons and unburned pyrolysis products). *Compartment fire behavior*: combustible gases; limits of flammability (lower explosive limit, upper explosive limit, ideal mixtures); ignition sources; fire gases; types of flame, e.g. colors, premixed, diffused.

- **Understand how fire develops and spreads within a compartment and how it can be extinguished**

  Compartment fire development: terminology, e.g. air-tract, under-pressure, over-pressure, neutral plane; stages of development (early, flashover, fully developed, decay); principle of thermal capacity and the concept of combustion inhibitors ('passives'); processes (smoldering fires, backdraft, fire-gas ignition).

  Compartment fire spread: factors involved (compartment construction, compartment size, fire loading, location of fire, changes in fire environment, ventilation); spread to adjacent compartments; effects of limited ventilation; effects of insufficient fuel.

  Extinguishing theory and methods: direct cooling; indirect cooling; gas cooling; with water (effects of steam); latent heat of fusion/vaporization.

- **Understand the methods used by firefighters to deal with and prevent fire development within a compartment**

  Effects: physiological effects (heat-stroke, heat syncope, heat exhaustion, dehydration); psychological effects (effects on knowledge, understanding and capability).

  Preventative and coping methods: training; self-assessment; understanding the limits and capabilities of Personal Protective Equipment (PPE); understanding the effects of humidity and hydration.

- **Understand the safety procedures relating to fire development within a compartment and how to implement them**

  Risk assessment: need for continual dynamic risk assessments.

  Safety procedures: checking the maintenance of the compartment; Personal Protective Equipment; trainer:learner ratio; pre-exercise safety brief; health monitoring of trainers and learners; branch techniques; movement within environment; environmental temperature monitoring.

Contingency plans: emergency plans (for withdrawal, extremes of weather, fatigue, physiological and psychological injury); provision of emergency first aid.

Unit One – Assessment criteria
Understand the principles of combustion and compartment fire behavior.

- Describe the causes of combustion;
- Explain the different types of combustion, the chemistry of combustion and its processes and products;
- Explain the principles of compartment fire behavior.

Understand how fire develops and spreads within a compartment and how it can be extinguished.

- Describe the key stages and processes in the development of a compartment fire;
- Explain the factors which affect the development and spread of a compartment fire;
- Explain the theory and methods of extinguishing compartment fires.

Understand the methods used by firefighters to deal with and prevent fire development within a compartment.

- Describe the physiological effects on firefighters during fire development within a compartment;
- Describe the psychological effects on firefighters during fire development within a compartment;
- Explain the various methods of preventing and managing these effects.

Understand the safety procedures relating to fire development within a compartment and how to implement them.

- Explain the need for continual dynamic risk assessment;
- Describe relevant safety briefs;
- Explain why appropriate emergency withdrawal and first aid procedures are needed during compartment fire training;
- Explain the safety procedures and how they would be implemented.

Delivery
This unit can be delivered through a combination of discussion-led and practical sessions. This is a theory-based unit. Therefore, the main delivery methods used will be formal lectures, presentations and guided group discussions. Syndicate work will also take place. These methods may be reinforced by using video presentations and computer simulations.

It is anticipated that learners will be fire and rescue service staff with a working knowledge of firefighting operations within the community. Learners should have the opportunity to use this existing knowledge. There are also a variety of experiments which can be carried out during the course of teaching this unit, to reinforce what is covered in the discussions. Examples of experiments which could aid the delivery of this unit are:

- Flask experiment with wood chips and polystyrene
- Candle and flame experiment
- Flammable range Bang Box experiment
- Gas aquarium experiment
- Fire growth demonstration/container session
- Window container session, describing the principles of backdraft prevention technique

The experiments should be delivered in a practical setting and could involve demonstrations in the classroom and in the open air. Learners should be given the opportunity to become involved with the running of these experiments, allowing them to reinforce the knowledge and understanding needed to manage experiments safely.

An example of an experiment which proves that 'smoke burns' is the decomposition of wooden chips in a glass container when heated by a Bunsen burner's flame. Smoke is emitted, which can then be ignited outside the glass container.

Learners will be expected to work in syndicate pairs or groups in order to meet the safety requirements, and to demonstrate knowledge, understanding and practical application of the unit content. Additionally, learners must undertake individual tasks within these syndicate pairs or groups, in order to develop their ideas and personal skills.

12.3 UNIT TWO – APPLICATION OF COMPARTMENT FIRE BEHAVIOR TRAINING

Learning outcomes
On completion of this unit a learner should:

1. Be able to demonstrate the appropriate tactical ventilation procedures and understand their beneficial effects;
2. Be able to demonstrate the appropriate extinguishing and cooling techniques prior to entry;
3. Be able to demonstrate the appropriate entry techniques, recognize the hazards and risks within the environment, and apply the appropriate tactics;
4. Be able to apply the appropriate procedures for operating a carbonaceous CFBT facility.

- **Be able to demonstrate the appropriate tactical ventilation procedures and understand their beneficial effects**
  Terminology: tactical ventilation (natural, forced, water-fogged); combustible gases; ignition sources; limits of flammability; over/under-pressure; neutral zone/plane; air inlet; exhaust vent; air route; air management; flashover; backdraft, e.g. ventilation-induced; fire-gas and cold smoke explosion; specific building locations, e.g. high-rise, basements, enclosed room; multi-room compartment fire behavior.
  Factors influencing use of tactical ventilation: location of fire; signs and symptoms of flashover and backdraft; fire-gas and cold smoke explosion; wind direction and strength; identification of appropriate air inlets and outlets; access to the structure and compartment; communications; method of fire attack; horizontal/vertical ventilation.

Tactical ventilation procedures: current best/good practice procedures (air management, natural ventilation, water–fogged ventilation).

Beneficial effects: removal of combustible gases (reducing/stopping risk of fire spread); lowered toxicity of the environment (improving survival rate for casualties, improving visibility, reducing subsequent water damage).

Risk assessment: appropriate Generic and Dynamic Risk Assessments (GRA/DRA).

- **Be able to demonstrate the appropriate extinguishing and cooling techniques prior to entry**

  Extinguishing techniques: direct and indirect; over-pressure and under-pressure.

  Cooling techniques: cooling and water-fogged ventilation techniques in compartments adjacent to the fire room compartment; cooling of the fire room compartment from the adjacent compartments; use of water spray to paint the wall linings to prevent further pyrolysis.

  Risk assessment: appropriate Generic and Dynamic Risk Assessments (GRA/DRA).

- **Be able to demonstrate the appropriate entry techniques, recognize the hazards and risks within the environment, and apply the appropriate tactics**

  Entry techniques: locating the fire; reading the stage of fire development; securing the area around the fire compartment before entry; cooling techniques to the fire compartment from adjacent compartment before entry; entry via a door; entry via a window.

  Hazards and risks within the environment: neutral zone/plane combustible gas layers' horizontal positions; smoke colors, e.g. black, gray; smoke temperatures, e.g. hot, cold; flame colors; position of flame fronts exiting the fire compartment; speed of the smoke layer movement.

  Emergency procedures: knowledge, understanding and application of current good practice emergency procedures.

- **Be able to apply the appropriate procedures for operating a carbonaceous CFBT facility**

  Carbonaceous systems: carbonaceous fuels, e.g. wood, fibrous materials; physical properties (flammability, toxicity, environmental issues); protocols for fuel load (compliance with GRA).

  Health monitoring procedures: e.g. for heat stress, for heat syncope, hydration levels, Personal Protective Equipment (PPE), firefighting equipment, communication systems, maintenance of the safety system of the facility.

  Risk assessments: application of knowledge and understanding relevant to Generic and Dynamic Risk Assessments (GRA/DRA).

Unit Two – Assessment criteria
Be able to demonstrate the appropriate tactical ventilation procedures and understand their beneficial effects

- Identify the factors to be considered when using tactical ventilation, and identify the benefits of this approach;
- Apply the appropriate tactical ventilation techniques;
- Apply appropriate dynamic risk assessments, based on the changing conditions of the fire environment.

Be able to demonstrate the appropriate extinguishing and cooling techniques prior to entry

- Apply the appropriate cooling techniques in compartments adjacent to the fire compartment;
- Apply the appropriate techniques to prevent further pyrolysis.

Be able to demonstrate the appropriate entry techniques, recognize the hazards and risks within the environment and apply the appropriate tactics

- Apply the appropriate entry techniques into a fire compartment, and the appropriate extinguishing techniques within the fire compartment;
- Identify the hazards and risks of the neutral zone/plane horizontal positions;
- Identify the hazards and risks in relation to the colors and types of smoke, the colors of flame and the position of flame fronts;
- Apply emergency procedures as appropriate.

Be able to apply the appropriate procedures for operating a carbonaceous CFBT facility

- Inform the users of the carbonaceous systems in use, and explain their physical properties;
- Apply loading protocols of the carbonaceous fuels in use;
- Apply the health monitoring procedures required during delivery of CFBT;
- Apply the appropriate generic risk assessment for a carbonaceous CFBT facility;
- Apply the appropriate dynamic risk assessment during the use of a carbonaceous CFBT facility, and apply emergency procedures if necessary.

Delivery
This unit is based on a balance of research, theory and practical exercises. Learners must have a sound theoretical knowledge of the fundamentals of compartment fire behavior before attempting this unit. The knowledge gained from studying case studies and real scenarios should be applied within a practice, purpose-built carbonaceous CFBT training simulator. Creating the required, safe, simulated environments will help learners to understand how these operational techniques and procedures can be applied.

Practical demonstrations should cover the use of current good practice equipment such as: the aquarium, the Bang Box, Bunsen burners and glass container with wooden chips inside, the single compartment chipboard box and the doll's house (a multi-compartment chipboard box).

Learners should use a range of methods to find out about policies, procedures and good practice relating to dealing with compartment fire to assist with their personal development. Examples of theses methods are textbooks, the brigade intranet, the Internet, technical journals and statutory instruments. They should also be encouraged to work individually, in syndicate pairs and in groups to enable them to think through and compare ideas, to share knowledge and understanding, to network and to assist in their personal development.

For good practice tactical ventilation procedures, tutors should refer learners to the *Fire Service Manual – Volume 4 Fire Service Training Guidance and Compliance Framework for CFBT* (HMSO, 2000).

Assessment

Assessment for this unit could take the form of training reports, debriefing presentations, videoed practical exercises, and professional discussions with the assessor, which could either be recorded on observation sheets or with audio recording equipment. Evidence should show depth and breadth of understanding, analysis and evaluation, an independent approach, intuition and perception, and an ability to apply the appropriate learning and development techniques.

Learners could be given a complete scenario (or several short scenarios) upon which their assessment is based. The scenarios must be developed in sufficient detail, to reflect the complexities of a real-life situation. Learners could produce a project, supported with answers to questions based on the scenario, or conduct a professional discussion with an assessor. Much of this unit is practical and, therefore, practical activities must be carried out, where appropriate. Simulated environments should be used. Evidence could be compiled through the use of observation sheets or video-recording equipment.

For the first learning outcome, learners must demonstrate the tactical ventilation procedures and be aware of beneficial effects listed in the unit content.

Learners must identify the factors to be considered when using tactical ventilation procedures and the benefits of this approach. This could be assessed by a practical demonstration or a presentation covering the appropriate unit content.

Learners must also demonstrate the tactical ventilation techniques appropriate for each situation. This could be achieved by using the CFBT facility to present demonstrations to other learners acting as trainees. Learners should also apply dynamic risk assessments within the changing fire environments.

For the second learning outcome, learners need to apply their knowledge and understanding of CFBT extinguishing and cooling techniques in a practical setting, within the fire compartment and in adjacent compartments. This can be done through facilitating CFBT simulated scenarios. Demonstrations could include:

- Original ignition stage to flashover/backdraft stage;
- Direct and indirect extinguishing;
- Over/under-pressure extinguishing techniques;
- Cooling and water-fogged ventilating techniques on smoke in compartments adjacent to the fire compartment;
- Cooling of the smoke in the fire room compartment from the adjacent compartments.

For the third learning outcome, learners need to identify the stage of a fire upon arrival, apply appropriate entry techniques, and use current good practice tactics to prevent further pyrolysis. Additionally, learners will need to show how they would implement the range of emergency procedures relevant to the carbonaceous CFBT facility. This should be done during the practical simulated scenarios undertaken in the carbonaceous CFBT facility. For the fourth learning outcome, learners must be able to use the appropriate procedures for operating the carbonaceous CFBT facility. Learners should describe the physical properties to users of the simulator and apply the loading protocols of the carbonaceous fuels used by the system. They should select various materials and apply the safety requirements related to the physical properties created during their decomposition.

Learners should also implement the health monitoring procedures required for CFBT trainers as identified in the Generic Risk Assessment (GRA). This could be assessed by direct observation of performance and supported by the use of questions and answers to confirm learner's knowledge and understanding of the procedures. Learners should be able to implement Generic and Dynamic Risk Assessments (GRA/DRA). For example, they should oversee the loading protocols in compliance with the GRA. Post scenario, learners could produce evaluation reports on the effectiveness of the Generic and Dynamic Risk Assessments. They could then discuss them with the tutors, and the providers of the carbonaceous CFBT facility, ensuring compliance and an understanding of *HSG 65 Successful Health and Safety Management* (HSE Book).

12.4 UNIT THREE – APPLICATION OF POSITIVE PRESSURE VENTILATION TRAINING

Learning outcomes
On completion of this unit a learner should:

1. Be able to demonstrate the appropriate Positive Pressure Ventilation (PPV) techniques;
2. Be able to demonstrate the appropriate entry and air control techniques, recognize the hazards and risks associated with using PPV techniques, and implement the appropriate emergency procedures;
3. Be able to apply the appropriate operating procedures for the positive pressure ventilation (PPV) training facility.

- **Be able to demonstrate the appropriate Positive Pressure Ventilation (PPV) techniques**
  Positive Pressure Ventilation (PPV) techniques: identifying the appropriate outlet and inlet locations and sizes; appropriate positioning of PPV; method of fire attack to extinguish fire within the fire compartment; air flow management; sequential ventilation of compartments adjacent to the fire room compartment; specific building locations (high-rise, basements and enclosed rooms).
  Hazards and risk identification and tactical responses: escaping hot combustible gases; provision of a covering water spray jet at the outlet (exit port); preference to open windows rather than breaking to create outlet (exit port); broken glass and debris; using aerial appliances to create high level outlet (exit port); ensuring that water spray jets are not directed into the created outlet (exit port).
- **Be able to demonstrate the appropriate entry and air control techniques, recognize the hazards and risks associated with using PPV techniques, and implement the appropriate emergency procedures**
  Entry and air control techniques: identification of location of the fire compartment; identification of wind direction and strength; identification of access for airflows created by PPV fans; current good practice for radio communications during operational use of PPV fans; performance of appropriate sequential ventilation around fire compartment.
  Hazards and risks: creating an appropriate size for the outlet (exit port); possibility of locally intensifying the fire; increasing potential for creating a

backdraft; and of creating a fire gas explosion/cold smoke explosion.

Emergency procedures: application of knowledge and understanding of current good practice emergency procedures.

- **Be able to apply the appropriate operating procedures for the Positive Pressure Ventilation (PPV) training facility**

Carbonaceous systems: carbonaceous fuels, e.g. wood, fibrous materials; physical properties (flammability, toxicity, environmental issues); protocols for fuel load (compliance with GRA).

Health-monitoring procedures: e.g. heat stress, heat syncope, hydration levels, Personal Protective Equipment (PPE), firefighting equipment, communication systems, maintenance protocols for the simulator.

Risk assessments: application of knowledge and understanding of safety systems relevant to GRA/DRA.

Unit Three – Assessment criteria
Be able to demonstrate the appropriate Positive Pressure Ventilation (PPV) techniques

- Apply appropriate sequential ventilation techniques in compartments adjacent to the fire compartment;
- Apply appropriate operational PPV techniques in specific locations;
- Apply appropriate inlet/outlet and airflow management techniques.

Be able to demonstrate the appropriate entry and air control techniques, recognize the hazards and risks associated with using PPV techniques, and implement the appropriate emergency procedures

- Apply appropriate entry and air control techniques into fire compartments;
- Identify the signs and symptoms of, and hazards and risks associated with, flashover and backdrafts;
- Apply the appropriate tactics to respond to flashover and backdrafts;
- Implement emergency procedures as appropriate.

Be able to apply the appropriate operating procedures for the Positive Pressure Ventilation (PPV) training facility

- Apply the appropriate generic risk assessment for the PPV training facility;
- Apply the health monitoring procedures required during delivery of PPV training;
- Apply the dynamic risk assessment used in the delivery of PPV training.

Delivery
Before they can safely attempt this unit, learners must:

- Have a thorough knowledge of the fundamental principles of compartment fire behavior;
- Be able to demonstrate good practice tactical ventilation procedures.

Learners should be putting theory into practice in the real work environment. This should involve using case studies in the classroom and practical scenarios in a current good practice purpose-built carbonaceous Positive Pressure Ventilation (PPV) training facility. Creating the required, safe, simulated environments will help learners to understand how these operational techniques and procedures are applied. Tutors should use a range of approaches in the delivery of this unit. For example, lectures, handouts, audiovisual aids, role play, and practical demonstration using the following current good practice equipment, e.g. PPV fans, breaking in tools and Personal Protective Equipment (PPE).

Learners should read widely about policies, procedures and good practice related to PPV training to assist with their personal development. Textbooks, the brigade intranet, the Internet, technical journals and statutory instruments are all useful resources. They should also be encouraged to work in syndicate pairs and groups, so as to think through and compare ideas, share knowledge and understanding, to network and to assist in their personal development.

Learners should be able to apply their knowledge and understanding of current good practice when using the PPV training facility.

Assessment

Assessment might take the form of training reports, presentations, practical exercises and video or audio evidence. Learners could be given scenarios upon which their assessment is based. The scenarios must go into a level of detail sufficient to reflect the complexities of a real-life situation. Much of this unit is practical, therefore evidence must be gathered, where appropriate, in a simulated environment.

Evidence should show depth and breadth of understanding, coherence, analysis, evaluation, independence, intuition and perception, and an ability to apply appropriate learning and development techniques.

For the first learning outcome, learners must apply appropriate sequential ventilation techniques in compartments adjacent to the fire compartment. They must also apply PPV techniques in specific structures and appropriate inlet/outlet and airflow management techniques. All of these could be assessed by direct observation of learners applying techniques during a simulated exercise(s). Tutors could also conduct a question and answer session with learners, to confirm their application of knowledge, and their understanding of appropriate techniques, for a range of situations. For the second learning outcome, learners need to apply appropriate entry and air control techniques into the fire compartment, identify the signs and symptoms of backdrafts and flashover, and the associated hazards and risks. This could be assessed through a practical session, followed by a one-to-one discussion with the tutor.

Lastly, learners will need to apply tactics to meet the needs of the incident and be able to implement emergency procedures as appropriate. This could be assessed by direct observation of learners and supported by questions and answers to confirm their application of knowledge and understanding relating to the scope of applications covered by the GRA. For the third learning outcome, learners must operate the PPV training facility using the appropriate procedures. Learners should be able to inform users of the physical properties of, and apply the appropriate loading protocols to, the carbonaceous fuels used by the system. They should select the various materials and apply the safety requirements related to the physical properties created during their decomposition.

Learners should also implement the health monitoring procedures required for PPV trainers as identified in the Generic Risk Assessment (GRA). This could be assessed by direct observation of performance and supported by the use of questions and answers to confirm learners' application, knowledge and understanding of the procedures.

Learners should be able to apply the Generic and Dynamic Risk Assessments (GRA/DRA). For example, they should oversee the loading protocols in compliance with the GRA. Post scenario, learners could provide evaluation reports of the effectiveness of the GRA/DRA, and discuss them with tutors and the providers of the PPV training facility, ensuring compliance and an understanding of *HSG 65 Successful Health and Safety Management* (HSE Book).

International CFBT qualifications

Edexcel supports **UK and international customers** with training related to BTEC qualifications. This support is available through a choice of training options offered in their published training directory or through customized training at your center. The support they offer focuses on a range of issues including:

- Planning for the delivery of a new program;
- Planning for assessment and grading;
- Developing effective assignments;
- Building your team and teamwork skills;
- Developing student-centered learning and teaching approaches;
- Building key skills into your program;
- Building in effective and efficient quality assurance systems.

The program of training they offer can be viewed on the Edexcel website (www.edexcel.org.uk).

BTEC is a registered trademark of Edexcel Limited
Guidance and units – Edexcel Level 3 BTEC Award and Certificate in Compartment Fire Behavior Training. 2007 Edexcel Limited 2007